DATE DUE

~~OC 2 9 '98~~		
~~NOV 2 5 2002~~		
~~FEB 12 2003~~		

DEMCO 38-296

This highly accessible and comprehensive book provides an introduction to the basic concepts and key circuits of radio-frequency systems, covering fundamental principles which apply to all radio devices, from wireless data transceivers on semiconductor chips to high-power broadcast transmitters.

Topics covered include filters, amplifiers, oscillators, matching networks, modulators, low-noise amplifiers, phase-locked loops, transmission lines, and transformers. The treatment of each subject emphasizes important physical insights while maintaining analytical rigor. Applications of radio-frequency systems are described in such areas as communications, radio and television broadcasting, radar, and radio astronomy.

The book contains many exercises, and assumes only a knowledge of elementary electronics and circuit analysis. It will be an ideal textbook for junior and senior courses in electrical engineering, as well as an invaluable reference for researchers and professional engineers in the area, or for those moving into the field of wireless communications.

RADIO-FREQUENCY ELECTRONICS

CIRCUITS AND APPLICATIONS

RADIO-FREQUENCY ELECTRONICS

CIRCUITS AND APPLICATIONS

JON B. HAGEN

National Astronomy and Ionosphere Center, Cornell University

CAMBRIDGE
UNIVERSITY PRESS

Y THE PRESS SYNDICATE OF THE UNIVERSITY OF CAMBRIDGE
g, Trumpington Street, Cambridge CB2 1RP, United Kingdom

UNIVERSITY PRESS
Building, Cambridge CB2 2RU, United Kingdom
40 West 20th Street, New York, NY 10011-4211, USA
10 Stamford Road, Oakleigh, Melbourne 3166, Australia

© Cambridge University Press 1996

First published 1996

Printed in the United States of America

Typeset in Times Roman

*A catalog record for this book is available from
the British Library*

Library of Congress Cataloging-in-Publication Data

Hagen. Jon B.
    Radio-frequency electronics: circuits and applications/Jon B. Hagen.
        p.   cm.
    ISBN 0-521-55356-3 (hardcover)
    1. Radio circuits. I. Title
    TK6560.H34 1996   95-48033
621.384'12–dc20   CIP

ISBN 0521 55356 3 hardback

CONTENTS

CONTENTS

CONTENTS

CONTENTS

PREFACE

This book was written to prepare the reader to analyze and design radio-frequency (RF) circuits. Developed as a text for a one-semester electrical engineering course at Cornell University, it can also be used for self-study and as a reference for practicing engineers. The discussions of systems, for example television and radio astronomy, complement the detailed analyses of the basic circuit blocks. In the discussions of these basic circuits, I have tried to convey an intuitive understanding from which mathematical analysis easily follows. The scope of topics is wide, and the level of analysis ranges from introductory to advanced. This seems to suit today's students who, though unfamiliar with radio-frequency circuits, are well prepared in engineering fundamentals and have good analytical skills. The only background assumed is basic engineering mathematics and physics, linear circuit analysis, and some elementary analog electronics. Many readers will have had more digital than analog experience, so the digital aspects of switching modulators and direct digital synthesizers are given only short explanations. On the other hand, some basic analog circuit elements such as transformers are now less commonly understood and are therefore reviewed in detail.

For helpful comments and suggestions I am grateful to many students and colleagues, especially Michael Davis, Paul Horowitz, Mario Ierkic, and Wesley Swartz.

Jon B. Hagen

Ithaca, NY
October 1996

INTRODUCTION

Consider the magic of radio. Portable, even hand-held, short-wave transmitters can reach thousands of miles beyond the horizon. Tiny microwave transmitters aboard space probes return data from across the solar system. And all at the speed of light. Yet before the late 1800s there was nothing to suggest that telegraphy through empty space would be possible even with mighty dynamos, much less with insignificantly small and inexpensive apparatus. The Victorians could extrapolate from experience to imagine flight aboard a steam-powered mechanical bird or space travel in a scaled-up Chinese skyrocket. But what experience would even have hinted at wire*less* communication? The key to radio came from theoretical physics. Maxwell consolidated the known laws of electricity and magnetism and added the famous displacement current term, $\partial D/\partial t$. By virtue of this term, a changing electric field produces a magnetic field, just as Faraday had discovered that a changing magnetic field produces an electric field. Maxwell's equations predicted that *electromagnetic waves* can break away from the electric currents that generate them and propagate independently through space with the electric and magnetic field components of the wave constantly regenerating each other.

Maxwell's equations predict the velocity of these waves to be $1/\sqrt{\varepsilon_0\mu_0}$ where the constants ε_0 and μ_0 can be determined by simple measurements of the static forces between electric charges and between current-carrying wires. The dramatic result is, of course, the experimentally known speed of light, 3×10^8 m/s. The electromagnetic nature of light is revealed. Hertz conducted a series of brilliant experiments in the 1880s in which he generated and detected electromagnetic waves with wavelengths very long compared to light. The utilization of Hertzian waves (the radio waves we now take for granted) to transmit information developed hand-in-hand with the new science of electronics.

Where is radio today? AM radio, the pioneer broadcast service, still exists along with FM, television, and two-way communication. Now radio also includes radar, surveillance, navigation and broadcast satellites, cellular telephones, remote control devices, and wireless data communications. Applications of radio frequency (RF) technology outside radio include microwave heaters, medical imaging systems, and cable television.

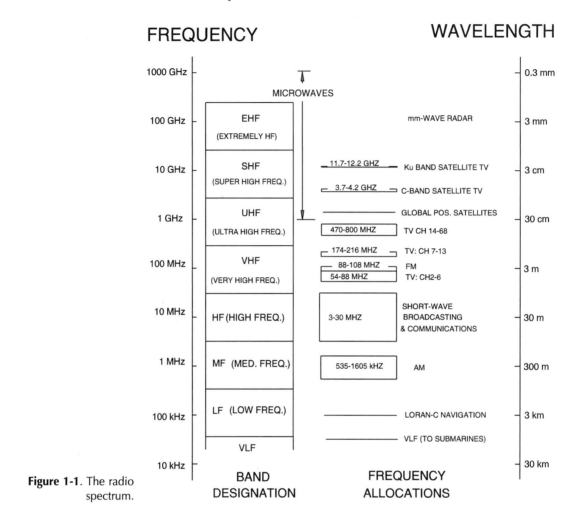

FREQUENCY

WAVELENGTH

Figure 1-1. The radio spectrum.

Radio occupies about eight decades of the electromagnetic spectrum, as shown in Figure 1-1.

RF CIRCUITS

The circuits discussed in this book generate, amplify, modulate, filter, demodulate, detect, and measure ac voltages and currents at radio frequencies. They are the blocks from which RF systems are designed. They scale up and down in both power and frequency. A six-section bandpass filter with a given passband shape, for example, might be large and water cooled in one application but subminiature in another. Depending on the frequency, this filter might be made of sheet metal boxes and pipes, of solenoidal coils and capacitors, or of piezoelectric mechanical resonators, yet the underlying circuit design remains the same. A class-C amplifier circuit might be a small section of an integrated circuit for a wireless data link or the largest part of a multimegawatt broadcast transmitter. Again, the design principles are the same.

NARROW-BAND NATURE OF RF SIGNALS

Note that most of the RF allocations have small fractional bandwidths, that is, the bandwidths are small compared to the center frequencies. The fractional bandwidth of the signal from any given transmitter is less than ten percent – usually much less. This means that the RF voltages throughout a radio system are very nearly sinusoidal. An otherwise purely sinusoidal RF "carrier" voltage must be *modulated* (varied in some way) to transmit information. Every type of modulation (audio, video, pulse, digital coding, etc.) works by varying the amplitude and/or the phase of the carrier. An unmodulated carrier has only infinitesimal bandwidth; it is a pure spectral line. Modulation always broadens the line into a spectral band, but the energy clusters around the carrier frequency. Oscilloscope traces of the RF voltages in a transmitter on a transmission line or antenna are therefore nearly sinusoidal. When modulation is present, the amplitude and/or phase of the sinusoid changes but only over many cycles. Because of this narrow-band characteristic, elementary sine wave ac circuit analysis serves for most RF work.

AC CIRCUIT ANALYSIS – A BRIEF REVIEW

The standard ac circuit theory that treats voltages and currents in linear networks is based on the linearity of the circuit elements. When a sinusoidal voltage or current generator drives a circuit, the resulting steady-state voltages and currents will all be perfectly sinusoidal and will have the same frequency as the generator. Normally we find the response of driven ac circuits by a mathematical artifice. We replace the given sinusoidal generator by a hypothetical generator whose time dependence is $e^{j\omega t}$ rather than $\cos(\omega t)$ or $\sin(\omega t)$. This source function has both a real and an imaginary part since $e^{j\omega t} = \cos(\omega t) + j\sin(\omega t)$. Such a nonphysical (because it is complex) source leads to a nonphysical (complex) solution. But the real and imaginary parts of the solution are separately good physical solutions that correspond to the real and imaginary parts of the complex source. The value of this seemingly indirect method of solution is that the substitution of the complex source converts the set of linear *differential* equations into a set of easily solved linear *algebraic* equations. When the circuit has a simple topology, as is often the case, it can be reduced to a single loop by combining obvious series and parallel branches. Several computer programs are available to find the currents and voltages in complicated ac circuits. Most versions of SPICE will do this steady-state ac analysis (which is much simpler than the transient analysis which is their primary function). Special linear ac analysis programs for RF and microwave work such as COMPACT, TOUCH-STONE, and MMICAD include circuit models for strip lines, waveguides, and other RF components. You can write a simple program to analyze ladder networks (see Problem 3) that will analyze most filters and matching networks.

IMPEDANCE AND ADMITTANCE

The coefficients in the algebraic circuit equations are functions of the complex *impedances* (V/I), or *admittances* (I/V), of the *RLC* elements. The voltage across an inductor is $L \, dI/dt$. If the current is $I_0 e^{j\omega t}$, then the voltage is $(j\omega L)I_0 e^{j\omega t}$. The impedance and admittance of an inductor are therefore respectively $j\omega L$ and $1/(j\omega L)$. The current into a capacitor is $C \, dV/dt$, so its impedance and admittance are $1/(j\omega C)$ and $j\omega C$. The impedance and admittance of a resistor are just R and $1/R$, respectively. Elements in series have the same current, so their total impedance is the sum of their separate impedances. Elements in parallel have the same voltage, so their total admittance is the sum of their separate admittances. The real and imaginary parts of impedance are called resistance and reactance while the real and imaginary parts of admittance (the reciprocal of impedance) are called *conductance* and *susceptance*.

SERIES RESONANCE

A capacitor and inductor in series have an impedance $Z_s = j\omega L + 1/j\omega C$. This can be written as $Z_s = j(L/\omega)(\omega^2 - 1/LC)$, so the impedance is zero when the (angular) frequency is $1/\sqrt{LC}$. At this *resonant frequency*, the *series LC* circuit is a perfect *short* circuit (Figure 1-2). Equal voltages are developed across the inductor and capacitor but they have opposite signs, and the net voltage drop is zero. At resonance and in the steady state there is no transfer of energy in or out of this combination. (Since the overall voltage is always zero, the power, IV is always zero.) However, the circuit does contain stored energy, which simply sloshes back and forth between the inductor and the capacitor. Note that this circuit, by itself, is a simple bandpass filter.

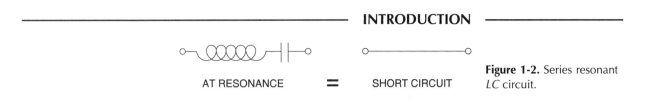

AT RESONANCE   =   SHORT CIRCUIT

Figure 1-2. Series resonant *LC* circuit.

PARALLEL RESONANCE

A capacitor and an inductor in parallel have an admittance $Y_p = j\omega C + 1/j\omega L$, which is zero when the (angular) frequency is $1/\sqrt{LC}$. At the resonant frequency, the *parallel LC* circuit is a perfect *open* circuit – a simple bandstop filter (Figure 1-3). Like the series *LC* circuit, the parallel *LC* circuit stores a fixed quantity of energy for a given applied voltage. These two simple combinations are important building blocks in RF engineering.

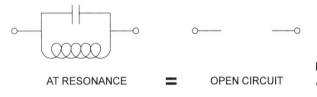

AT RESONANCE   =   OPEN CIRCUIT

Figure 1-3. Parallel resonant *LC* circuit.

NONLINEAR CIRCUITS

Many important RF circuits, including mixers, modulators, and detectors, are based on nonlinear circuit elements such as diodes and saturated transistor switches. Here we cannot use the linear $e^{j\omega t}$ analysis but must use time domain analysis. Usually the nonlinear elements can be replaced by simple models to explain the circuit operation. Full computer modeling can be used for accurate circuit simulations.

PROBLEMS

1. A generator has a source resistance r_s and an open circuit rms voltage V_0. Show that the maximum power available from the generator is given by $P_{max} = V_0^2/(4r_s)$ and that this maximum power will be delivered when the load resistance, R_L, is equal to the source resistance, r_s.

2. A passive network, for example a circuit composed of resistors, inductors, and capacitors, is placed between a generator with source resistance r_s and a load resistor R_L. The *power response* of the network (with respect to these resistances) is defined as the fraction of the generator's maximum available power that reaches the load. If the network is *lossless*, that is, contains no resistors or other dissipative elements, its power response function can be found in terms of the impedance Z_{in}, seen looking into the network with the load connected. Show that the expression for the power response of the lossless network is given by

$$P(\omega) = \frac{4r_s R}{(R + r_S)^2 + X^2}$$

where $R = \text{Re}(Z_{in})$ and $X = \text{Im}(Z_{in})$.

3. Most filters and matching networks take the form of the ladder network shown below. Write a program that reads a circuit file specifying the series and shunt elements and finds the power response function as defined in Problem 2. (This problem will be of use in many later problems and will be expanded in scope several times.)

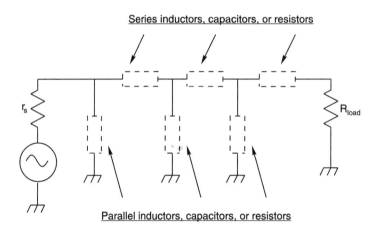

Series inductors, capacitors, or resistors

Parallel inductors, capacitors, or resistors

Hints: One approach is to begin from the load resistor and calculate the input impedance as the elements are added, one by one. When all the elements are in place, the formula in Problem 2 gives the power response – as long as none of the elements are resistors. The process is repeated for every desired frequency.

A better approach, which is no more complicated and which allows resistors, is the following: Assume a current of $1 + j0$ amperes is flowing into the load resistor. The voltage at this point is therefore $R_L + j0$ volts. Move to the left one element. If this is a series element, the current is unchanged but the voltage is higher by IZ, where Z is the impedance of the series element. If the element is a shunt element, the voltage remains the same but the input current is increased by VY where Y is the admittance of the shunt element. Continue adding elements, one at a time, updating the current and voltage. When all the elements are accounted for, you have the input voltage and current, and could calculate the total input impedance of the network terminated by the load resistor. Instead, however, go one more step, treating the source resistance, r_s, as just another series impedance. This gives you the voltage of the source generator, from which you can calculate the maximum power available from the source. Since you already know the power delivered to the load, $(I)^2 R_L$, you can find the power response. Repeat this process for every desired frequency.

The ladder elements (and the start frequency, stop frequency, and frequency increment if you like) can be treated as data, that is, they can be located together in a block of the program or in a file so they can be changed easily. For now the program only needs to deal with six element types: series and parallel inductors, capacitors, and resistors. Each element in the circuit file must therefore have an identifier such as "PL," "SL," "PC," "SC," "PR," and "SR" or 1, 2, 3, 4, 5, 6, or whatever, plus the value of the component in henrys, farads, or ohms. Organize the circuit file so that it begins with the element closest to R_L and ends with some identifier such as "EOF" (for "end of file") or some distinctive number.

An example of this program, written in Microsoft QBasic, is shown here, together with some output data from an example circuit file (line number 600). Use this circuit and data to check your own program. You will want to add some headings for the print-out, and maybe graphing capability.

```
'Qbasic program to calculate transmission through an RCL ladder network
RSOURCE= 1000: RLOAD = 50
FOR F=1E6 TO 2E6 STEP 5E4
  OMEGA=2 * 3.14159 * F
  IR=1: II=0:VR= IR*RLOAD: VI=II*RLOAD 'assume 1 amp into load.
  READ TYPE$ ' "PC" is parallel capacitor, "SL" is series inductor, etc.
    DO UNTIL TYPE$ = "EOF" 'EOF denotes end of circuit file.
    READ VALUE
    IF TYPE$ = "PC" THEN B= OMEGA * VALUE: G=0: GOSUB 400 'to update I.
    IF TYPE$ = "SC" THEN X= -1/(OMEGA * VALUE):R=0: GOSUB 500 'to update V.
    IF TYPE$ = "PL" THEN B= -1/(OMEGA * VALUE):G=0: GOSUB 400 'to update I.
    IF TYPE$ = "SL" THEN X= OMEGA * VALUE:R=0 GOSUB 500 'to update V.
    IF TYPE$ = "PR" THEN G= 1/VALUE: B=0: GOSUB 400 'to update I.
    IF TYPE$ = "SR" THEN R= VALUE: X=0: GOSUB 500 'to update V.
    READ TYPE$
  LOOP
  R=RSOURCE: X=0: GOSUB500 'to get generator voltage.
  'calculate fraction of maximum possible power transfer.
  FRAC= (1^2*RLOAD)/((VR*VR+VI*VI)/(4*RSOURCE))
  PRINT F; FRAC; 10/LOG(10)*LOG(FRAC) 'freq, frac & frac in decibels
  RESTORE 600 'rewind data.
NEXT F
END
400 'subroutine to update real and imaginary parts of I.
  IR=IR+(VR*G - VI*B): II=II+(VI*G + VR*B): RETURN
500 'subroutine to update real and imaginary parts of V.
  VR=VR+(IR*R - II*X): VI=VI+(II*R + IR*X): RETURN
  'example circuit file: a 463 pF parallel capacitor and
  'a 23.1 microhenry series inductor
  600 DATA SL,23.1E-6,PC,463E-12,EOF
```

Program output

Frequency (MHz)	Fraction	dB
1.00	0.4179	−3.789
1.05	0.4601	−3.372

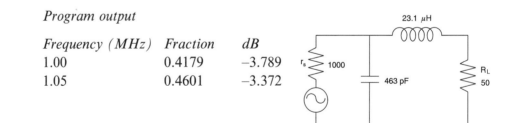

IMPEDANCE MATCHING I

Matching normally means the use of a lossless (nonresistive) network between an ac (here RF) source and a load in order to maximize the power transferred to the load. An antenna tuner, for example, is a device to match an antenna to a transmitter. The same circuit, if it is built into the transmitter, might be called an output tuner. In the direct current (dc) circuit of Figure 2-1, maximum power is transferred when the load resistance is equal to the source resistance. (You can verify that making the load resistance equal the source resistance maximizes the current × voltage product of the load.)

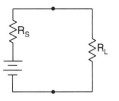

Figure 2-1. Maximum power is transferred to R_L when $R_L = R_s$.

TRANSFORMER MATCHING

In the case of an alternating current (ac) source, a transformer can make the load resistance match the source resistance (and vice versa), as shown in Figure 2-2. The ac situation often has a complication: the source and/or load may be reactive, that is, have an unavoidable built-in reactance. An example of a reactive load is an antenna; many antennas are purely resistive at only one frequency. Above this resonant frequency they usually look like a resistance in series with an inductor, and below the resonant

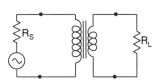

Figure 2-2. Transformer converts R_L to R_s for maximum power transfer.

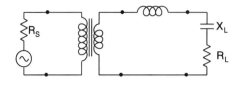

Figure 2-3. Series reactor makes the load a pure resistance.

frequency they look like a resistance in series with a capacitor. An obvious way to deal with this is first to cancel the reactance to make the load and/or source impedance purely resistive and then use a transformer to match the resistances. In the circuit of Figure 2-3, an inductor cancels the reactance of a capacitive (but not purely capacitive) load. If we are working at 60 Hz we would say the inductor corrects the power factor of the load.

From the standpoint of the load, the matching network converts the source impedance, $R_s + j0$, into the complex conjugate of the load impedance. And in general, when a matching device is used between two devices, each device will look into an impedance that is the complex conjugate of its own impedance. Whenever the source and/or load has a reactive component, the match will be frequency-dependent, that is, away from the design frequency the match will not be perfect. In fact, with reactive sources and/or reactive loads, *any* lossless matching circuit will be frequency-dependent – a filter of some kind – whether we like it or not.

L-NETWORKS

More often than not, matching circuits use no transformers (i.e. no coupled inductors). Figure 2-4 shows a two-element L-network (a rotated letter "L") that will match a source to a load resistor whose resistance is smaller than the source resistance. The trick is to put a reactor, Xp, in *parallel* with the *larger* resistance. We will consider a specific example: $R_s = 1000$ ohms and $R_L = 50$ ohms. The impedance of the left-hand side is given by

$$Z_{\text{left}} = R_{\text{left}} + jX_{\text{left}} = \frac{1,000jX_{\text{p}}}{1,000 + jX_{\text{p}}} = \frac{(1,000jX_{\text{p}})(1,000 - jX_{\text{p}})}{(1,000 + jX_{\text{p}})(1,000 - jX_{\text{p}})}$$

$$= \frac{1,000^2 jX_{\text{p}} + 1,000X_{\text{p}}^2}{1,000^2 + X_{\text{p}}^2}. \tag{2-1}$$

We can pick the value of X_{p} so that the real part of Z_{left} will be 50 ohms, that is, equal to the load resistance. Using Equation (2-1), we find that

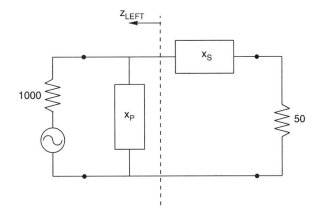

Figure 2-4. Two reactors (X_p, in parallel; X_s, in series) in an L-network match R_L to R_s.

$X_p^2 = 52,632$, so we can pick either $X_p = 229$ (an inductor) or $X_p = -229$ (a capacitor). The left-hand side now has the correct equivalent series resistance, 50 ohms, but it is accompanied by an equivalent series reactance, X_{left}, given by the imaginary part of Equation (2-1). We can cancel X_{left} by inserting a series reactor, X_s, equal to $-X_{left}$. Figure 2-5 shows the matching circuits that result when X_p is an inductor and when X_p is a capacitor.

The final step is to find the values of L and C that produce the specified reactances at the given frequency. For the circuit of Figure 2-5b, $\omega L = 218$. Suppose the design frequency is 1.5 MHz ($\omega = 2\pi \times 1.5 \times 10^6$), near the top of the AM broadcast band. Then $L = 23.1\,\mu H$ and $C = 462\,pF$. Note that the values of the two reactors are completely determined by the source and load resistances. Except the choice of which element is to be an inductor and which is to be a capacitor, there are no free parameters in this two-element matching circuit. The match is perfect at the design frequency, but away from that frequency we must accept the resulting frequency response. The frequency responses (fractional power reaching the load versus frequency) for the two circuits of Figure 2-5 are

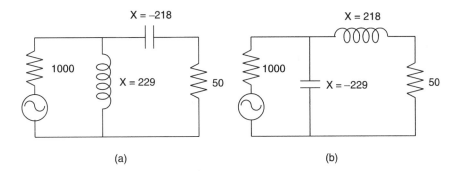

(a) (b)

Figure 2-5. The two realizations for the L-network of Figure 2-4.

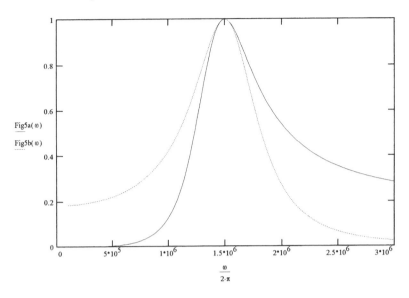

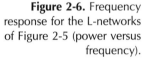

Figure 2-6. Frequency response for the L-networks of Figure 2-5 (power versus frequency).

plotted in Figure 2-6. Note that around the design frequency, that is, around the resonant peak, the curves are virtually identical. Otherwise, the complete cut-off at very low frequencies of Figure 2-5a and the complete cutoff at very high frequencies of Figure 2-5b can be predicted from inspection of the circuits.

QUICK DESIGN PROCEDURE FOR L-NETWORKS

If you remember only that the parallel reactance goes across the larger resistance you will be able to repeat the steps used above and design *L*-networks. But if you are doing these things often it is worth memorizing the following "*Q*-factor" for *L*-network design:

$$Q_{\mathrm{EL}} = \sqrt{\frac{R_{\mathrm{high}}}{R_{\mathrm{low}}} - 1}. \tag{2-2}$$

You can verify (see Problem 6) that the ratios $R_{\mathrm{high}}/X_{\mathrm{p}}$ and $X_{\mathrm{s}}/R_{\mathrm{low}}$ are both equal to this factor, Q_{EL}. Remember the definition of Q_{EL} and these ratios immediately give you the L-network reactance values. You can also verify that, when Q_{EL} is large, the two elements in an L-network have nearly equal and opposite reactances, that is, together they resonate at the design frequency. In this case the magnitude of the reactances is given by the geometric mean of R_{high} and R_{low} (especially easy to remember).

When the ratio of the source resistance to the load resistance is much different from unity, an L-network produces a narrow-band match, that is, the match will be good only very close to the design frequency. Conversely, when the impedance ratio is close to unity, the match is wide. The width of any resonance phenomenon is described by a factor, *the effective Q* (or *circuit Q* or just *Q*), which is equal to the center frequency divided by the two-sided 3 dB bandwidth (difference between the half power points). Equivalently, Q_{eff} is the reciprocal of the fractional bandwidth. When an ideal voltage generator drives a simple *RLC* series circuit, Q_{eff} is given by X/R, where X is either X_L or X_C at the center frequency (since they are equal). The *L*-network matching circuit is equivalent to a simple series *RLC* circuit, but Q_{EL} is twice Q_{eff} because the nonzero source resistance is also the series. The matching circuit makes the effective source resistance equal to the load resistance so the total series resistance of the loop is twice the load resistance. As a result, the fractional bandwidth is given by $1/Q_{\text{eff}} = 2/Q_{\text{EL}}$. In many applications the bandwidth of the match is important, and the match provided by the L-network (which is completely determined by the source and load resistances) may be too narrow or too wide. When matching an antenna to a receiver, for example, one wants a narrow bandwidth so that signals from strong nearby stations will not overload the receiver. In another situation the signal produced by a modulated transmitter might have more bandwidth than the L-network would pass. Networks described below solve these problems.

HIGHER *Q*: PI AND T-NETWORKS

Higher *Q* can be obtained with back-to-back L-networks, each one transforming down to a center impedance that is lower than either the generator or the source resistance. The resulting pi network is shown in Figure 2-7. With the simple L-networks we had $Q_{\text{EL}} = \sqrt{19} = 4.4$. In this pi network both the 1000 ohm source and the 50 ohm load are matched down to a center impedance of 10 ohms (a free parameter). The bandwidth is equiva-

Figure 2-7. Pi network (back-to-back L-networks) provides higher *Q*.

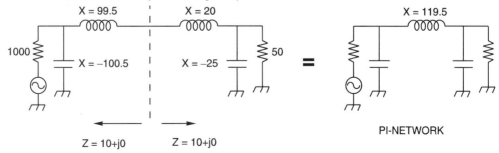

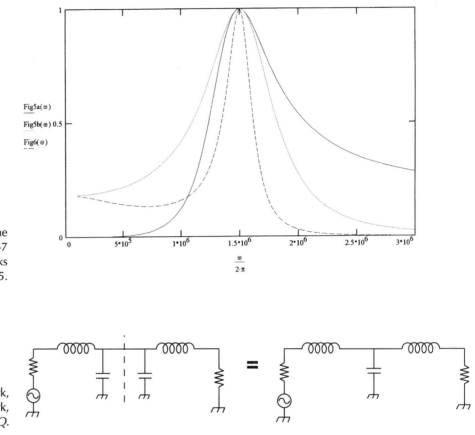

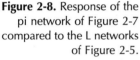

Figure 2-8. Response of the pi network of Figure 2-7 compared to the L networks of Figure 2-5.

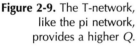

Figure 2-9. The T-network, like the pi network, provides a higher Q.

lent to that of an L-network with $Q_{\text{EL}} = 11.95$. When $R_{\text{high}} \gg R_{\text{low}}$, the pi network has a bandwidth equivalent to that of an L-network with $Q_{\text{EL}} = \sqrt{R_{\text{high}}/R_{\text{center}}}$. Again, the fractional bandwidth is given by $1/Q_{\text{eff}} = 2Q_{\text{EL}}$. The response of this pi network is shown in Figure 2-8 together with the responses of the L-networks of Figure 2-5.

You can guess that we could just as well have used "front-to-front" L-networks, each one transforming up to a center impedance that is *higher* than both the source or the load. This produces the T-network of Figure 2-9. Note that both the pi network and the T network have a free parameter (the center impedance), which gives us some control over the frequency response while still providing a perfect match at the center or design frequency.

LOWER Q – THE DOUBLE L-NETWORK

In a double L-network (Figure 2-10) the first stage transforms to an impedance between the source and load impedances. The second stage takes it

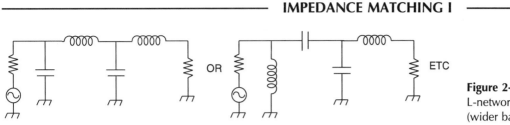

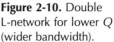

Figure 2-10. Double L-network for lower Q (wider bandwidth).

the rest of the way. The process can, of course, be in smaller steps with any number of cascaded networks. A long chain of L networks forms an artificial transmission line that tapers in impedance to produce a frequency-independent match. Real transmission lines (i.e. lines with distributed L and C) are sometimes physically tapered to provide this kind of impedance transformation. A tapered transmission line is sometimes called a transformer, since, like the transformer in Figure 2-2, it provides frequency-independent matching.

EQUIVALENT SERIES AND PARALLEL CIRCUITS

To design the L-network we used the fact that a two-element parallel XR circuit, where $1/Z = 1/R_p + 1/(jX_p)$, has an equivalent series circuit, where $Z = R_s + jX_s$. Conversion between equivalent series and parallel representations is used so often it is worth a few more words. If you are given, for example, an antenna or a black box with two terminals and you make measurements at a *single frequency* you can only determine whether the box is "capacitive," that is, equivalent to an R, C combination or is "inductive," that is, equivalent to an R, L combination. Suppose it is capacitive. Then you can represent it equally well as a series circuit where $Z = R_s + 1/(j\omega C_s)$ or as a parallel circuit where $1/Z = 1/R_p + j\omega C_p$. As long as you are working only at (or never very far from) the single frequency, either representation is equally valid, even if the box contains a complicated circuit with discrete resistors, capacitors, inductors, transmission lines, metallic and resistive structures, and so on. If you measure the impedance at more than one frequency you might determine that the box does indeed contain a simple parallel RC or series RC circuit or that its impedance variation at least resembles that of a simple parallel circuit more than it resembles that of a simple series circuit.

LOSSY REACTORS AND EFFICIENCY OF MATCHING NETWORKS

So far we have considered networks made of ideal inductors and capacitors. Real components, however, are lossy due to the finite conductivity of metals, lossy dielectrics or magnetic materials, and even radiation. Power dissipated in nonideal components is power that does not reach the load so, with lossy components, we must consider the efficiency of a matching network. As explained above, a lossy reactor can be modeled as an ideal L or C together with either a series or parallel resistor. Normally we can make the approximation that the values of L or C and the value of the associated resistor are constant throughout the band of interest. Let us consider the efficiency of the L network that uses a series inductor and a parallel capacitor. We will assume that the loss in the capacitor is negligible compared to the loss in the inductor. (This is very often the case with lumped components.) We will model the lossy inductor as an ideal inductor in series with a resistor of value r_s. The ratio of the inductive reactance, X_L, to this resistance value is the *quality factor*, Q_U, where the subscript denotes "*unloaded Q*" or *component Q*. (Less series resistance certainly implies a higher quality component.) Note that this resistance, like the inductor, is in series with the load resistor so the same current, I, flows through both. The power delivered to the load is $I^2 R_L$, and the power dissipated in r_s is $I^2 r_s$. Using the relations $X_s = Q_{EL} R_L$ and $Q_U = X_s/r_s$, we find the efficiency of the match is given by

$$\eta = \text{efficiency} = \frac{\text{power out}}{\text{power in}} = \frac{I^2 R_L}{I^2 R_L + I^2 r_s} = \frac{1}{1 + Q_{EL}/Q_U}. \qquad (2\text{-}3)$$

Efficiency is maximized by maximizing the ratio Q_U/Q_{EL}, that is, the ratio of unloaded Q to loaded Q. If we model the lossy inductor as a parallel LR circuit and define the unloaded Q as r_p/X_p we will get the same expression for efficiency (see Problem 7). Likewise, if the loss occurs in the capacitor we will also get this expression as long as we define the unloaded Q of the capacitor again as parallel resistance over parallel reactance or as series reactance over series resistance. When the load resistance is very different than the source resistance the effective Q of an L-network will be high, so, for high efficiency, the unloaded Q of the components must be very high. The double L-network, with its lower loaded Qs, can be used to provide higher efficiency.

Q-FACTOR SUMMARY

Loaded Q, the *Q*-factor associated with *circuits* can be either high or low, depending on the application. Narrow-band filters have high loaded Q. Wide-band matching circuits have low loaded Q. Loaded Q is therefore not a measure of quality. *Unloaded Q*, however, which specifies the losses in *components*, is indeed a measure of quality since lowering component losses always increases circuit efficiency.

1. A nominal 47 ohm 1/4 W carbon resistor with 1.5″ wire leads is measured at 100 MHz to have an impedance of $48 + j39$ ohms. Find the component values for (a) an equivalent series RL circuit and (b) an equivalent parallel RL circuit.

2. (a) Design an L-network to match a 50 ohm generator to a 100 ohm load at a frequency of 1.5 MHz. Let the parallel element be an inductor. Use your circuit analysis program (see Chapter 1, Problem 3) to find the frequency response of this circuit from 1 to 2 MHz in steps of 50 kHz.

(b) Same as Problem 2(a), but let the parallel element be a capacitor.

3. Design a double-L matching network for the generator, source, and frequency of Problem 2(a). For maximum bandwidth, let the intermediate impedance be the geometric mean of the source impedance and the load impedance, that is, $\sqrt{50 \times 100}$. Use your circuit analysis program (see Chapter 1, Problem 3) to find the response, as in Problem 2.

4. Suppose the only inductors available for building the networks of Problems 2(a) and 3 have a Q_U (unloaded Q) of 100 at 1.5 MHz. Assume the capacitors have no loss. Calculate the efficiencies of the matching networks at 1.5 MHz. Check your results using your circuit analysis program (see Chapter 1, Problem 3).

5. The diagram below shows a network that allows a 50 ohm generator to feed two loads (which might be antennas). The network divides the power such that the top load receives twice as much power as the bottom load. The generator is matched, that is, it sees 50 ohms. Find the values of X_{L_1}, X_{L_2}, and X_C. *Hint*: transform each load first with an L-section network and then combine the two networks into the circuit shown.

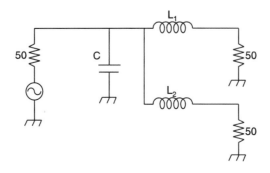

6. Verify the prescription given for calculating the values of an L-network: $X_p = \pm R/Q$ and $X_s = \pm rQ$ where $R > r$ and $Q = \sqrt{R/r - 1}$.

7. At a single frequency, a lossy inductor can be modeled as a lossless inductor in series with a resistance or as a lossless inductor in parallel with a resistance. Convert the series combination r_s, L_s to its equivalent parallel combination r_p, L_p, and show that Q_U defined as X_s/r_s is equal to Q_U defined as r_p/X_p.

LINEAR AMPLIFIERS

3

In this chapter we will look at conventional resistance-controlled linear amplifiers that are commonly used as dc amplifiers, audio amplifiers, video amplifiers, and RF amplifiers. We will see that, while the basic operating principle remains the same, the circuit configurations are determined by the nature of the load, the nature of the signal, and whether high efficiency is needed. For now we will concentrate on the topology of the output circuit or "business end" of these amplifiers. The term "linear" means that the output signal should be linearly proportional to the input signal over some continuous input range, that is, the output should be a scaled reproduction of the input.

SINGLE-LOOP AMPLIFIERS

A basic amplifier circuit is a loop formed by the dc supply, the active device (transistor or tube), and the load as shown in Figure 3-1. The manually adjustable resistor (rheostat) in the left-hand circuit represents an active device (the transistor in the right-hand circuit). It throttles or "valves" the current, thereby controlling the voltage drop across the load resistor, R_L. Since the active device is used as a variable resistor it *must* dissipate power (heat) except when the output voltage is at zero or at the full supply voltage. The dc supply voltage is chosen to be the maximum voltage ever needed at the load. The efficiency of the amplifier depends only on the waveform, that is, the efficiency is not a function of the transistor characteristics. The input or "drive signal" is applied to the

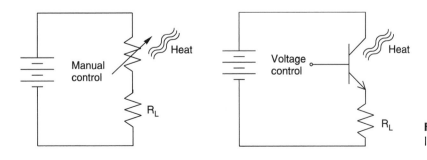

Figure 3-1. Basic single-loop amplifier.

base of the transistor to control its resistance. (For the moment we will not specify the second or return connection for the drive.) The power delivered to the load is greater than the drive power (usually much greater). The ratio of these powers is known as the gain. In RF work, "gain" always means power gain.

THE EMITTER FOLLOWER

When the return connection for the drive signal is made at the bottom of the load resistor, we get the amplifier shown in Figure 3-2. Because only the collector is at an ac ground point this circuit is classified as a common-collector amplifier. (The collector is, of course, not at dc ground, but the battery provides a ground connection for ac.) It is also known as an *emitter follower* because, if R_L is not too small, the transistor will adjust its current flow to make the instantaneous emitter voltage almost identical to the instantaneous base voltage. Here "almost identical" means a small dc offset (the average emitter voltage will be about 0.7 V less than the average base voltage) and a slight reduction in signal amplitude (the amplitude of the ac signal produced at the emitter will be maybe 1 percent less than the amplitude of the signal applied at the base. (See Problem 4.)

The emitter follower can deliver positive voltages to the load over a continuous range from zero to the supply voltage but it cannot furnish negative voltage at all. The load resistor and the transistor form a voltage divider in which the resistance of the transistor controls the voltage drop across the load. This works if the load actually is a resistor, but does not work at all if the load is a capacitor, for example a beam deflection electrode in an oscilloscope cathode ray tube. Despite these

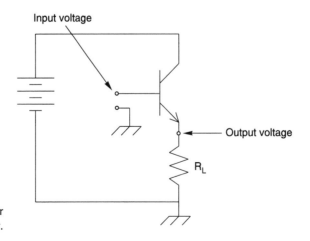

Figure 3-2. The emitter follower.

limitations, there are situations for which the single-loop amplifier is quite adequate. Consider an electronically regulated oven. The load resistor is the heating element. The transistor can change the magnitude of the voltage across this resistor, and there is no need to change the polarity. Moreover, in this application, the emitter follower will be as efficient as any other resistance-controlled circuit working from a single power supply.

COMMON-EMITTER AND COMMON-BASE AMPLIFIERS

Before looking at circuits that overcome the limitations of the single-loop circuit we should mention two common variations.

A common-emitter amplifier is shown in Figure 3-3. Here the load and the transistor have changed positions. Since the dc supply, the transistor, and the load are still in series, the basic operation and efficiency are the same as with the emitter follower. But with the emitter tied to ground, the drive voltage is placed directly across the base–emitter junction, and the transistor current will be a nonlinear (exponential) function of the drive voltage. An external circuit could be used to generate an inverse exponential (logarithmic) drive signal to linearize the amplifier. This happens when a current source is used instead of a voltage source to drive the common-collector power amplifier. (The linearity of the overall amplifier will depend on β, the small-signal current gain, remaining constant over large excursions.) The common-emitter amplifier has more power gain than the emitter follower.

A common-base amplifier is shown in Figure 3-4. In this circuit the drive current flows through the main loop. Gain is obtained because, while the driver and load have essentially the same current (the base current might be only 1 percent as large as the collector-emitter current) the voltage swing at the collector, determined by the supply voltage, is much

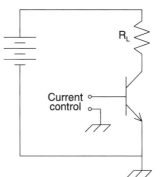

Figure 3-3. Common-emitter amplifier.

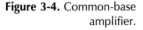

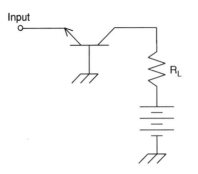

Figure 3-4. Common-base amplifier.

greater than the voltage swing at the emitter, determined by the near short-circuit base–emitter junction.

ONE TRANSISTOR, TWO SUPPLIES

If the amplifier has to supply output voltages of either polarity and also has to handle arbitrary waveforms, it will require a circuit with two power supplies (or a single "floating" power supply as we will see later). The two-loop circuit shown in Figure 3-5 still uses only one active device. Here, R_E is a pull-down resistor. It pulls the output toward the negative supply when the transistor lets it. The average current or *bias current* in R_E depends on the average output voltage. For some signals, such as audio, the average voltage is zero. In this case the values of V_{neg} and R_E are chosen so that the maximum negative swing can be equal to the maximum positive swing. For maximum efficiency this makes $V_{neg} = 2V_{pos}$ and $R_E = R_L$. You can verify, however, that the maximum efficiency is poor, only 1/12. Biased amplifiers are known as class-A amplifiers, and are commonly used in small signal applications where power dissipation is always low and efficiency is not a prime consideration. This biased amplifier circuit could drive a purely capacitive load since R_E provides a discharge path. The maximum negative load current is limited, however, by the value of R_E.

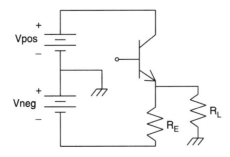

Figure 3-5. A dual-supply, single-transistor amplifier.

Making R_E small increases the current capacity but also increases the bias current and hurts the efficiency.

TWO TRANSISTORS, TWO SUPPLIES

The two-transistor *push–pull* configuration shown in Figure 3-6 provides output voltage of both polarities and can achieve high efficiency. This two-loop, two-transistor circuit consists of the original npn emitter follower plus a complementary (pnp) emitter follower, instead of a pull-down resistor, to supply negative voltage. The top transistor pulls the load positive. The lower transistor pulls it negative, that is, pushes it away from positive. Any parasitic shunt capacitance present in the load can be charged or discharged quickly by overdriving the appropriate transistor until the desired output voltage is reached. This requires feedback; the driver has to know when to apply the extra drive. But even without an external feedback loop, the follower circuits provide inherent negative feedback; if the load voltage (emitter voltage) does not immediately follow the drive voltage (base voltage) the emitter-to-base voltage increases, turning up the conduction of the transistor. The push–pull circuit is *the* circuit of choice for arbitrary waveforms; the efficiency is as high as possible for a linear circuit (78% for a sine wave of maximum amplitude). For capacitive loads, the slew rate, dV/dt is not limited by the value of a pull-down resistor (or the resistive component of the load). The push–pull amplifier normally runs as class B, which means that when the output voltage is zero both transistors are turned off and there is no dissipation. For positive current the top device is active and the bottom device is off. For negative current the situation is reversed. It is important that the crossover be smooth and continuous. Sometimes push–pull amplifiers are run as class AB, where a small bias current ensures a smooth crossover. Another advantage of the complementary push–pull circuit is that the drive signals for the two transistors are identical except for a small dc bias. This advantage is

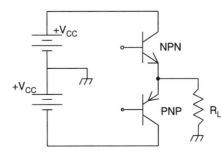

Figure 3-6. Complementary (pnp/npn) push–pull amplifier.

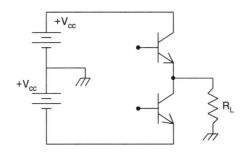

Figure 3-7. Totem pole push–pull amplifier.

lost when two identical transistors are used to build the asymmetric "totem pole" amplifier shown in Figure 3-7. Here the two transistors must be driven out-of-phase, that is, the drive signals on the two bases must have opposite polarities. Worse yet, the top transistor is an emitter follower that follows its drive *voltage*, while the bottom transistor is in a common-emitter configuration and follows its drive *current*. Separate drivers are needed to achieve linearity with this awkward circuit. We will see below that transformers permit building a symmetric push–pull amplifier with two identical devices (two npn transistors, two pnp transistors, or two tubes).

There is a push–pull amplifier that uses a single power supply. This so-called bridge amplifier is shown in Figure 3-8a. This three-loop circuit has no direct connection between the power supply and the load; either the supply or the load must "float," that is, have no ground connection. In this circuit, one pair of transistors can be operated as on–off switches. The half-bridge circuit of Figure 3-8b replaces two transistors with resistors; as with the full-bridge circuit, the load can be supplied with current of either polarity, and no bias current needs to flow through the load. But the maximum voltage swing is cut in half, and efficiency suffers because of the necessary bias current through the bridge resistors.

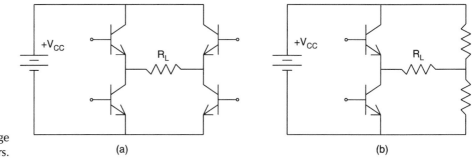

Figure 3-8. Bridge amplifiers.

(a)

(b)

AC AMPLIFIERS

While the push–pull configuration is the most efficient all-purpose amplifier, amplifiers can be simplified when they are restricted to ac signals (audio), and can be simplified even more when restricted to narrowband ac signals (RF).

AUDIO AMPLIFIERS

Audio signals, having no dc component, are symmetric about zero volts, so ac coupling can be used to eliminate the need for a negative power supply. The circuit at the left in Figure 3-9 is a single-supply version of the single-transistor class-A amplifier shown earlier. The maximum efficiency is still only 1/12. But if we replace the pull-down resistor with a choke (inductor), as shown in the right-hand figure, we eliminate the power dissipation in the pull-down element. The choke allows the output to go negative as well as positive. No dc blocking capacitor is needed (if the choke has negligible dc resistance). There must be sufficient bias current through the choke to keep the transistor always on for the continuous control needed in linear operation. You can calculate that the maximum efficiency of this current is 50%; the choke improves the efficiency by a factor of six. In the most common arrangement of this circuit (the common-emitter configuration) the load is placed above the transistor, as shown in Figure 3-10. The version at the right shows that a blocking capacitor is needed if one end of the load must be grounded. (This is also the most common circuit topology for small signal amplifiers; when efficiency and large swings are unimportant, a resistor often takes the place of the choke.) As we have seen, current drive is called for to achieve linearity if the emitter is tied directly to ground. At the expense of some power loss, however, the emitter can be tied to ground through a resistor to allow the emitter voltage to follow the drive (base) voltage. Then the emitter current, collector current, and output voltage are also linearly

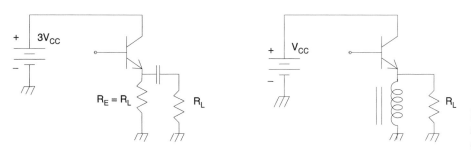

Figure 3-9. Single-ended audio amplifiers (common collector).

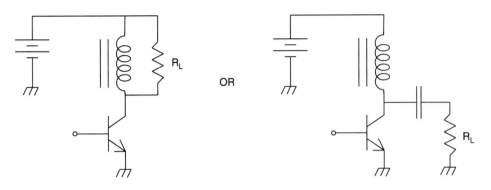

OR

Figure 3-10. Single-ended audio amplifiers (common emitter).

proportional to the input voltage. This technique of linearizing a common emitter amplifier is known as "emitter degeneration" or "series feedback."

Often the impedance of a given load is wrong for maximum power transfer, and the choke and blocking capacitor are replaced by a transformer, as shown in Figure 3-11.

All of these choke-coupled or transformer-coupled class-A amplifiers provide a peak-to-peak collector swing that is twice V_{cc}. (Note that the simplest model of a transformer, the "ideal transformer," would predict a maximum peak-to-peak swing of only V_{cc}, since R_L and the transformer would be replaced by only the transformed resistor.) If a transformer is needed for impedance transformation, a second identical active device can be added together with a center-tapped transformer to make the symmetric push–pull amplifier shown in Figure 3-12. The push–pull circuit shown in Figure 12, like the transformerless push–pull circuits, can be operated in class B for high efficiency. Tube-type high-fidelity amplifiers use this transformer circuit. Transistor high-fidelity amplifiers are usually transformerless and use the complementary push–pull arrangement. They can operate with a single power supply by using capacitor coupling to the loudspeaker. (Replace the bridge resistors in the circuit of Figure 3-8b with capacitors.)

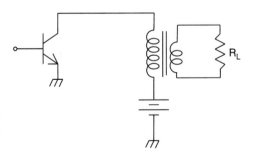

Figuer 3-11. Single-ended transformer-coupled audio amplifier.

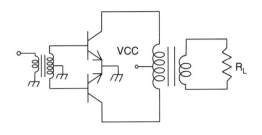

Figure 3-12. Transformer-coupled symmetric push–pull amplifier.

A NOTE ON EFFICIENCY   It might seem that the maximum efficiency of the class-B amplifier (78%) is only slightly better than the maximum efficiency of the class-A amplifier (50%). But these maximum efficiencies apply only when the amplifier is delivering a sine wave of maximum amplitude. For speech and music, the average power is much less than the maximum power. The class-B amplifier has little dissipation when the signal is low, but the class-A amplifier, with its continuous bias current, always draws power equal to twice the maximum output power. (A class-A amplifier rated for 25 W output would consume a continuous 50 W from its supply, while a class-B amplifier of equal power rating would consume, on average, only a few watts.)

RF AMPLIFIERS

All the ac considerations for audio amplifiers apply also to RF amplifiers. But, because RF amplifiers are narrow-band ac amplifiers, there are a couple of interesting tricks. Both small-signal RF amplifiers and RF power amplifiers often look like the class-A audio circuit with the choke in the collector, as shown in Figure 3-13. At high frequencies the parallel-output capacitance of the transistor (shown by dotted lines) could effectively short circuit the output side of the amplifier. But over a narrow frequency range this parasitic capacitance can be "cancelled out" with a parallel resonating inductance. (The dc supply is an ac short, so that the

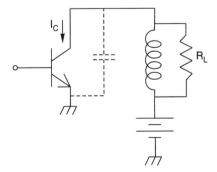

Figure 3-13. Single-ended RF amplifier.

inductor is effectively in parallel with the output capacitance.) Parasitic input capacitance can be similarly resonated away. While this RF amplifier has the same topology as the single-ended class-A audio amplifier, the resonant tuning circuit lets it be run as an efficient class-B power amplifier. (Remember that class-B audio amplifiers and class-B dc amplifiers must use the push–pull "double-ended" arrangement.) Here the flywheel effect* of the resonant circuit makes class-B operation possible with a single-ended circuit. To see how this works, consider the collector voltage and current waveforms shown in Figure 3-14 (for maximum signal conditions). On the second half of the cycle the collector voltage is decreasing. The transistor is pulling the voltage down (I_c is positive), drawing power from the supply. But on the first half of the cycle, the voltage is increasing. Negative current is needed, but there is no complementary circuit to provide the push. Instead, the flywheel effect of the resonant circuit carries the voltage through the positive half cycles. This is like pushing a swing: the pushes you provide can be always in the same direction; the pendulum effect carries the swing symmetrically through the other side of the cycle. (Of course you have to push twice as hard if you push only in one direction.) The flywheel effect must be strong enough to maintain the desired sinusoidal waveform over the positive half cycles when the transistor is off, that is, the loaded Q of the resonant circuit cannot be too low. Additional capacitance in parallel with the output capacitance of the transistor provides the energy storage needed to raise the Q as needed. (The inductor must then have a correspondingly lower value in order that the circuit always resonates at the operating frequency.)

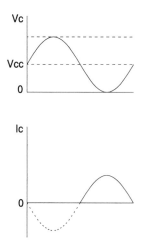

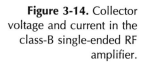

Figure 3-14. Collector voltage and current in the class-B single-ended RF amplifier.

*Energy in a parallel (or series) LC circuit oscillates back and forth cyclically between the inductor and the capacitor, with a sinusoidal current. This is analogous to the mechanical energy stored in the (cyclical) motion of a rotating flywheel.

A NOTE ON MATCHING A POWER AMPLIFIER TO ITS LOAD

Examination of the circuit of a power amplifier shows how large the peak output voltage can be – usually the supply voltage, V_{cc}. The maximum power, P_{max}, into a load, R, will therefore be $V_{cc}^2/(2R)$. If we need a given maximum power, P_{max}, and have a given load, R, then the value of V_{cc} should be $V_{cc} = \sqrt{2RP_{max}}$. If the amplifier must work down to dc, V_{cc} must be just this value too for the amplifier to have the desired maximum power. (Of course V_{cc} could be higher, but with the penalty of reduced efficiency – see Problem 1.) For ac amplification, however, we can transform the given load impedance to the desired value, $V_{cc}^2/(2P_{max})$, by means of a transformer (for audio or RF) or a tuned matching network (for RF only). Note that the transformer or matching network transforms the given load impedance to the value $V_{cc}^2/(2P_{max})$ and not to some impedance that characterizes the transistor or tube. For power amplifiers, the tube or transistor is chosen by its ability to withstand the applied voltages and currents (and to work properly at the maximum required frequency). The small-signal impedances of the device play no role in the design of the output matching circuit of a power amplifier.

PROBLEMS

1 (a) Analyze the class-B amplifier shown below to show that the efficiency is $\pi/4$ when the amplifier is driving the load resistor with a sine wave whose peak-to-peak amplitude is $2V_{cc}$ (a full-power sine wave).

Hint: Remember that in class-B operation the bottom transistor is off when the output voltage is positive and the top transistor is off when the output voltage is negative. Since the operation is symmetric, it is sufficient to consider just the half cycle, when the top transistor is on. For this half cycle, find the average power dissipated in the load (the average of IV_L) and the average power delivered by the supply (the average of IV_{cc}).

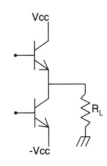

(b) What is the efficiency when the amplitude of the sine wave is only a fraction, α, of the maximum amplitude?

(c) What is the efficiency when the output is a square wave of maximum amplitude?: (*Answer:* 100%.)

2. The class-A amplifier shown below is operating at maximum power, applying a 24 V peak-to-peak sine wave to the load resistor. Assume the choke has zero dc resistance and enough inductance to block any ac current and that the capacitor has enough susceptance to prevent any ac voltage drop.

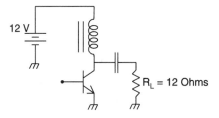

(a) Draw the waveform of the collector voltage. *Hint:* Remember that there can be no dc voltage drop across the choke.

(b) Draw the waveform of the collector current. *Hint:* Remember that there is no ac current through the choke.

(c) What power is drawn from the supply under the maximum signal sine wave condition?

(d) Show that the efficiency under this maximum signal sine wave condition is 50%.

(e) What power is drawn from the supply if the signal is zero?

3. An ideal push–pull amplifier does not have the current limitation of the emitter follower, so it can drive capacitive loads at high frequencies. But what about inductive loads? Suppose a load has an unavoidable series inductance but large voltages at high frequencies must be produced across the resistive part of the load. How does this impact the amplifier design?

4. Justify the statements made about the voltage gain (about 99%) and the offset (about 0.7 V) of the emitter follower. Use the relation between the emitter current and base-to-emitter voltage of a (bipolar) transistor, $I \approx I_s \exp[(V_b - V_e)/.026]$. To get a value for I_s, assume that $I = 10\,\text{mA}$ when $V_b - V_e = 0.7\,\text{V}$. Remember that in the emitter follower, $V_e = I_e R$.

Assume a reasonable value for R such as 1000 ohms and find V_e for several values of V_b.

5. Find the power gain of the emitter follower. Use the fact that the input current (base current) is less than the emitter current by a factor $1/(\beta + 1)$ where β is the current gain of the transistor (typically of the order of 100). Remember that the output voltage is essentially the same as (follows) the input voltage.

6. The emitter follower amplifier shown below has a load that includes an unavoidable parallel capacitance.

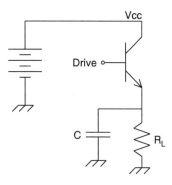

(a) What is the maximum peak-to-peak voltage that can be delivered to the load at low frequencies (where the capacitor can be neglected)?

(b) At what frequency will a sine wave output signal of half the maximum amplitude become distorted? *Hint*: Express the emitter current as the sum of the resistor current and capacitor current and note that distortion will occur if this current ever should be negative (the transistor can only supply positive current). *Answer*: $\omega = \sqrt{3}/(RC)$.

4 FILTERS I

Bandpass filters are key elements in radio circuits, where they are used, for example, to determine the pass band selectivity of radio receivers. Here we will discuss lumped element filters made of inductors and capacitors. We will first look at low-pass filters and then see how they serve as prototypes for conversion to bandpass filters. We begin with the well-established low-pass filter prototypes – Butterworth, Chebyshev, Bessel, and others. These low-pass prototypes are simple *LC* ladder networks with series inductors and shunt capacitors (Figure 4-1).

An *n*-section low-pass filter has *n* components (capacitors plus inductors). The end components can be either series inductors, as shown in Figure 4-1, or shunt capacitors, or one of each. Since they contain no (intentional) resistance, these filters are reflective filters; outside the pass band, it is mismatch that keeps power from reaching the load. The ladder network can be redrawn as a cascade of voltage dividers, as in Figure 4-2. At high frequencies the division ratio increases, so the load is increasingly isolated from the source. For frequencies well above the cutoff, each circuit element contributes 6 dB of (power) attenuation per octave (20 dB per decade). Within the pass band, an ideal low-pass filter provides a perfect match between the load and the source. Filters with many sections

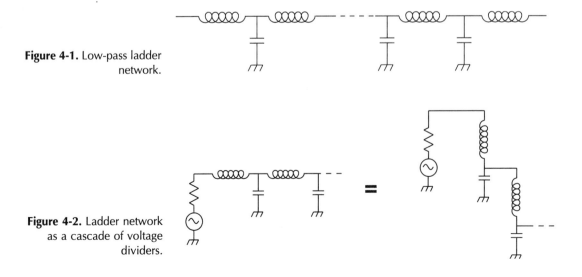

Figure 4-1. Low-pass ladder network.

Figure 4-2. Ladder network as a cascade of voltage dividers.

approach this ideal. When the source and load impedances have no reactance (built-in or parasitic) it is theoretically possible to have a perfect match across a wide band.

PROTOTYPE LOW-PASS FILTERS

The Butterworth filter is maximally flat, that is, it is designed so that at zero frequency the first $n - 1$ derivatives with respect to frequency of the power transfer function are zero. The final condition (needed to determine the values of n elements) is the specification of the cutoff frequency f_0, often specified as the 3 dB or half-power frequency. The frequency response of the Butterworth filter turns out to be

$$\left|\frac{V_{\text{out}}}{V_{\text{in}}}\right|^2 = \frac{1}{1 + (f/f_0)^{2n}}. \tag{4-1}$$

While it is the flattest filter, the Butterworth filter does not have skirts as sharp as those of the Chebyshev filter. The trade-off is that the Chebyshev filters have some passband ripple. The design criterion for the Chebyshev filter is that these ripples all have equal depth. The response is given by

$$\left|\frac{V_{\text{out}}}{V_{\text{in}}}\right|^2 = \frac{1}{1 + (V_{\text{r}}^{-2} - 1)\cosh^2[n\,\cosh^{-1}(f/f_0)]} \tag{4-2}$$

where V_{r} is the height of the ripple valley (in voltage) above the baseline.
    You will find tables of filter element values in many handbooks and textbooks. A set of tables from Matthaei et al. [1] is given in Appendix 4-1 at the end of this chapter. These tables are for normalized filters, that is, the cutoff frequency* is 1 radian/s ($1/2\pi$ Hz). The value of the nth component is g_n farads or henrys, depending on whether the filter begins with a capacitor or with an inductor. The proper source impedance is $1 + j0$ ohms. This is also the proper load impedance except for the even-order Chebyshev filters, where it is $1/g_{n+1} + j0$ ohms. Figure 4-3 shows plotted power responses of a Butterworth filter and several Chebyshev filters.

---

*The cutoff frequency for the Butterworth filters is the half-power (3 dB) point. For an n dB Chebyshev filter it is the highest frequency for which the response is down by n dB. (See Figure 4-3.)

Power

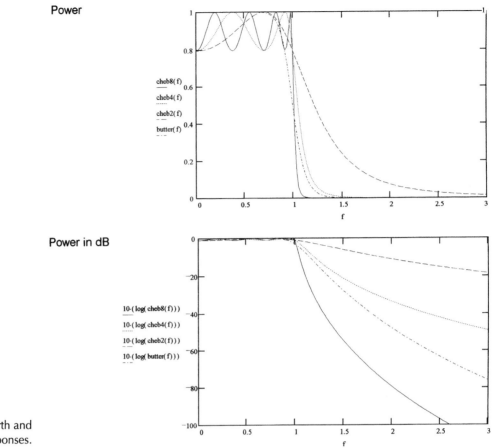

Power in dB

Figure 4-3. Butterworth and Chebyshev responses.

A LOW-PASS FILTER EXAMPLE

As an example, we will look at the three-section Butterworth low-pass filter. From Table A4-1, the filter has values of 1 H, 2 F, and 1 H (Figure 4-4a) or (1 F, 2 H and 1 F (Figure 4-4b). The (identical) responses for these two filters are given in Table 4-1 and plotted in Figure 4-5. Note that they work as advertised; the 3 dB point is at 0.159 Hz.

Suppose we need a three-section Butterworth that is 5 kHz wide and works between a 50 ohm generator and a 50 ohm load. We can easily find the element values by scaling the prototype. The values of the inductors are just multiplied by 50 (we need 50 times the reactance) and divided by $2\pi \times 5,000$ (we need to reach that reactance at 5 kHz, not 1 radian/s). Similarly, the capacitor values are divided by 50 and divided by $2\pi \times 5,000$. Figure 4-6 shows the circuit resulting from scaling the values of Figure 4-4b. The response of the scaled filter is shown in Table 4-2 and Figure 4-7.

TABLE 4-1

FREQUENCY RESPONSE FOR FILTERS OF FIGURE 4–4

Frequency (Hz)	Power	dB
0.00	1.000	−0.0
0.0321	0.000	−0.0
0.0640	0.996	−0.02
0.095	0.955	−0.20
0.1270	0.792	−1.01
0.1590	0.500	−3.01
0.1910	0.251	−6.00
0.2230	0.117	−9.31
0.2540	0.056	−12.5
0.2860	0.029	−15.4
0.3180	0.015	−18.1

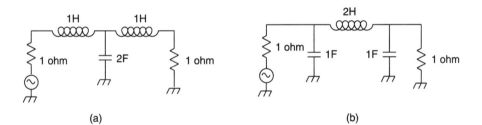

(a)   (b)

Figure 4-4. Equivalent three-section Butterworth low-pass filters.

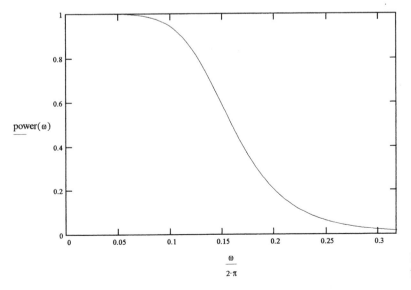

$\dfrac{\omega}{2\cdot\pi}$

Figure 4-5. Plotted response for the filters of Figure 4-4.

TABLE 4-2

RESPONSE OF THE SCALED LOW-PASS FILTER

Frequency (Hz)	Power	dB
0	1.000	−0.0
1,000	1.000	−0.0
2,000	0.996	−0.02
3,000	0.956	−0.20
4,000	0.793	−1.01
5,000	0.500	−3.01
6,000	0.251	−6.00
7,000	0.117	−9.31
8,000	0.056	−12.5
9,000	0.029	−15.4
10,000	0.015	−18.1

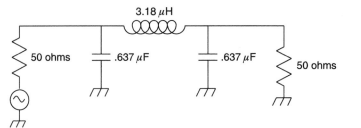

Figure 4-6. Filter of Figure 4-4b. Converted to 50 ohms and a 5 kHz cutoff frequency.

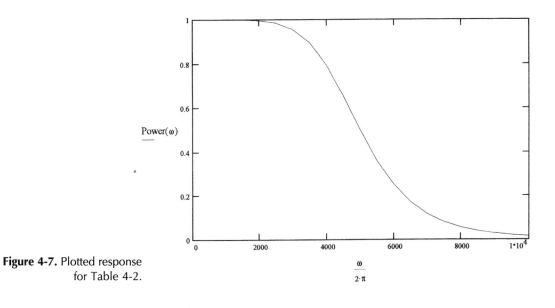

Figure 4-7. Plotted response for Table 4-2.

CONVERSION TO BANDPASS FILTERS

Here we will see how to convert low-pass filters into bandpass filters. Remember how the low-pass filters work: as frequency increases, the series arms (inductors), which are short circuits at dc, begin to pick up reactance. Likewise, the shunt arms (capacitors), which are open circuits at dc, begin to pick up susceptance. Both effects impede the signal transmission, as we have seen. To convert these low-pass filters in the most direct way to bandpass filters, we can replace the inductors by series LC combinations and the capacitors by parallel LC combinations. The series combinations are made to resonate (have zero impedance) at the center frequency of the desired bandpass filter, just as the inductors had zero impedance at dc, the "center frequency" of the prototype low-pass filter. It is important to note that as we move away from resonance, a series LC arm picks up reactance at twice the rate of the inductor alone. This is easy to see. The reactance of the series arm is given by

$$X_{\text{series}} = \omega \text{L} - 1/(\omega \text{C}). \tag{4-3}$$

At ω_0, $X = 0$, so

$$\omega_0 \text{L} = 1/(\omega_0 \text{C}) \tag{4-4}$$

and

$$d\text{X}/d\omega = \text{L} + 1/(\omega_0^2 \text{C}) = 2\text{L}. \tag{4-5}$$

As we move off resonance, the inductor *and* the capacitor provide equal contributions to the reactance. Likewise, the parallel LC circuits, which replace the capacitors in the prototype low-pass filter, pick up susceptance at twice the rate of their capacitors. With this in mind let us convert our 5 kHz low-pass filter into a bandpass filter. Suppose we want the center frequency to be 500 kHz and the bandwidth to be 10 kHz. As we move up from the center frequency, the series arms must pick up reactance at the same rate that the inductors picked up reactance in the prototype low-pass filter. Similarly, the shunt arms must pick up susceptance at the same rate that the capacitors picked up susceptance in the prototype. This will cause the bandpass filter to have the same shape above the center frequency as the prototype had above dc. If the 3dB point of the prototype filter was 5 kHz, the upper 3 dB point of the bandpass filter will be at 5 kHz above the center frequency. The bandpass filter, however, will have a mirror image response as we go below the center frequency. (Below the center frequency the reactances and susceptances change sign but the response remains the same.)

Let us calculate the component values. As we leave the center frequency, the series circuits will get equal amounts of reactance from the L and the C as explained above. Therefore the series inductor values should be exactly half what they were in the low-pass prototype. Note: no matter how high we make the center frequency, the values of the inductors are reduced only by a factor of two from those of the scaled low-pass filter. The series capacitors are chosen to resonate at the center frequency with the new (half-value) series inductors. The values of the parallel arms are determined similarly; the parallel capacitors must have half the value they had in the prototype low-pass filter. Finally, the parallel inductors are chosen to resonate with the new (half-value) parallel capacitors. These simple conversions yield the bandpass filter shown in Figure 4-8. The response of this bandpass filter is given in Table 4-3 and Figure 4-9.

While this theoretical filter works perfectly (since its components are lossless), the component values are impractical; typical real components with these values would be too lossy to achieve the calculated filter

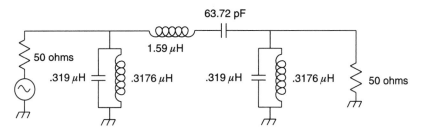

Figure 4-8. Bandpass filter.

TABLE 4-3

RESPONSE OF THE SCALED LOW-PASS FILTER

Frequency (kHz)	Power	dB
490	0.014	−18.1
492	0.053	−12.8
494	0.241	−6.19
496	0.785	−1.05
498	0.996	−0.018
500	1.000	−0.00
502	0.996	−0.018
504	0.801	−0.966
506	0.260	−5.84
508	0.059	−12.9
510	0.016	−17.9

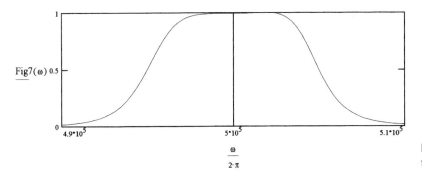

Figure 4-9. Plotted response for Table 4-3.

shape. When a bandpass filter is to have a large fractional bandwidth (bandwidth divided by center frequency) this direct conversion from low pass to bandpass can be altogether satisfactory. It is when the fractional bandwidth is small, as in this example, that the direct conversion gets into trouble.* We will see later that the problem is solved by transforming the prototype low-pass filters into somewhat more complicated bandpass circuits known as *coupled resonator filters*. Those filters retain the desired shape (Butterworth, Chebyshev, etc.) and can serve, in turn, as prototypes for filters made from quartz or ceramic resonators and for filters made with resonant irises (thin plates that partially block a waveguide).

BIBLIOGRAPHY

1. G. L. Matthaei, L. Young, and E. M. T. Jones (1964) , *Microwave Filters, Impedance-Matching Networks and Coupling Structures*. New York: McGraw Hill (reprinted by Artech House, Boston). (Contains fully developed designs comparing measured results with theory (spectacular fits even at microwave frequencies) and has an excellent introduction and review of the theory.)
2. D. G. Fink (1975), *Electronic Engineers' Handbook*. New York: McGraw-Hill, Section 12. (This section,"Filters, coupling networks, and attenuators," by M. Dishal contains an extensive list references.)

*The component problem with the straightforward low-pass-to-bandpass conversion is that the values of the series inductors are very different than the values of the parallel inductors. (The same is true of the capacitors, but high-Q capacitors can usually be found.) In the above example, the inductors differ by a factor of about 5,000, and it is normally impossible to find high-Q components over this range. (Low-Q inductors, of course, make the filter lossy and, if not accounted for, distort the bandpass shape.) The inductors in coupled resonator filters are all of about the same value. If a high-Q inductor can be found, the coupled resonator filter is designed for whatever impedance calls for that value of inductor and then transformers or matching sections are used at each end to convert to the desired impedance.

APPENDIX 4.1.

Component Values for Normalized Low-pass Filters (From Matthaei et al. [1])

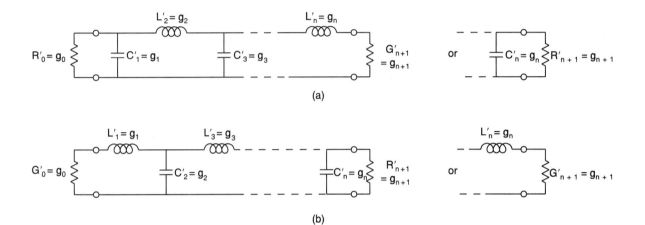

(a)

(b)

<table>
<tr><td colspan="2">**TABLE A4-1**</td></tr>
</table>

ELEMENT VALUES FOR BUTTERWORTH (MAXIMALLY FLAT) LOW-PASS FILTERS (the 3 dB point is at $\omega = 1$ radian/second)

Value of n	g_1	g_2	g_3	g_4	g_5	g_6	g_7	g_8	g_9	g_{10}	g_{11}
1	2.000	1.000									
2	1.414	1.414	1.000								
3	1.000	2.000	1.000	1.000							
4	0.7654	1.848	1.848	0.7654	1.000						
5	0.6180	1.618	2.000	1.618	0.6180	1.000					
6	0.5176	1.414	1.932	1.932	1.414	0.5176	1.000				
7	0.4450	1.247	1.802	2.000	1.802	1.247	0.4450	1.000			
8	0.3902	1.111	1.663	1.962	1.962	1.663	1.111	0.3902	1.000		
9	0.3473	1.000	1.532	1.879	2.000	1.879	1.532	1.000	0.3473	1.000	
10	0.3129	0.9080	1.414	1.782	1.975	1.975	1.782	1.414	0.9080	0.3129	1.000

TABLE A4-2

ELEMENT VALUES FOR CHEBYSHEV LOW-PASS FILTERS (For a filter with N-dB ripple, the last N-dB point is at $\omega = 1$ radian/second)

Value of n	g_1	g_2	g_3	g_4	g_5	g_6	g_7	g_8	g_9	g_{10}	g_{11}
0.01 dB ripple
1	0.0960	1.0000
2	0.4488	0.4077	1.1007
3	0.6291	0.9702	0.6291	1.0000
4	0.7128	1.2003	1.3212	0.6476	1.1007
5	0.7563	1.3049	1.5773	1.3049	0.7563	1.0000
6	0.7813	1.3600	1.6896	1.5350	1.4970	0.7098	1.1007
7	0.7969	1.3924	1.7481	1.6331	1.7481	1.3924	0.7969	1.0000
8	0.8072	1.4130	1.7824	1.6833	1.8529	1.6193	1.5554	0.7333	1.1007
9	0.8144	1.4270	1.8043	1.7125	1.9057	1.7125	1.8043	1.4270	0.8144	1.0000
10	0.8196	1.4369	1.8192	1.7311	1.9362	1.7590	1.9055	1.6527	1.5817	0.7446	1.1007
0.1 dB ripple
1	0.3052	1.0000
2	0.8430	0.6220	1.3554
3	1.0315	1.1474	1.0315	1.0000
4	1.1088	1.3061	1.7703	0.8180	1.3554
5	1.1468	1.3712	1.9750	1.3712	1.1468	1.0000
6	1.1681	1.4039	2.0562	1.5170	1.9029	0.8618	1.3554
7	1.1811	1.4228	2.0966	1.5733	2.0966	1.4228	1.1811	1.0000
8	1.1897	1.4346	2.1199	1.6010	2.1699	1.5640	1.9444	0.8778	1.3554
9	1.1956	1.4425	2.1345	1.6167	2.2053	1.6167	2.1345	1.4425	1.1956	1.0000
10	1.1999	1.4481	2.1444	1.6265	2.2253	1.6418	2.2046	1.5821	1.9628	0.8853	1.3554
0.2 dB ripple
1	0.4342	1.0000
2	1.0378	0.6745	1.5386
3	1.2275	1.1525	1.2275	1.0000
4	1.3028	1.2844	1.9761	0.8468	1.5386
5	1.3394	1.3370	2.1660	1.3370	1.3394	1.0000
6	1.3598	1.3632	2.2394	1.4555	2.0974	0.8838	1.5386
7	1.3722	1.3781	2.2756	1.5001	2.2756	1.3781	1.3722	1.0000
8	1.3804	1.3875	2.2963	1.5217	2.3413	1.4925	2.1349	0.8972	1.5386
9	1.3860	1.3938	2.3093	1.5340	2.3728	1.5340	2.3093	1.3938	1.3860	1.0000
10	1.3901	1.3983	2.3181	1.5417	2.3904	1.5536	2.3720	1.5066	2.1514	0.9034	1.5386

Value of n	g_1	g_2	g_3	g_4	g_5	g_6	g_7	g_8	g_9	g_{10}	g_{11}
0.5 dB ripple											
1	0.6986	1.0000									
2	1.4029	0.7071	1.9841								
3	1.5963	1.0967	1.5963	1.0000							
4	1.6703	1.1926	2.3661	0.8419	1.9841						
5	1.7058	1.2296	2.5408	1.2296	1.7058	1.0000					
6	1.7254	1.2479	2.6064	1.3137	2.4758	0.8696	1.9841				
7	1.7372	1.2583	2.6381	1.3444	2.6381	1.2583	1.7372	1.0000			
8	1.7451	1.2647	2.6564	1.3590	2.6964	1.3389	2.5093	0.8796	1.9841		
9	1.7504	1.2690	2.6678	1.3673	2.7239	1.3673	2.6678	1.2690	1.7504	1.0000	
10	1.7543	1.2721	2.6754	1.3725	2.7392	1.3806	2.7231	1.3485	2.5239	0.8842	1.9841
1.0 dB ripple											
1	1.0177	1.0000									
2	1.8219	0.6850	2.6599								
3	2.0236	0.9941	2.0236	1.0000							
4	2.0991	1.0644	2.8311	0.7892	2.6599						
5	2.1349	1.0911	3.0009	1.0911	2.1349	1.0000					
6	2.1546	1.1041	3.0634	1.1518	2.9367	0.8101	2.6599				
7	2.1664	1.1116	3.0934	1.1736	3.0934	1.1116	2.1664	1.0000			
8	2.1744	1.1161	3.1107	1.1839	3.1488	1.1696	2.9685	0.8175	2.6599		
9	2.1797	1.1192	3.1215	1.1897	3.1747	1.1897	3.1215	1.1192	2.1797	1.0000	
10	2.1836	1.1213	3.1286	1.1933	3.1890	1.1990	3.1738	1.1763	2.9824	0.8210	2.6599
2.0 dB ripple											
1	1.5296	1.0000									
2	2.4881	0.6075	4.0957								
3	2.7107	0.8327	2.7107	1.0000							
4	2.7925	0.8806	3.6063	0.6819	4.0957						
5	2.8310	0.8985	3.7827	0.8985	2.8310	1.0000					
6	2.8521	0.9071	3.8467	0.9393	3.7151	0.6964	4.0957				
7	2.8655	0.9119	3.8780	0.9535	3.8780	0.9119	2.8655	1.0000			
8	2.8733	0.9151	3.8948	0.9605	3.9335	0.9510	3.7477	0.7016	4.0957		
9	2.8790	0.9171	3.9056	0.9643	3.9598	0.9643	3.9056	0.9171	2.8790	1.0000	
10	2.8831	0.9186	3.9128	0.9667	3.9743	0.9704	3.9589	0.9554	3.7619	0.7040	4.0957

Value of n	g_1	g_2	g_3	g_4	g_5	g_6	g_7	g_8	g_9	g_{10}	g_{11}
3.0 dB ripple											
1	1.9953	1.0000									
2	3.1013	0.5339	5.8095								
3	3.3487	0.7117	3.3487	1.0000							
4	3.4389	0.7483	4.3471	0.5920	5.8095						
5	3.4817	0.7618	4.5381	0.7618	3.4817	1.0000					
6	3.5045	0.7685	4.6061	0.7929	4.4641	0.6033	5.8095				
7	3.5182	0.7723	4.6386	0.8039	4.6386	0.7723	3.5182	1.0000			
8	3.5277	0.7745	4.6575	0.8089	4.6990	0.8018	4.4990	0.6073	5.8095		
9	3.5340	0.7760	4.6692	0.8118	4.7272	0.8118	4.6692	0.7760	3.5340	1.0000	
10	3.5384	0.7771	4.6768	0.8136	4.7425	0.8164	4.7260	0.8051	4.5142	0.6091	5.8095

PROBLEMS

1. Design a five-element low-pass filter with a Chebyshev 0.5 dB ripple shape. Let the input and output impedances be 100 ohms. Use parallel capacitors at the ends. The bandwidth (from dc to the last 0.5 dB point) is to be 100 kHz. (This is the bandwidth convention for the Chebyshev tables included in Appendix 1.) Use the tables to find the values of the prototype 1 ohm, 1 radian/s filter and then alter these values for 100 ohms and 100 kHz.

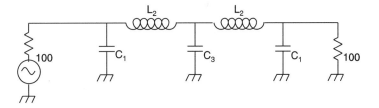

2. Use the results of Problem 1 to design a five-element bandpass filter with a Chebyshev 0.5 dB ripple shape. Let the input and output impedances remain at 100 ohms. The center frequency is to be 5 MHz and the total bandwidth (between outside 0.5 dB points) is to be 200 kHz.

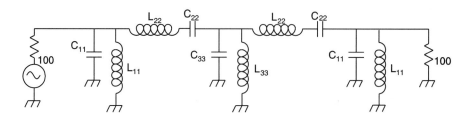

3. Convert the filter of Problem 2 to operate at 50 ohms by adding an L-section matching network at each end.

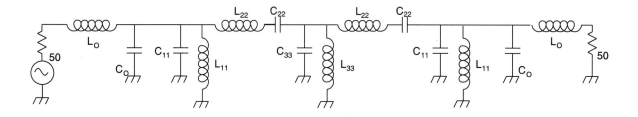

Test the filter of Problem 3 using your ladder network analysis program (see Chapter 1, Problem 3), sweeping from 4.5 to 5.5 MHz in steps of 20 kHz.

4. The one-section bandpass filter shown below uses a single parallel resonator. In its prototype low-pass filter, the resonator is a single shunt capacitor.

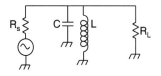

Show that the frequency response of this filter is given by

$$\frac{P}{P_{\max}} = \frac{1}{1 + Q^2[(f/f_0) - (f_0/f)]^2}$$

where f_0 is the resonant frequency of the *LC* combination and Q is defined as $R/(\omega_0 L)$, where R is the parallel combination of R_s and R_L

5. High-pass filters are derived from low-pass filters by changing inductors to capacitors and vice versa and replacing the component values in the prototype low-pass tables by their reciprocals. (A 2 F capacitor, for example, would become a 0.5 H inductor.) The prototype high-pass response at ω will be equal to the prototype low-pass response at $1/\omega$. Convert the low-pass filter of Figure 4-4b into a high-pass filter. (*Answer:* 1 H, 0.5 F, and 1 H.) Next, scale it to have a cutoff frequency of 5 kHz and to operate at 50 ohms. Finally, convert the scaled filter into a band-stop filter with a stop band 10 kHz wide, centered at 500 kHz.

6. Enhance your ladder network analysis program (see Chapter 1, Problem 3) to calculate not just the amplitude response of a network but also the phase response (phase angle of the output voltage minus phase angle of the input voltage). Calculate the phase response of the Butterworth filter in Figure 4-4a. *Note:* Ladder networks belong to a class of networks ("minimum phase networks") for which the amplitude response uniquely determines the phase response and vice versa. Later we will encounter "allpass" filters which are not in this class; phase varies with frequency while amplitude remains constant.

5 FREQUENCY CONVERTERS

A common operation in RF electronics is frequency translation, meaning that all the signals in a given frequency band are shifted to a higher frequency band or to a lower frequency band. Every spectral component is shifted by the same amount. Cable television boxes, for example, shift the selected cable channel to a low VHF channel (normally channel 3 or 4). Nearly every radio receiver and television receiver uses the *superheterodyne* principle, where the desired channel is first shifted to a standard intermediate frequency band or "IF." Most of the amplification is done in the fixed IF band, with the advantage that nothing in this major portion of the receiver needs to be adjusted when a different station or channel is selected. The same principle can be used in frequency agile transmitters; it is often easier to shift an already modulated signal than to generate it from scratch at any required frequency. Frequency translation is also called conversion, and even more often called "mixing." Several common mixer circuits are described below.

THE IDEAL MULTIPLIER AS A MIXER

A mixer combines the input signal, which is to be shifted, with a reference signal whose frequency is equal to the desired shift. In a radio receiver, the reference signal is known as the local oscillator or "LO" signal, because it is supplied by an oscillator that is part of the receiver. Mixers, in order to produce new frequencies, must necessarily be nonlinear because linear circuits can change only the amplitudes and phases of a set of superposed sine waves. In audio work, "mixing" means addition (a linear superposition that produces no new frequencies). In RF work, however, mixing means multiplication; RF mixers form *products* of the input signals and the LO signal. This nonlinear combination (multiplication) produces new signals at the shifted frequencies. Let us consider first an ideal multiplier used as a mixer (Figure 5-1). The output voltage from the multiplier is the product of the two input voltages. In Figure 5-2 a sine wave input signal is multiplied by an LO that is 1.455 times higher in frequency. These multiplicands are shown in the top graph. The bottom graph shows their product, which can be seen to contain frequencies both higher and lower than

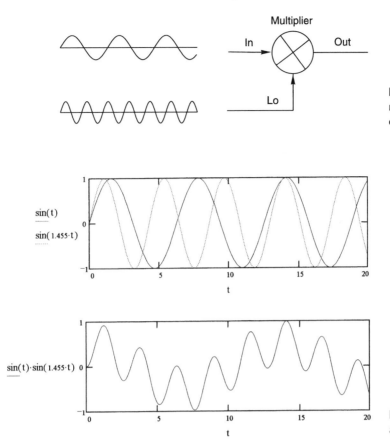

Figure 5-1. Voltage multiplier as a frequency converter (mixer).

Figure 5-2. Multiplier input and output waveforms.

the original frequencies. A standard trigonometric identity shows that the output consists of just two frequencies: an up-shifted signal at $\omega_L + \omega_R$ and a down-shifted signal at $\omega_L - \omega_R$:

$$\sin(\omega_R t)\sin(\omega_L t) = 1/2[\cos(\omega_R - \omega_L)t - \cos(\omega_R + \omega_L)t]. \qquad (5\text{-}1)$$

If we replace the single RF signal by $V_1 \sin(\omega_1 t) + V_2 \sin(\omega_2 t)$, a signal with two spectral components, the output will be

$$\begin{aligned} &V_1 \cos[(\omega_1 - \omega_L t)] - V_1 \cos[(\omega_1 + \omega_)] + V_2 \cos[(\omega_2 -_L t)] \\ &- V_2 \cos[(\omega_2 + \omega_L t)]. \end{aligned} \qquad (5\text{-}2)$$

Just as this linear combination of two signals is faithfully copied into both an up-shifted band and a down-shifted band, any linear combination, that is, any spectral distribution, of signals will be faithfully copied into these shifted output bands. In this sense a mixer is a linear device. Generally, only one of the bands is desired; an appropriate bandpass filter will reject the other. Ideal analog multipliers have not been common in RF electronics, but at least one, the Gilbert cell transconductance multiplier, is now available in fast versions intended as mixers.

SWITCHING MIXERS

If the LO is a square wave rather than a sine wave, the mixer output will contain not only the fundamental up-shifted and down-shifted outputs but also components at offsets corresponding to the third, fifth and all other odd harmonics of the LO frequency, that is, at offsets found in the Fourier decomposition of the square wave. These new components are usually very easy to filter out, so there is no disadvantage in using a square wave LO. In fact, there is an advantage; since the multiplier multiplies the input signal only by either $+1$ or -1 it can be replaced by an SPDT switch that connects the output alternately to the input signal and to the negative of the input signal. This equivalence is shown in Figure 5-3. The phase inversion needed for the bottom side of the switch can easily be done with a center-tapped transformer, and the switching can be done with two transistors, one for the high side and one for the low side. In the circuit of Figure 5-4 the switches are FETs (field effect transistors). (Mixers based on transistors are called active mixers.) A second transformer provides the LO phase inversion so that one FET is on while the other is off.

We could just as well have taken the signal from the center tap and used the FETs to ground one end of the secondary and then the other. With this arrangement, shown in Figure 5-5, it is easier to provide the drive signals to the transistors since they are not floating.

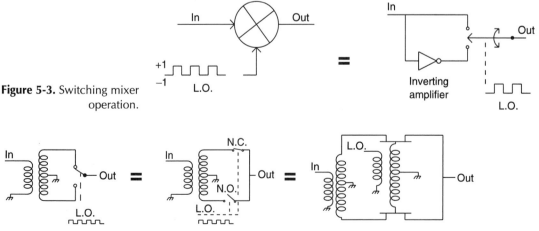

Figure 5-3. Switching mixer operation.

Figure 5-4. Active switching mixer.

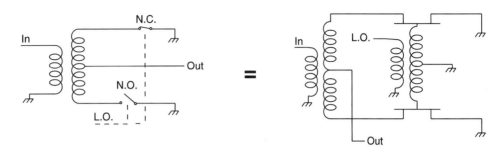

Figure 5-5. Alternate active switching mixer.

Diodes are more commonly used as the switching elements in the circuit described above. This passive switching mixer circuit is shown below in Figure 5-6. Voltage from the LO transformer alternately drives the top diode pair and the bottom pair into conduction. The LO signal is made large enough that the conducting diodes have very low impedance (small depletion region) and the nonconducting diodes have a very large impedance (wide depletion region). The end of the input transformer connected between the on diodes is effectively connected to ground through the secondary of the LO transformer. Note that current uses both sides of the LO transformer on the way to ground, so no net flux is created in that transformer and it has zero impedance for this current. This circuit is usually drawn in the form shown in Figure 5-7, and is referred to as a double balanced diode ring mixer.

All the switching mixers shown above are "double balanced" with respect to their port-to-port isolation. No LO energy appears at the RF or IF ports, and no RF, except the mixing products, appears at the IF port. A balanced mixer is desirable, for example, when it is the first element in receiver. An unbalanced mixer would allow LO energy to reach the antenna, and the radiation could cause interference to other receivers

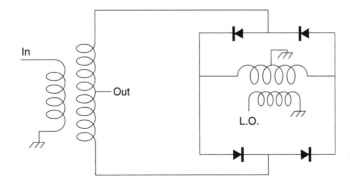

Figure 5-6. Mixer using diodes as switches.

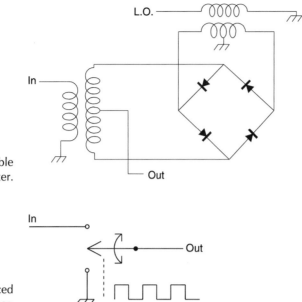

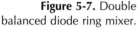

Figure 5-7. Double balanced diode ring mixer.

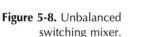

Figure 5-8. Unbalanced switching mixer.

(and could also reveal the position of the receiver). Simpler mixers, discussed below, do not have this balance. An unbalanced switching mixer is shown in Figure 5-8. It multiplies the signal by a square wave that goes from +1 to 0 (rather than +1 to −1), which is just a +1/2 to −1/2 square wave together with a bias of 1/2. The square wave term produces the up-shifted and down-shifted bands as before, but the bias term allows the RF input to get through, unshifted, to the output.

GENERAL NONLINEAR DEVICE MIXER

Finally we come to the simplest mixers which use a single nonlinear device. A diode is, of course, nonlinear, and many mixers use just a single diode. The RF and LO voltages are simply added, and the sum is applied to the diode. The current, a nonlinear function of the applied voltage, will contain mixing products at frequencies $Nf_{\text{RF}} \pm Mf_{\text{LO}}$, where N and M are simple integers. A transistor, if driven by the sum of the RF and LO voltages, will likewise have a current containing these mixing products, especially if the LO signal causes large nonlinear excursions. Sometimes a dual-gate FET is used as a mixer; the LO voltage is applied to one gate and the RF voltage is applied to the other. This provides some isolation between the LO and RF (which is provided automatically in a balanced mixer such as the diode ring mixer).

DIODE MIXER

Figure 5-9 is a hypothetical circuit for a single-diode mixer. The first operational amplifier (op amp) is used to sum the RF and LO voltages. That sum is applied to the diode. The input of the second op amp is a virtual ground, so the full sum voltage is applied across the diode. The second op amp is used as a current-to-voltage converter; it produces a voltage proportional to the current in the diode. This op amp circuit is intended to emphasize that the *sum* of the RF and LO signals is applied to the diode. Commonly used circuits use passive components, and the summing is not always obvious. Diodes are actually exponential devices; the current versus applied voltage is given by

$$I = I_s[\exp(V/V_{th}) - 1] \tag{5-3}$$

where $V_{th} = V_{thermal} = kT/e = 26\,\text{mV}$. In a small-signal situation, that is, when $V \ll 26\,\text{mV}$, we can expand the exponential to get the output of the above mixer:

$$V_{out} = I_s R[V/V_{th} + (V/V_{th})^2/2! + (V/V_{th})^3/3! + \ldots]. \tag{5-4}$$

Since $V = V_{RF} + V_{LO}$, the first term will give feed-through (no balance) at the RF and LO frequencies. The second term (the square law term) will produce the desired up-shifted and down-shifted sidebands because the square of $V_{LO} + V_{RF}$ contains the cross-*product*, $2V_{LO}V_{RF}$. This term also produces bias terms and double-frequency components. The third-order term will give outputs at the third harmonics of the RF and LO frequencies and at $2\omega_{RF} + \omega_{LO}$, $2\omega_{RF} - \omega_{LO}$, $2\omega_{LO} + \omega_{RF}$, and $2\omega_{LO} - \omega_{RF}$. Normally these products are far removed from the desired output band and can be filtered out. If the input voltage is small enough we do not have to continue the expansion. For larger signals, however, the next term (fourth order) gives undesirable products within the desired output band. To see how this happens, consider an input signal with two components, $A_1 \cos(\omega_1 t)$ and $A_2 \cos(\omega_2 t)$. One of the fourth-power output terms will be, except for a constant,

$$\cos(\omega_L t)[A_1 \cos(\omega_1 t) + A_2 \cos(\omega_2 t)]^3. \tag{5-5}$$

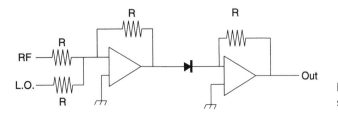

Figure 5-9. Hypothetical single-diode mixer circuit.

This expression contains a term

$$\cos(\omega_L t)3A_1 \cos^2(\omega_1 t)A_2 \cos(\omega_2 t)$$
$$+ 3/2 A_1 A_2 \cos(\omega_L t)[1 - \cos(2\omega_1 t)\cos(\omega_2 t)] \tag{5-6}$$

which in turn, contains

$$\cos(\omega_L t)[\cos[(2\omega_1 + \omega_2)t] + \cos[(2\omega_1 - \omega_2)t] =$$
$$1/2\{\cos[(\omega_L + 2\omega_1 + \omega_2)t] + \cos[(\omega_L - 2\omega_2)t]\}. \tag{5-7}$$

When ω_1 and ω_2 are close to each other, $2\omega_1 + \omega_2$ and $2\omega_1 - \omega_2$ are nearby and can lie within the desired output band. In a radio receiver, this means that two very strong signals will create a mixing product at a nearby, that is, in-band, frequency that will interfere with reception of any weak signal at that frequency.

We will see later that multiplication, the basis of mixing, is also the operation needed to modulate the amplitude of a carrier, that is, to produce AM modulation. Multiplication, mixing and (AM) modulation are all the same basic operation.

PROBLEMS

1. Sometimes two multipliers, two phase shifters, and an adder are used to build a mixer that has only one output band (a so-called single-sideband mixer). The design for an *upper*-sideband mixer, for example, follows directly from the identity

$$\cos(\omega_{RF} t)\cos(\omega_{LO} t) - \sin(\omega_{RF} t)\sin(\omega_{LO} t) = \cos[(\omega_{RF} + \omega_{LO})t].$$

Draw a block diagram for this upper sideband mixer.

2. The diode ring switching mixer also works when the LO and RF ports are interchanged. Explain the operation in this case.

3. The conversion gain of a mixer is defined as the ratio of the output power (the IF signal) to the input power (the RF signal). Calculate the conversion gain (actually a loss, since it is less than unity) of an ideal switching mixer. *Hint*: The square wave LO signal contains a sine wave whose frequency is equal to the desired shift, but it also contains sine waves at all odd multiples of this frequency. Only the fundamental component causes power to be shifted to the desired IF frequency.

4. Consider a situation where two signals of the same frequency but with a phase difference, θ, are separately mixed to a new frequency. Suppose identical mixers are used and that they are driven with the same LO signal. Show that the phase difference of the shifted signals is still θ.

5. In RF engineering, considerable use is made of the trigonometry identities $\cos(a + b) = \cos a \cos b - \sin a \sin b$ and $\sin(a + b) = \sin a \cos b + \cos a \sin b$. Prove these identities, either using geometric constructions or using the identity $e^{jx} = \cos x + j \sin x$.

6 RADIO RECEIVERS

THE BASIC REQUIREMENTS

Three basic specifications for any kind of radio receiver are *amplification*, *sensitivity*, and *selectivity*, that is, does a weak signal at the antenna terminals from a desired station produce a sufficiently strong and uncorrupted output (audio, video, or data) and does this output remain satisfactory in the presence of strong signals at nearby frequencies? *Sensitivity* is often specified as the minimum detectable input signal, for example 1 μV across the 50 ohm input jack of the receiver, or as the noise power added by the receiver, referred to the input. *Selectivity* is normally determined by one main bandpass-determining filter, and might be specified as "3 dB down at 2 kHz from center frequency and 20 dB down at 10 kHz from the center frequency." (Receiver manufacturers usually do not specify the exact shape, such as "2 dB ripple Chebyshev filter of eight sections with 3 dB points separated by 4 kHz.")

AMPLIFICATION

Let us consider how much amplification an ordinary radio receiver must have. One milliwatt of audio power into a typical earphone produces a sound level some 100 dB above the threshold of hearing. A barely discernible audio signal can therefore be produced by -100 dBm (100 dB below 1 mW or 10^{-13} W). Let us specify that a receiver, for comfortable earphone listening, must provide 50 dB more than this minimum level or 10^{-8} W. You can see that, with efficient circuitry, the batteries in a portable receiver could last a very long time! (Sound power levels are surprisingly small; you radiate only about 1 mW of acoustic power when shouting and about 1 nW when whispering.) How much signal power arrives at a receiver? A simple wire antenna could, in fact, intercept 10^{-8} W of RF power at a distance of about 20,000 km from a 10 kW radio station at 1 MHz having an omni-directional transmitting antenna, so let us first consider "self-powered" receivers.

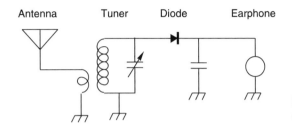

Figure 6-1. Self-powered crystal set receiver.

CRYSTAL SETS

The earliest radios, crystal sets, were self-powered. A crystal diode rectifier recovered the modulation envelope, converting enough of the incoming RF power into audio power to drive the earphone. A simple *LC* tuned circuit served as a bandpass filter to select the desired station, and could also serve as an antenna matching network. The basic crystal set receiver is shown in Figure 6-1.

The considerations given above show that the self-powered receiver can have considerable range. But when the long-wire antenna is replaced by a compact but very inefficient loop antenna and the earphone is replaced by a loudspeaker, considerable amplification is needed. In addition, we will see later that the diode detector, when operated at low signal levels, has a square law characteristic. For proper envelope detection of an AM signal, the signal applied to a diode detector must have a high level, several milliwatts. The invention of the vacuum tube provided the needed amplification. Receivers normally contain both RF and audio amplifiers. RF amplification provides enough power for proper detector operation, while subsequent audio amplification provides the power to operate loudspeakers.

TRF RECEIVERS

The first vacuum tube radios were simply crystal sets with RF preamplification and audio postamplification, as described above. These TRF (tuned RF) sets* had individual tuning adjustments for each of several

---

*Early radios were called "radio sets" because they were literally a set of parts including one or more tubes and batteries (or maybe just a crystal detector), inductors, "condensers", resistors, and headphones. Many of these parts were individually mounted on wooden bases, and, together with "hook-up" wires, would spread out over a table top.

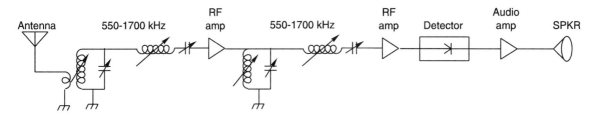

Figure 6-2. TRF receiver.

cascaded RF amplifier stages. Changing stations required the user to adjust every dial (often with the aid of a tuning chart or graph).

Figure 6-2 shows a hypothetical TRF receiver with cascaded amplifiers and bandpass filters. Note that all the inductors and capacitors should be variable in order to tune the center frequency of the bandpass filters and also maintain the proper bandwidth, about 10 kHz for AM. (In a practical circuit, the bandpass filters would use a coupled resonator design rather than the straightforward low-pass-to-bandpass conversion design shown here.)

THE SUPERHETERODYNE RECEIVER

The disadvantages of the TRF set were the cost and inconvenience of having many tuning adjustments. Most of these adjustments were eliminated with the invention of the superheterodyne circuit by Armstrong in 1917. Armstrong's circuit consists of a fixed-tuned, that is, single-frequency, TRF receiver preceded by a frequency converter (mixer and local oscillator) so that the signal from any desired station can be shifted to the frequency of the TRF receiver. This frequency is known as the intermediate frequency or *IF*. The superheterodyne is still the circuit used in nearly every radio, television, and radar receiver. Among the few exceptions are some toy walkie-talkies, microwave receivers used in radar-operated door openers, and some highway speed trap radar detectors. Figure 6-3 shows the classic broadcast band "superhet." Selectivity is provided by fixed-tuned bandpass filters in the IF amplifier section. The detector here is still a diode, that is, the basic crystal set. (Later we will analyze this detector carefully and consider alternative detector circuits.) All the RF gain can be contained entirely in the fixed-tuned IF amplifier. We will see later that there are sometimes reasons for having some amplification ahead of the mixer as well. Figure 6-3 also serves as the block diagram for FM broadcast receivers (except that the IF frequency is usually 10.7 MHz, and an FM detector is used in place of the envelope

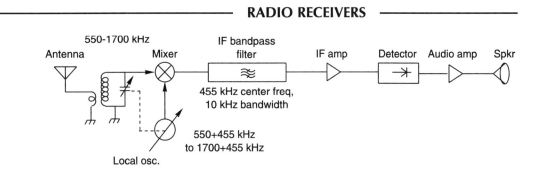

Figure 6-3. Standard superheterodyne receiver for the AM broadcast band.

detector) and for television receivers (except that the IF center frequency is about 45 MHz, and this basic radio block is followed by video- and sound-processing circuits).

NOTE There was indeed a *heterodyne* receiver that preceded the super-heterodyne. Invented by the radio pioneer Reginald Fessenden, the hetero-dyne receiver converted the incoming RF signal directly to audio. This design is still occasionally built under the name "direct conversion receiver." The combination of a front-end IF converter followed by a heterodyne back-end receiver (rather than a TRF back-end receiver) is referred to as a superheterodyne with a product detector. (See Chapter 27.)

IMAGE REJECTION

The superheterodyne receiver has some disadvantages of its own. With respect to signals at the input to the mixer, the receiver will simultaneously detect signals at the desired frequency and also any signals present at an undesired frequency known as the *image frequency*. To see this, consider a practical example. Suppose we have a conventional AM receiver with an IF frequency of 455 kHz and suppose the local oscillator is set at 1,015 kHz in order to receive a station broadcasting at 1,015−455 = 560 kHz. (The frequency of 560 kHz is near the lower edge of the AM broadcast band.) All the mixers we have considered will also produce a 455 kHz IF signal from any input signal present at 1,470 kHz, that is, 455 kHz above the local oscillator. If the receiver has no RF filtering before the mixer and if there happens to be a signal at 1,470 kHz, it will be detected along with the desired 560 kHz signal. A bandpass filter ahead of the mixer is needed to pass the desired frequency but suppress signals at the image frequency. Note that in this example (the most common AM receiver design), this antiimage bandpass filter must be tunable and, for the receiver to have single-dial tuning, the tuned filter must always "track" exactly 455 kHz below the local oscillator (LO) frequency. In this example, the tracking requirement is not difficult to satisfy; since the image frequency is more than an octave above the desired frequency, the simple one-section filter shown in Figure 6-2 can be fairly broad and still provide adequate image rejection, maybe 20 dB. (Note, though, that 20 dB is inadequate if a signal at the image frequency is 20 dB stronger than the signal at the desired frequency.)

What if a receiver with this same 455 kHz IF frequency is also to cover the short-wave bands? The worst image situation occurs at the highest frequency, 30 MHz, where the image is only about 3% higher in frequency than the desired frequency. A filter 20 dB down at only 3% from its center frequency will need to have many sections, all of which must be tuned simultaneously with a mechanical multisection variable capacitor or voltage-controlled varicaps. As explained above, the center frequency of the filter must track with a 455 kHz offset from the LO frequency in order for the desired signal to fall within the narrow IF pass band. Image rejection is not simple when the IF frequency is much lower than the input frequency.

SOLVING THE IMAGE PROBLEM

A much higher IF frequency can solve the image problem. If the AM broadcast band receiver discussed above were to have an IF of 10 MHz rather than 455 kHz, the LO could be tuned to 10.560 MHz to tune in a station at 560 kHz. The image frequency would be 20.560 MHz. As the radio is tuned up to the top end of the AM broadcast band, 1,700 kHz, the image frequency increases to 21.700 MHz. In this case, a fixed-tuned bandpass filter, wide enough to cover the entire broadcast band, can be placed ahead of the receiver to render the receiver insensitive to images. This system is shown in Figure 6-4. Only the LO needs to be changed to tune this receiver. Of course, the 10 MHz IF filter must still have a narrow 10 kHz pass band to establish the basic selectivity of the receiver.

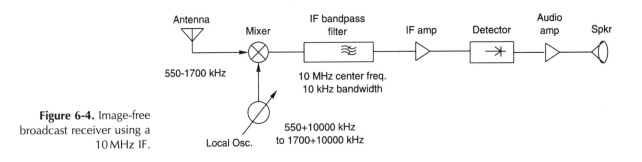

Figure 6-4. Image-free broadcast receiver using a 10 MHz IF.

The circuit of Figure 6-4 is entirely practical, although it is more expensive to make the necessary narrow-band filters at higher frequencies; quartz crystal resonator elements usually take the place of lumped *LC* elements. If the input band is wider, for example 3–30 MHz for a short-wave receiver, the IF frequency would have to be much higher, and narrow-band filters are impractical. (Even at 10 MHz, a bandwidth of 10 kHz implies a fractional bandwidth of only 0.1 percent.) It is still possible to solve the image problem with a high IF in a *double-conversion* super-heterodyne receiver, as described below.

DOUBLE CONVERSION SUPERHETERODYNE RECEIVER

Figure 6-5 shows how a second frequency converter takes the first IF signal at 10 MHz and converts it down to 455 kHz, where it can be processed by the standard IF section of the receiver of Figure 6-2. The 10 MHz first IF filter can be wider than the ultimate pass band. Suppose, for example, that it has a bandwidth of 500 kHz. The second LO frequency is at 10.455 MHz, so the second mixer would produce an image from a signal at 10.910 MHz. But note that our first IF filter cuts off at 10 MHz + 0.500 kHz/2 = 10.25 MHz, so there will not be any signals at 10.910 MHz.

This system has its own special disadvantages: the receiver usually cannot be used to receive signals in the vicinity of its first IF, as it is difficult to avoid direct feed-through into the IF amplifier. Multiple conversions require multiple local oscillators, and various sum and difference frequency combinations inevitably are produced by nonlinearities, and show up as spurious signals known as "birdies."

Present practice for communications receivers is to use double or triple conversion with a first IF frequency at, say, 40 MHz. The front-end image filter is usually a 30 MHz low-pass filter. In as much as modern crystal filters can have a fairly small bandwidth even at 40 MHz, the output of the first IF section can be mixed down to a second IF with a much lower

Figure 6-5. Double-conversion superheterodyne receiver.

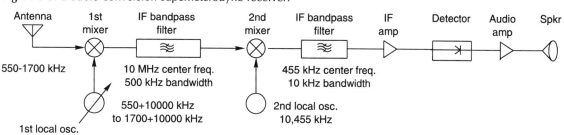

frequency. Sometimes triple conversion is necessary when the final IF frequency is very low, such as 50 kHz. The use of first IF frequencies in the VHF region requires very stable LOs, but crystal oscillators and frequency synthesizers provide the necessary stability. (Oscillator phase noise was a problem in the first generation of receivers with synthesized LOs; the oscillator sideband noise was converted into the pass band by strong signals near the desired signal but nominally outside the pass band.)

AUTOMATIC GAIN CONTROL

Nearly every receiver has some kind of automatic gain control (AGC) to adjust the gain of the RF and/or IF amplifiers according to the strength of the input signal. Without this feature, the receiver will overload; overdriven amplifiers go nonlinear ("clip"), and the output is distorted as well as too loud. The output sound level of an FM receiver does not nominally vary with signal level, but overloading the IF amplifier stages will still produce distortion, and FM receivers also need AGC. Television receivers need accurate AGC to maintain the correct contrast level. Any AGC circuit is a feedback control system. In simple AM receivers the diode detector provides a convenient dc output that can control the bias current (and hence gain) of the RF amplifiers.

NOISE BLANKERS

Many receivers, including most television receivers, have a noise blanker circuit to reduce the effects of impulse noise such as the spiky noise produced by auto ignition systems. Here the interfering pulses are of such short duration that the IF stages can be gated off briefly while the interference is present. The duty cycle of the receiver remains high, and the glitch is all but inaudible (or invisible). An important consideration is that the gating must be done before the bandwidth is made very narrow because narrow filters elongate pulses.

DIGITAL SIGNAL PROCESSING IN RECEIVERS

In principle, all the filtering and detection in a receiver could be done digitally, at least after some initial amplification and bandwidth limiting. Digital processing of the IF signal makes it possible to realize any desired filter amplitude and phase response. Good performance requires sufficient processing power (available in general-purpose digital signal processor

(DSP) chips) and high-speed high-resolution analog-to-digital converters. The proposed standard for advanced television (ATV) requires substantial digital processing at the receiver, not only for obviously digital tasks (decoding, etc.) but also for signal processing, such as adaptive multipath signal cancellation.

BIBLIOGRAPHY

1. *Handbook for Radio Amateurs*, 71st Edition, The American Radio Relay League, Newington, Connecticut, 1994. (A full five pounds of practical circuits, explanations, and construction information.)
2. W. Gosling (1986), *Radio Receivers*. New York: Peter Peregrinus. (Good contemporary discussion of receivers.)
3. U. Rohde and T. Bucher (1988), *Communications Receivers, Principles & Design*. New York: McGraw-Hill. (A whole course in itself.)

PROBLEMS

1. The FM broadcast band extends from 88 to 108 MHz. Standard FM receivers use an IF frequency of 10.7 MHz. What is the required tuning range of the local oscillator?

2. Why are airplane passengers asked not to use radio receivers while in flight?

3. Two sinusoidal signals that are different in frequency, if simply added together, will appear to be a signal at a single frequency but amplitude modulated. This "beat" phenomenon is used, for example, to tune two guitar strings to the same frequency. When they are still at slightly different frequencies, the sound seems to pulsate slowly at a rate equal to their frequency difference. Show that

$$\sin[(\omega_0 - \delta\omega)t] + \sin[(\omega_0 + \delta\omega)t] = A(t)\sin(\omega_0 t)$$

where

$$A(t) = 2\cos(\delta\omega t).$$

(The ear responds to the intensity of the sound, i.e., the square of the amplitude, so the perceived pulsation rate is $2\delta\omega$, the difference frequency. Note that new frequencies are generated in the non-linear detection process, not in the linear addition operation.)

4. Using an AM receiver in an environment crowded with many stations, you will sometimes hear an annoying high-pitched 10 kHz tone together with the desired audio. If you rock the tuning back and forth the pitch of this tone does not change. What causes this?

5. When tuning an AM receiver, especially at night, you may hear "heterodynes" or whistling audio tones that change frequency as you slowly tune the dial. What causes this? Can it be blamed on the receiver? (*Answer*: Yes.)

6. With modern components and digital control we could build good TRF radios. What advantages would such a radio have over a super-heterodyne receiver? What would be the disadvantages?

CLASS-C AND CLASS-D AMPLIFIERS

Class-C and class-D RF power amplifiers are so nonlinear (output signal versus input signal) that they might better be called synchronized sine wave generators. They consist of a power supply, at least one switching element (tube, transistor, . . .), and an *LCR* circuit. The "*R*" is the load, often the radiation resistance of an antenna, and the circuit is resonant at the operating frequency. The high efficiency obtained with these amplifiers is important for applications ranging from the smallest battery-operated transmitters to megawatt broadcast transmitters and industrial heaters. The output amplitude, although a nonlinear function of the input amplitude, is linearly proportional to the power supply voltage. These amplifiers can therefore be amplitude modulated by simply varying the supply voltage.

CLASS-C AMPLIFIERS

An idealized class-C amplifier is shown in Figure 7-1. The active device (transistor or tube) is modeled here as a switch having a constant on-resistance, r, and an infinite off-resistance. This model is a fairly good representation of a power field effect transistor (FET). The switch is closed for less than half the RF cycle, during which time the power supply deli-

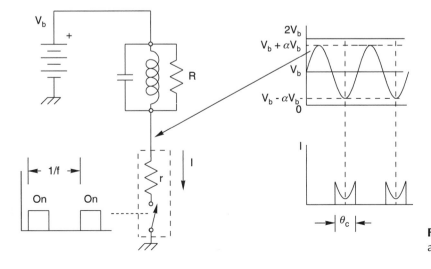

Figure 7-1. Class-C amplifier operation.

vers a pulse of current to keep the *RLC* circuit pumped up. The conduction angle, θ_c, is usually of the order of 120° (see Figure 7-1). Normally the *LC* circuit has a high Q (at least 5), so its flywheel action minimizes distortion of the sine wave caused by the abrupt pull-down of the switch and by the damping between pulses. Note that it is also possible to pulse the switch only every other cycle or every third cycle, and so on. When operated this way, the circuit is a class-C doubler or tripler – a frequency multiplier. The drive is shown as a rectangular pulse, but is often a sine wave, biased so that conduction takes place just around the positive tips.

SIMPLIFIED ANALYSIS OF CLASS-C OPERATION

Class-C amplifiers are normally run in saturation, meaning that the switch, when on, always has its lowest possible resistance (which should be much less than the load resistance). The amplitude of the drive signal is made more than enough to put the switch into its fully on or fully off state. Here we will analyze the circuit of Figure 7-1 to find the output voltage (and therefore power) and the efficiency. Referring to Figure 7-1, θ_c is the conduction angle, V_b is the supply voltage, r is the on resistance of the switch, and αV_b is the peak voltage of the sine wave. We can find α as follows. The input power (the power supplied by the battery) must be equal to the sum of the output power (the power dissipated in R) plus the power dissipated in the switch resistance r. These terms are given respectively by the average of the battery voltage times the current, $(\alpha V_b)^2/(2R)$, and the average of I^2r. The power equation becomes

$$\frac{1}{2\pi}\int_{-\theta_c/2}^{\theta_c/2} V_b\left(\frac{V_b - \alpha V_b \cos\theta}{r}\right) d\theta = \frac{(\alpha V_b)^2}{2R}$$

$$+ \frac{1}{2\pi}\int_{-\theta_c/2}^{\theta_c/2}\left(\frac{V_b - \alpha V_b \cos\theta}{r}\right) r\, d\theta. \tag{7-1}$$

Rearranging this equation we get

$$\frac{1}{\pi r/R}\int_{0}^{\theta_c/2}(1 - \alpha\cos\theta)(\alpha\cos\theta)d\theta - \frac{\alpha^2}{2} = 0. \tag{7-2}$$

Carrying out the integral in Equation (7-2) and solving for α, we find

$$\alpha(\theta_c, r/R) = \frac{2\sin(\theta_c/2)}{\theta_c/2 + (\sin\theta_c)/2 + \pi r/R}. \tag{7-3}$$

The quantity α^2 (proportional to the output power) is plotted in Figure 7-2 for five values of r/R, the ratio of the switch resistance to the load resistance. The middle value, $r/R = 0.1$, is typical in actual practice. Maximum power is produced for a conduction angle of 180°, that is, if the switch is closed during the entire negative voltage loop. If the conduction angle exceeds 180°, the incursions into the positive loop extract energy from the tuned circuit, and the power is reduced. Note that α^2, and therefore α, can be greater than unity, especially when the conduction angle is 180°. We will see below, however, that much higher efficiency is obtained for conduction angles substantially less than 180°. The efficiency is given by the output power divided by the power supplied by the battery:

$$\eta(\theta_c, r/R) = \frac{(\alpha V_b)^2/2R}{1/2\pi \int_{-\theta_c/2}^{\theta_c/2} V_b[(V_b - \alpha V_b \cos\theta')/r]d\theta'\}}. \qquad (7\text{-}4)$$

Evaluating the integral in Equation (7-4) we find

$$\eta(\theta_c, r/R) = \frac{(\pi r/R)\alpha^2}{\theta_c - 2\alpha \sin(\theta_c/2)} \qquad (7\text{-}5)$$

where α is given by Equation (7-3). This expression for efficiency is plotted in Figure 7-3 for the same five values of r/R. For $r/R = 0.1$, the efficiency is a maximum at about 90°.

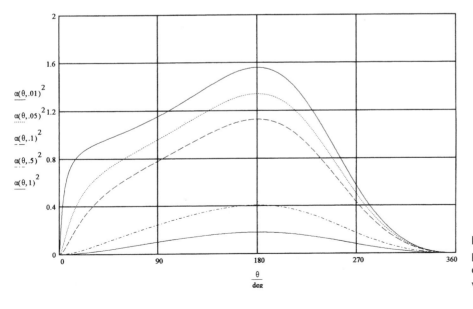

Figure 7-2. Class-C output power (α^2) versus conduction angle for five values of r/R.

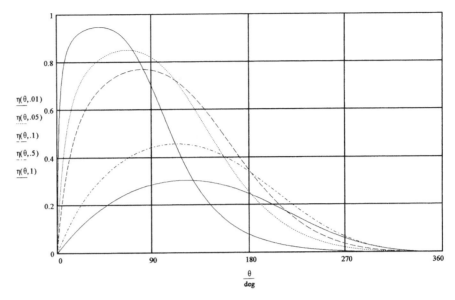

Figure 7-3. Class-C efficiency versus conduction angle for five values of r/R.

GENERAL ANALYSIS OF A CLASS-C OPERATION WITH A NONIDEAL TUBE OR TRANSISTOR

The simplified analysis of Figures 7-2 and 7-3 will not give accurate results for class-C amplifiers made with tubes or bipolar transistors, because these devices do not have the simple constant on-resistance characteristic of an FET. Nevertheless, their nonlinear characteristics are specified graphically on data sheets, and accurate class-C analysis can be done numerically. The method is basically the same as the simplified analysis; one assumes the resonant LC circuits at the input and output have enough Q to force the input and output waveforms to be sinusoidal. For this analysis, the device characteristics are plotted in a "constant-current" format. In the case of a tube, curves of constant plate current are plotted on a graph whose axes are plate voltage and grid voltage. Sinusoidal plate and grid voltages are assumed, and a numerical integration of plate current × plate voltage, averaged through one complete cycle, gives the power dissipated in the device. Power supply voltage × *average* current gives the total input power. The difference is the power delivered to the load. The designer selects a device and a power supply voltage and assumes trial waveforms (bias points and sine wave amplitudes). Usually several trial designs are needed, for example to maximize output power with the given device or to maximize efficiency for a specified output power.

DRIVE CONSIDERATIONS

Class-C amplifiers using vacuum tubes* nearly always drive the control grid positive when the tube is conducting. The grid draws current and dissipates power. Data sheets include grid current curves, so that the designer can use the procedure outlined above to verify that the chosen operating cycle stays within both the maximum plate dissipation rating and the maximum grid dissipation rating. When tetrodes are used, a third analysis must be done to calculate the screen grid dissipation.

SERIES-FED AND SHUNT-FED CIRCUITS

Figure 7-4 shows two common circuit configurations. The series-fed circuit of Figure 7-4a is equivalent to that of Figure 7-1 except that a transistor replaces the hypothetical switch. Note that the circuits for class-C amplifiers are the same as those of class-A and class-B "real amplifiers." The only differences are that the class-C amplifier has the control element (base, gate, or grid) biased beyond the cutoff so that the device conducts for less than half the cycle, and, when the device is on, it is fully on (saturated). The shunt-fed circuit of Figure 7-4b allows one end of the load to be connected to ground. An RF choke connects the supply to the transistor, and a blocking capacitor keeps dc off the tank circuit. The RF choke has a large inductance, so the current through it is essentially constant. The switch pulls pulses of charge from the blocking capa-

Figure 7-4. Series-fed (a) and shunt-fed (b) amplifiers.

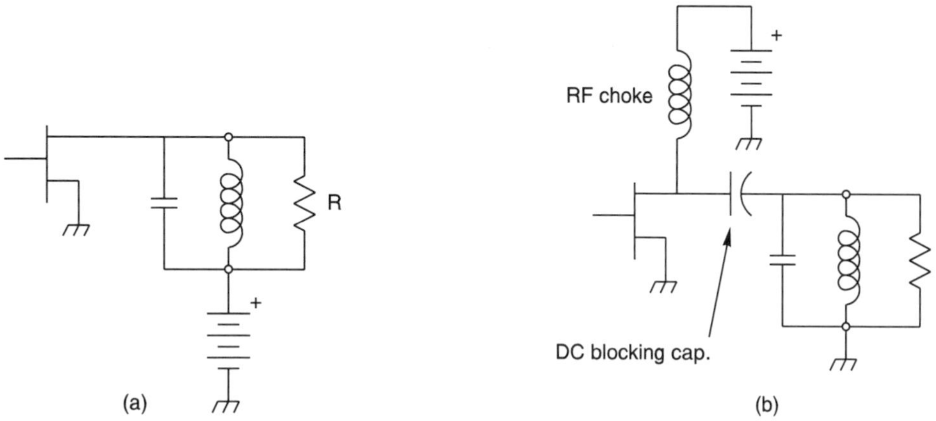

*A triode vaccum tube is analogous to an npn transistor. The plate, control grid, and cathode of the tube correspond respectively to the collector, base, and emitter of the transistor. Tetrode tubes have an additional grid, the screen grid, between the control grid and the plate. The screen grid is usually run at a fixed bias voltage, and forms a useful electrostatic shield between the control grid and the plate. (See [3].)

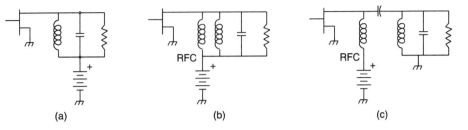

Figure 7-5. Equivalence of series-fed and shunt-fed circuits.

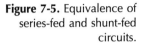

citor. This charge is replenished by the current through the choke. Figure 7-5 shows, from right to left, the equivalence of the shunt-fed and series-fed circuits. The RF choke placed in parallel with the inductor in Figure 7-5b does not affect the circuit operation. Usually the choke has so much inductance it hardly changes the value of L, but in any case, the parallel combination can be made to have the original value of L. The choke provides an alternate dc path to the switch so the original RLC tank circuit can be isolated from dc by the large-value blocking capacitor in Figure 7-5c. (Note that the series-fed and shunt-fed equivalence holds for amplifiers of class A, B, or C and for large-signal or small-signal operation.)

THE CLASS-C AMPLIFIER AS A VOLTAGE MULTIPLIER

An important property of the saturated class-C amplifier is that the peak voltage of the output RF sine wave is directly proportional to the supply voltage ($V_{pk} = \alpha V_b$). In the standard saturated operation, the proportionality constant, α, is about 0.9. The class-C amplifier is therefore equivalent to a voltage multiplier, which forms the product of a nearly unit amplitude input sine wave times the power supply voltage. Modulating (varying) the power supply voltage of a class-C amplifier is the classic method used in AM transmitters. (Note that this property of a class-C amplifier would be considered a defect for an op amp – poor power supply rejection.)

HIGH-POWER CLASS-C AMPLIFIERS

Tubes are available with maximum plate dissipations of up to more than 1 MW. Typical class-C amplifiers have efficiencies of 75–85 percent, so a single large tube can produce in excess of 6 MW of RF power. The highest power used in radio broadcasting is about 2 MW, but much higher power amplifiers (or class-C oscillators) are found in industrial heating applications such as curing plywood and welding. AM and FM broadcasting stations have traditionally used class-C amplifiers for high efficiency and, in the case of AM, for the linear plate modulation characteristic. Newer AM transmitters use solid state amplifiers, either class-C or class-D circuits (discussed below). High-power solid state transmitters combine the power of many low-power ($\approx 1\,\text{kW}$) modules.

MODIFIED CLASS-C AMPLIFIERS FOR HIGHER EFFICIENCY

A basic problem with the switching action in the class-C amplifier is that the switch shunts the supply voltage directly across a parallel *LRC* circuit. The capacitor acts as an instantaneous short; since the voltages do not match, it wants infinite current, but the maximum current is limited by the resistance of the switch. Therefore, during the pulse, or at least until the voltage difference is negligible, there must be a nonzero voltage times current product, and the switch *must* dissipate energy. Sometimes additional reactive elements are included in the output circuit to excite selected harmonic currents and deliberately distort the plate voltage. In particular, the bottom of the voltage waveform can be flattened somewhat so that the voltage times current product in the switch is reduced and the circuit efficiency is increased. Raab [2] refers to this as *class-F* operation.

CLASS-D AMPLIFIERS

The standard class-C amplifier can, in principle, approach 100% efficiency, but only as the duty cycle approaches zero and the pulse current approaches infinity. Class-D amplifiers can, in principle, achieve 100% efficiency with comfortable duty cycles and currents. At least two switches are required, but neither is forced to support simultaneously both voltage and current.

SERIES RESONANT CLASS-D AMPLIFIER

The series resonant class-D concept is shown in Figure 7-6a. An SPST switch produces a square wave. An *LC* filter lets the fundamental sine wave component reach the load, *R*. The bottom of the switch could be connected to a negative supply, but ground will work since the output is ac coupled. A real circuit is shown in Figure 7-6b; two transistors form the switch. This two-transistor circuit is push–pull, that is, the transistors are driven out-of-phase so that when one is on the other is off. Let us find the voltage on the load and calculate the efficiency. Since the capacitor also acts as a dc block, we can consider this square wave to be symmetric about zero, swinging from $-V_{\text{b}}/2$ to $+V_{\text{b}}/2$. The series *LC* extracts the fundamental sinusoidal component of the square wave. Recall how to find Fourier components: our square wave can be decomposed as $A_1 \sin(\omega t) + A_3 \sin(3\omega t) + \dots$. We are after the first coefficient, A_1, which

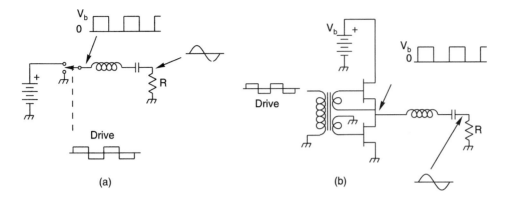

Figure 7-6. Class-D series amplifier.

can be found by calculating the average value of the product of the square wave times $\sin(\omega t)$. This calculation is shown in Figure 7-7. Since $A_1 = 2V_b/\pi$, the output power, $A_1^2/(2R)$, is given by $2V_b^2/(\pi^2 R)$. (We do not often get to write "$\pi^2 R$"!) As with the class-C amplifier analysis, we will assume the FETs have constant on-resistance, r. The current through the load passes through one or the other of the switches, so the ratio of output power to switch power is $I^2R/(I^2r)$, and the efficiency is $\eta = R/(R+r)$. You can see that, in as much as $r \ll R$, the efficiency can approach 100%. Note also that the class-D amplifier, like the class-C amplifier, is a voltage multiplier, because its output sine wave amplitude is directly proportional to the supply voltage.

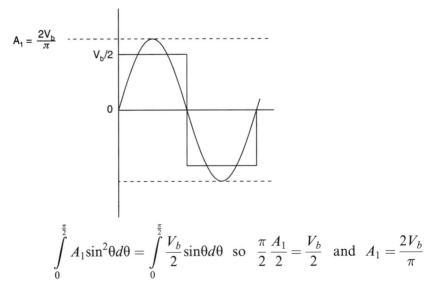

Figure 7-7. Extraction of the fundamental sine wave component of a square wave.

$$\int_0^{\frac{\pi}{2}} A_1\sin^2\theta\, d\theta = \int_0^{\frac{\pi}{2}} \frac{V_b}{2}\sin\theta\, d\theta \quad \text{so} \quad \frac{\pi}{2}\frac{A_1}{2} = \frac{V_b}{2} \quad \text{and} \quad A_1 = \frac{2V_b}{\pi}$$

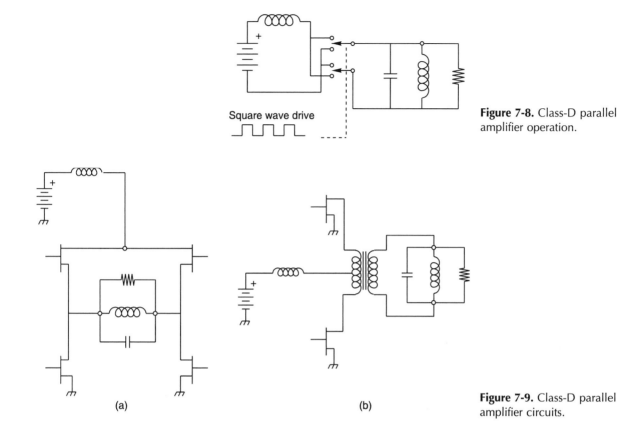

Figure 7-8. Class-D parallel amplifier operation.

Square wave drive

(a)   (b)

Figure 7-9. Class-D parallel amplifier circuits.

PARALLEL RESONANT CLASS-D AMPLIFIER

The class-D amplifier of Figure 7-6 is a square wave source driving a series *RLC* circuit. It is also possible to have a square wave current source driving a parallel *RLC* circuit. In the circuit of Figure 7-8, a large inductor provides the constant current. A DPDT switch commutates the load. Two practical versions of this circuit are shown in Figure 7-9. The circuit on the left uses four transistors in a bridge, while the circuit on the right uses a transformer and gets by with only two transistors.

WHICH CIRCUIT TO USE: CLASS C OR CLASS D?

The class-D amplifier has the attraction of potentially unlimited efficiency. Nevertheless, the class-C amplifier, especially the improved efficiency designs (class F) have respectable efficiencies. The class-C amplifier requires only one active device and can operate at the upper frequency limit of the tube or transistor. The class-D amplifier can be difficult to get running. (With any totem pole circuit, care must be taken to avoid letting the upper and lower devices conduct simultaneously, even for brief instants.) Class D requires that the devices operate as ideal switches, so the devices usually limit class-D operation to relatively low frequencies. AM broadcast transmitters can use class D. FM broadcast transmitters (VHF) use class C. At microwave frequencies even class-C operation is difficult; conduction angles are large, and a nominal class-C amplifier might actually operate in class A.

BIBLIOGRAPHY

1. H. L. Krauss, C. W. Bostian and F. H. Raab (1980), *Solid State Radio Engineering*, New York: John Wiley.
2. F. H. Raab (1975) "High efficiency amplification techniques", *IEEE Circuits and Systems* 7: 3–11.
3. *Care and Feeding of Power Grid Tubes*. Varian-Eimac Division Laboratory Staff, San Carlos, California, 1967.

PROBLEMS

1. Suppose the simple series class-D amplifier of Figure 7-6b is driving a 50 ohm resistive load at a frequency of 1 MHz. The values of the inductor and capacitor are 9.49 μH and 2.67 nF (equal and opposite reactances at 1 MHz). This resonant circuit passes the 1 MHz component of the square wave and greatly reduces the harmonics. By what factor is the 3 MHz (i.e. third harmonic) power delivered to the load lower than the fundamental (1 MHz) power? *Hints*: In a square wave, the amplitude of the third harmonic is one-third the amplitude of the fundamental. Remember that at 1 MHz, $X_L = X_C$ while at 3 MHz X_L is increased by a factor of three and X_C is decreased by a factor of three.

2. A single-tube class-C amplifier with 75% efficiency is providing 500 kW of continuous-wave output power. The supply voltage is 60 kV and the conduction angle is 90°. What is the average current drawn from the power supply? (*Answer*: 11.1 A.) What is the average current when the tube is on? (*Answer*: 44.4 A.)

3. Explain why the class-C amplifier will have low efficiency if the drive pulses do not saturate (fully turn on) the tube or transistor.

4. The current pulses shown in Figure 7-1 correspond to the case where the sine wave peak voltage is less than V_b, that is, $\alpha < 1$. Sketch the form of the current pulses when $\alpha > 1$. (Assume here that the switch can conduct current in either direction.)

TRANSMISSION LINES

8

We draw circuit diagrams with lumped components connected by lines that represent zero-length wires. But all real wires have some length and their distributed capacitance and inductance can be complicated circuits in their own right. When a circuit has dimensions that are not small compared to the wavelength, these effects become important. On the other hand, when interconnections are made with proper transmission lines, these effects are entirely predictable and can be accounted for in the design. In this chapter, we will have in mind the common two-conductor lines such as coaxial cable and twin lead.

FUNDAMENTALS

Every increment of a transmission line contributes series inductance and shunt capacitance; the ladder network shown in Figure 8-1 becomes a real transmission line in the limit that δC and δL go to zero. (*Note*: for some situations, such as baseband telephony and digital data transmission, the model must also include series and shunt resistance. At radio frequencies, however, the series reactance is usually much greater than the series resistance, and the parallel reactance is usually much less than the parallel resistance, so both resistances can be neglected.) A piece of transmission line placed ahead of some impedance Z will, in general, produce a modified impedance Z', as shown in Figure 8-2. The only impedance not modified is a resistance whose value is equal to the *characteristic impedance* of the line, Z_0. This impedance is given by $Z_0 = \sqrt{L/C} + j0$, where L and C are the inductance and capacitance per unit length. It is sufficient to know either L or C, as they are related by $LC = 1/v_{\text{phase}}^2 = \varepsilon/c^2$ where v_{phase} is the velocity of propagation, ε is the dielectric constant (relative to

Figure 8-1. Transmission line model – a ladder network of infinitesimal LC sections.

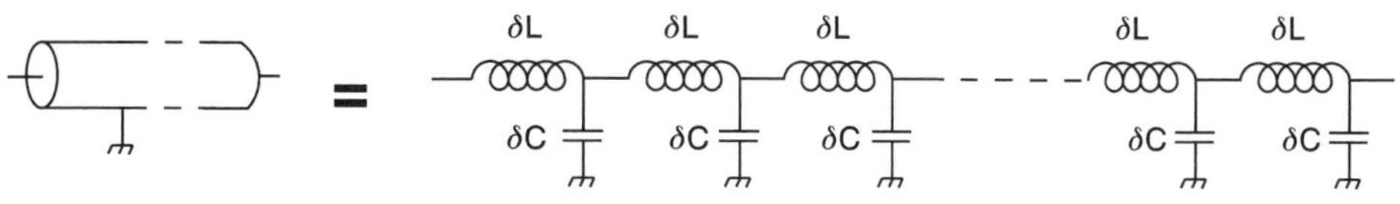

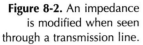
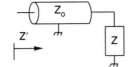

Figure 8-2. An impedance is modified when seen through a transmission line.

vacuum), and c is the speed of light. (This relation between L and C holds for any two-conductor structure with translational symmetry in one direction, for example an unlikely transmission line consisting of a square inner conductor inside a triangular outer conductor.)

How "real" is characteristic impedance? If we connect an ordinary dc ohmmeter to the end of a piece of 50 ohm cable will it indicate 50 ohms? Answer: In principle, yes, if the piece is *very* long so that the reflection does not arrive back at the meter while we make the measurement. Otherwise, a short at the far end of the cable produces a negative reflected voltage that cancels the applied voltage, and we will read zero ohms. An open circuit at the far end produces positive reflected voltage whose negative current (negative simply because the reflected wave travels backward) cancels the applied current, and we will read infinite ohms. But with a pulse generator and an osilloscope you can easily make an ohmmeter that is fast enough to determine Z_0.

DETERMINATION OF CHARACTERISTIC IMPEDANCE AND PROPAGATION VELOCITY

To see that $Z_0^2 = L/C$, consider the circuit shown in Figure 8-3, where we have added a very small piece (δx meters) of cable to a perfect termination, that is, a resistor of Z_0 ohms. Let us see what the relation between L and C must be in order that the impedance remain equal to Z_0. This circuit looks just like the L-section matching that can (and must) transform a resistance into a *different* resistance. But here, because the δL and δC are infinitesi-

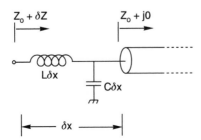

Figure 8-3. Adding another infinitesimal section must leave Z_0 unchanged.

mal, we will show that the change in impedance, δZ_0 will be zero if $Z_0^2 = L/C$.

Remember that L and C are the inductance and capacitance per unit length. From inspection of the circuit we can write

$$Z_0 + \delta Z = j\omega L\,\delta x + \frac{Z_0[1/(j\omega C\,\delta x)]}{Z_0 + 1/(j\omega C\,\delta x)} = j\omega L\,\delta x + \frac{Z_0}{j\omega C\,\delta x Z_0 + 1}.$$
$$(8\text{-}1)$$

Simplifying the right-hand term in the limit that δx goes to zero produces

$$Z_0 + \delta Z \to Z_0 + j\omega\,\delta x(L - Z_0^2 C). \qquad (8\text{-}2)$$

Thus, δZ will be zero if $Z_0 = \sqrt{L/C}$, that is, the impedance will remain constant as the line is lengthened. (You can confirm that interchanging the order of the shunt capacitance and the series inductance does not change this result.)

A similar argument shows how a wave propagates down the line. Here we apply an input $e^{j\omega t}$, and find the voltage drop across an incremental length of line (Figure 8-4). Since we already know the input impedance is Z_0, the input current must be V/Z_0, and the drop across the inductor, δV, is $(V/Z_0)(j\omega L\,\delta x)$. But this is just the differential equation

$$\frac{dV}{dx} = -j\omega\frac{L}{Z_0}V = -j\omega\sqrt{LC}\,V. \qquad (8\text{-}3)$$

The solution to this familiar differential equation is

$$V = V_f e^{-j\omega\sqrt{LC}x} = V_f e^{-jkx} \qquad (8\text{-}4)$$

where $k = \omega\sqrt{LC} = \omega/v_{\text{phase}}$. We have launched this wave in the positive x direction because, if we include the time dependence, the voltage is

$$V = V_f e^{j(\omega t - kx)}. \qquad (8\text{-}5)$$

Any point of constant phase moves according to $\omega t - kx = $ constant, so $dx/dt = \omega/k = 1/\sqrt{LC} = v_{\text{phase}}$. This phase velocity is independent of ω; there is nominally no dispersion. (Common transmission lines, however,

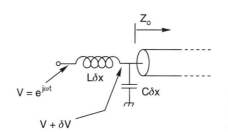

Figure 8-4. Finding the change in voltage, δV, for a change in length, δx.

use insulating material whose dielectric constant is somewhat frequency-dependent, so they do exhibit some dispersion.)

MODIFICATION OF AN IMPEDANCE BY A TRANSMISSION LINE

Now we can get on with finding how a piece of line modifies an impedance (see Figure 8-2). A wave traveling in the positive x direction will be proportional to $e^{j(\omega t - kx)}$ where $k = 2\pi/\lambda = \omega/v_{\text{phase}}$. Likewise, a wave traveling to the left (the negative x direction) is proportional to $e^{j(\omega t + kx)}$. A superposition of these two waves is a general (single-frequency) solution for the voltage on the line. Consider a piece of line of length l whose right-hand end, at $x = 0$, is connected to some impedance Z_L (L for "load"). Assume that some source produces an incident wave traveling to the right, e^{-jkx}, and that Z_L causes a reflected wave, ρe^{jkx}, traveling to the left. At any point x, the voltage on the line is $V(x) = e^{-jkx} + \rho e^{jkx}$. The corresponding current is $I(x) = 1/Z_0(e^{-jkx} - \rho e^{jkx})$ as shown in Figure 8-5.

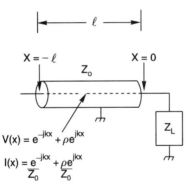

Figure 8-5. Determination of the voltage and current at the input of a transmission line.

The minus sign occurs because the reflected wave travels in the negative x direction. At the right-hand end $(x = 0)$, Ohm's law requires that $V(0)/I(0) = Z_L$. This will give us ρ:

$$\frac{V(0)}{I(0)} = Z_L = \frac{(1 + \rho)}{(1 - \rho)/Z_0}$$

so

$$\frac{Z_L}{Z_0} = \frac{(1 + \rho)}{(1 - \rho)}$$

and

$$\rho = \frac{(Z_L - Z_0)}{(Z_L + Z_0)}. \tag{8–6}$$

We now use this expression for ρ to get $V(-l)/I(-l)$, which is what we are after, that is, the input impedance at a point l to the left of the load:

$$
\begin{aligned}
Z' &= \frac{e^{-jk(-l)} + \rho e^{jk(-l)}}{e^{-jk(-l)}/Z_0 - \rho e^{jk(-l)}/Z_0} \\
&= Z_0 \frac{(Z_L + Z_0)e^{jkl} + (Z_L - Z_0)e^{-jkl}}{(Z_L + Z_0)e^{jkl} - (Z_L - Z_0)e^{-jkl}} \\
&= Z_0 \frac{Z_L + jZ_0 \tan(kl)}{Z_0 + jZ_L \tan(kl)}.
\end{aligned}
\tag{8-7}
$$

This important result (the last line) is not hard to remember; it has no minus signs and is symmetric. Just remember

$$(1 + j \tan)/(1 + j \tan).$$

Once you have written this down, you will remember how to put in the coefficients. Some important special cases are:

1. If $Z_L = Z_0$ then $Z' = Z_0$ as expected.
2. If $Z_L = 0$ (a short) then $Z' = jZ_0 \tan(kl)$, a pure reactance, which is inductive for $kl < \pi/2$, then capacitive, and so on.
3. If $Z_L = $ infinity (an open circuit) then $Z' = Z_0/j \tan(kl)$, which is capacitive for $kl < \pi/2$, then inductive, and so on.

Note that "inductive" does not mean a lumped inductor because $Z_0 \tan(kl) = Z_0 \tan(\omega l/v_{\text{phase}})$ is not proportional to ω (except for small l, where a shorted piece of line is a good approximation to a lumped inductor). Likewise, a short open-ended line is a good approximation to a lumped capacitor.

We have seen how a transmission line of length l transforms an impedance Z into Z' via Equation (8-7). Note that the transformed impedance depends on the length of the cable (unless $Z = Z_0$). Sometimes we have to make an impedance measurement through a cable. For example, it is inconvenient to measure the impedance of an antenna at its feedpoint, which may be high in the air, so most often it is measured through its feedline. Solving for Z_{ant} (the antenna impedance) in terms of Z' (the measured impedance) gives

$$Z_{\text{ant}} = Z_0[Z' - jZ_0 \tan(kl)]/[Z_0 - jZ' \tan(kl)]. \tag{8-8}$$

This inverse operation is similar to the "de-embedding" process used to deduce the characteristics of a transistor after it has been measured in a mounting fixture.

PROBLEMS

1. A common 50 ohm coaxial cable, RG214, has a shunt capacitance of 30.8 pF/foot. Calculate the series inductance per foot and the propagation velocity.

2. (a) Use the "tan tan" formula to show that a short length, δx, of transmission line, open circuited at the far end, behaves like a capacitor, that is, that it has a positive susceptance directly proportional to frequency. Express the value of this capacitor in terms of the capacitance/unit length of the cable *Hint: $\tan\theta \approx \theta$ for small θ.*

(b) Show that a short length, δx, of a transmission line, short circuited at the far end, behaves like an inductor, that is, it has a negative susceptance inversely proportional to frequency. Express the value of this inductor in terms of the inductance/unit length of the cable.

3. (a) Find a formula for the characteristic impedance of a lossy cable where the loss can be due to a series resistance per unit length R, as well as a parallel conductance per unit length G (R represents the ohmic loss of the metal conductors while G represents dielectric loss.)

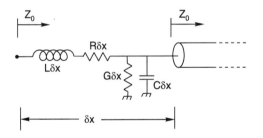

Hint: You can generalize the result for the lossless cable by simply replacing L by $L + R/j\omega$ and C by $C + G/j\omega$.

(b) Find the formula for the propagation constant k of this lossy cable. (*Hint:* Apply the substitutions given above to the formula $k = \omega\sqrt{LC}$.) What distance (in wavelengths) is required to reduce by $1/e$ the power of a signal at frequency ω_1 if $R/\omega_1 = 0.01L$? (Assume $G = 0$.)

4. If the (sinusoidal) voltage, V, and current, I, at the right-hand end of a transmission line are given, find the corresponding voltage, V', and current, I', at the left-hand end.

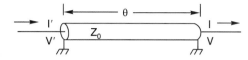

Hint: Assume the (complex) voltage on the line is given by $V(\phi) = V_F e^{-j\phi} + V_R e^{j\phi}$. The current is then related to the voltage by $Z_0 I(\phi) = V_F e^{-j\phi} - V_R e^{j\phi}$. Let $\phi = 0$ at the right-hand end. Show that $V_F = (V + IZ_0)/2$ and $V_R = (V - IZ_0)/2$. Then show that, at the left-hand end, where $\phi = \theta$, that $V' = V \cos\theta + IZ_0 j \sin\theta$ and $I' = I \cos\theta + j \sin\theta \, V/Z_0$.

5. Use the results of Problem 4 to upgrade your ladder network analysis program (see Chapter 1, Problem 3) to handle another type of element, a series transmission line. Three parameters are necessary to specify the line. These could be the characteristic impedance, the physical length, and the velocity of propagation. For convenience in later problems, however, let the three parameters be the characteristic impedance (Z_0), the electrical length (θ_0) in degrees for a particular frequency, and that frequency (f_0). A 50 ohm cable that has an electrical length of 80° at 10 MHz would appear in the circuit file as "TL, 50, 80, 10E6." For any frequency f the electrical length is then $\theta = \theta_0 f / f_0$.

6. A transmission line of 50 ohms is connected in parallel with an equal length transmission line of 75 ohms, that is, at each end the inner conductors are connected and the outer conductors are connected. The cables have equal phase velocities. Show that the characteristic impedance of this composite transmission line is given by $(50 \times 75)/(50 + 75)$, that is, the characteristic impedances add like parallel resistors.

7. In the circuit shown below, the impedance Z is modified by a transmission line in parallel with a lumped impedance Z_1, which could be R, C, or L or a network. Find an expression for Z'.

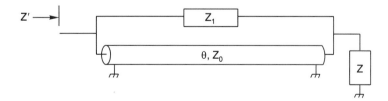

Hint: Extend the argument used in the text to find Z' for a cable without a bridging lumped element. Assume a forward and reverse wave in the cable with amplitudes 1 and ρ. The voltage on the cable is then $V(x) = e^{j\omega t}(e^{-jkx} + \rho e^{jkx})$ and the current is $I(x) = Z_0^{-1} e^{j\omega t} (e^{-jkx} - \rho e^{jkx})$. The current into Z is the sum of the current from the cable and the current from Z_1, while the current into the circuit is the sum of the current into the cable and the current into Z_1.

IMPEDANCE MATCHING II 9

IMPEDANCES SPECIFIED BY REFLECTION COEFFICIENT

To find how an impedance is modified when viewed through a length of transmission line of characteristic impedance Z_0, we first found the reflection coefficient at the end of the line where the given impedance is attached. (The reflection coefficient is the ratio of the voltage of the reflected wave to the voltage of the incident wave.) Ohm's law required that $\rho = (Z - Z_0)/(Z + Z_0)$. We can regard this ρ as characterizing the given impedance, as the relation can be inverted:

$$Z = R + jX = Z_0(1 + \rho)/(1 - \rho). \tag{9-1}$$

In antenna and microwave work, where transmission lines are commonly used as interconnections, it is convenient to use ρ rather than Z and to work in the complex ρ plane. The reflection coefficient for the given impedance when seen through a length l of transmission line is just

$$\rho' = \rho e^{-j2kl}. \tag{9-2}$$

This is easy to see. When we add a length of cable, the phase of the incident wave is delayed by kl getting to the end of the cable, and the reflected wave is delayed by the same kl getting back again. The effect of a cable is therefore to rotate the complex number ρ clockwise by $2kl$ to give ρ', the modified reflection coefficient. (Since $e^{j\omega t}$ rotates counterclockwise, a time delay is a clockwise displacement.) Very often one does further operations on ρ without having to transform back to $Z = R + jX$. The equations above show the one-to-one relation (mapping) between points in the ρ-plane and their corresponding R- and X-values. Note that the ρ-plane is a complex plane but that it is *not* the $R + jX$ plane. Let us look at a few special points in the ρ-plane:

1. The center of the plane, $\rho = 0$, corresponds to no reflected wave, so this point represents the impedance $Z_0 + j0$.
2. The magnitude of ρ must be less than or equal to 1 for passive impedances. (Otherwise the reflected wave has more power than the incident wave.)
3. The point $\rho = -1 + j0$ corresponds to $Z = 0$, a short circuit.
4. The point $\rho = 1 + j0$ corresponds to $Z = \infty$, an open circuit.

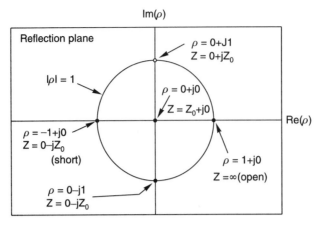

Figure 9-1. Impedances mapped into the reflection plane.

5. Points on the circle $|\rho| = 1$ correspond to pure reactances. All points inside this circle map to impedances with nonzero R.

6. The point $\rho = 0 + j1$ corresponds to an inductance, $Z = 0 + jZ_0$. All points in the top half plane are "inductive," that is, $Z = R + j|X|$ or, equivalently, $Y = G - j|B|$.

7. The point $\rho = 0 - j1$ corresponds to a capacitance, $Z = 0 - jZ_0$. All points in the bottom half plane are "capacitive," that is, $Z = R - j|X|$ or, equivalently, $Y = G + j|B|$.

These special cases of mapping of Z into ρ are shown in Figure 9-1. In this ρ-plane, if you plot $\rho(R, X)$ where R is a constant and X varies, you will get a circle centered on the real axis and tangent to the line $\text{Re}(\rho) = 1$. For every value of R there is one of these "resistance circles." The resistance circle for $R = 0$ is the unit circle in the ρ-plane. The resistance circle for $R = \infty$ is a circle of zero radius at the point $\rho = 1 + j0$. Likewise, if you plot $\rho(R, X)$, where X is a constant and R varies, you will get "reactance circles" centered on the line $\text{Re}(\rho) = 1$ and tangent to the line $\text{Im}(\rho) = 0$. These circles are shown in Figure 9-2.

If you now trim the circles to leave only the part within the $|\rho| = 1$ circle (corresponding to passive impedances, that is, impedances whose real part is positive) you are left with a useful piece of graph paper, the famous Smith chart, shown in Figure 9-3.

We will see that graphing on the reflection plane (Smith chart) allows one to follow an impedance as it is changed through successive operations such as adding a series piece of transmission line or adding a series L or a series C. But first, note that we could just as well have plotted ρ as a function of G and B. As you might guess, this produces "G-circles" and "B-circles," as shown in Figure 9-4. Including G- and B-circles on the chart

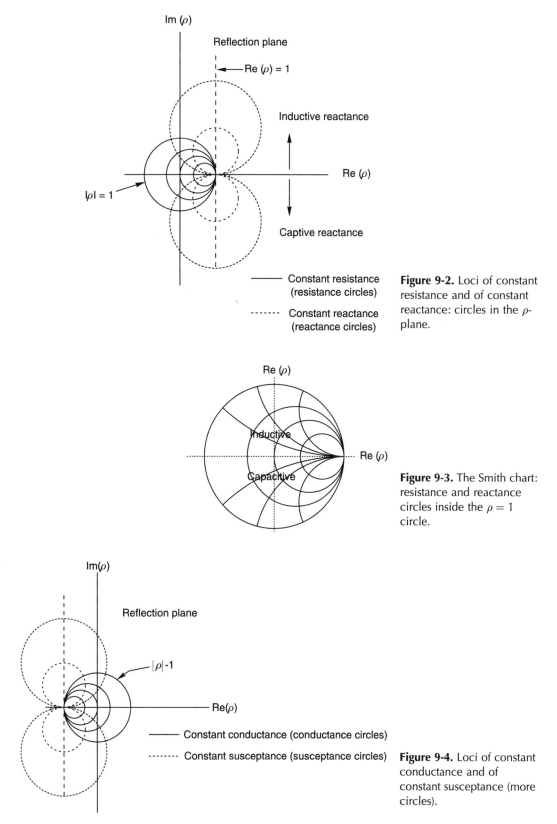

Figure 9-2. Loci of constant resistance and of constant reactance: circles in the ρ-plane.

Figure 9-3. The Smith chart: resistance and reactance circles inside the $\rho = 1$ circle.

Figure 9-4. Loci of constant conductance and of constant susceptance (more circles).

will let us deal with parallel as well as series elements. This full-blown Smith chart, which gets very dense, is shown in Figure 9-5.

Designing a matching network becomes an exercise in moving from a given ρ to a desired ρ' in the reflection plane. Working graphically, it is often easy to find the strategy to design the simplest matching circuit. Element values can be read from the chart, making it something of a calculator, but now it is more common to machine-calculate these values. The graphical presentation of impedances, however, helps to visualize a problem, and is worth learning. Take a look at a catalog of radio frequency or microwave transistors and you will see that their input and output impedances are plotted in the ρ-plane, that is, on the Smith chart.

Sometimes the Smith chart is scaled for a specific Z_0 (such as 50 or 75 ohms). Other charts are normalized; for example, the $R = 1$ circle would be the 50 ohm circle if we are dealing with 50 ohm cable. Remember that, because the Smith chart is just the complex reflection plane, impedances of constant R or constant X map to circles on the chart. Constant B and constant G also map into circles. We have seen that in the reflection plane it is particularly easy to find the new reflection coefficient, ρ' that results from connecting a length of transmission line to a point whose impedance Z corresponds to ρ. The Smith chart lets us go from Z to ρ to ρ' to Z', as follows:

1. Given $Z = R + jX$ (or $Y = G + jB$) use the resistance and reactance (or conductance and susceptance circles) to locate ρ on the chart.
2. Rotate that point clockwise around the origin by twice the phase length of the line to find the point ρ'.
3. Use the circles to read R' and X' (or G' and B').

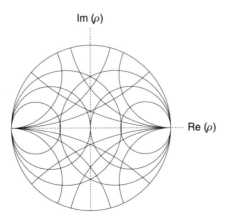

Figure 9-5. Smith chart with R-, X-, G-, and B-circles.

Let us revisit the 1,000 ohm-to-50 ohm matching problem from Chapter 2 and design several matching networks with the aid of the Smith chart.

I. The starting impedance, 1,000 ohms, and the final target impedance, 50 ohms, are indicated on the chart shown in Figure 9-6. Also shown is the circle for $R = 50$ ohms. We can use a transmission line to reach the 50 ohm circle. This line moves us along the dashed circle. Now we have $R = 50$ ohms, but X is capacitive. A series inductor will cancel the capacitive reactance, taking us to $Z = 50 + j0$ (the center of the chart).

II. Another solution would be to use a longer piece of cable to circle most of the chart, hitting the 50 ohm circle in the top half of the plane (Figure 9-7). At this point we have $Z = 50 + jX$, where X is positive (inductive). We can add a series capacitor to cancel this X and again arrive at $Z = 50 + j0$.

III. So far we have only used series elements. Let us travel around to the $G = 1/50$ circle. Then we can add a shunt element to reach the center of the chart. The first intersection of the $G = 1/50$ circle is in the lower half-plane (capacitive), so from this point we would use a shunt inductor. Instead of a lumped inductor we might use a shorted length of transmission, as shown in Figure 9-8, to make a matching circuit using only transmission line elements.

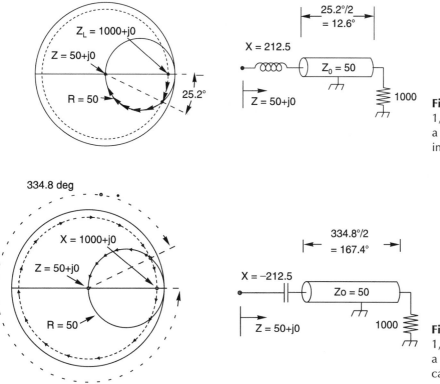

Figure 9-6. Conversion from 1,000 ohms to 50 ohms via a transmission line and inductor.

Figure 9-7. Conversion from 1,000 ohms to 50 ohms via a transmission line and a capacitor.

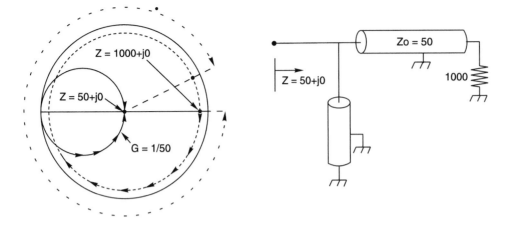

Figure 9-8. Conversion from 1,000 ohms to 50 ohms via series and shunt transmission lines.

IV. Here is a solution (Figure 9-9) that uses no transmission line. We draw in the $G = 1/1,000$ circle. If we apply shunt reactance we can move along this circle. Let us pick a shunt inductance that will move us upward along the G-circle to the 50 ohm circle. We now have $R = 50$ ohms, but there is inductive reactance. As in the above example, we can now cancel the inductance reactance with a series capacitor. This is just our original L-network.

V. If we had used shunt capacitance rather than shunt inductance, we would have moved downward to the 50 ohm circle (Figure 9-10). The remaining series capacitance can be cancelled with an inductor. This produces the L-network where the positions of the L and C are reversed.

In these examples our final impedance was at the center of the chart ($Z = 50 + j0$), but you can see that these techniques allow us to transform any point on the chart (i.e. any impedance) into any other point on the chart (any other impedance).

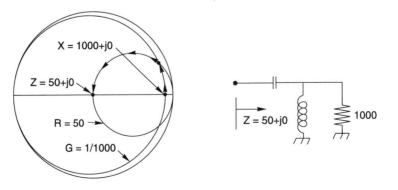

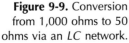

Figure 9-9. Conversion from 1,000 ohms to 50 ohms via an *LC* network.

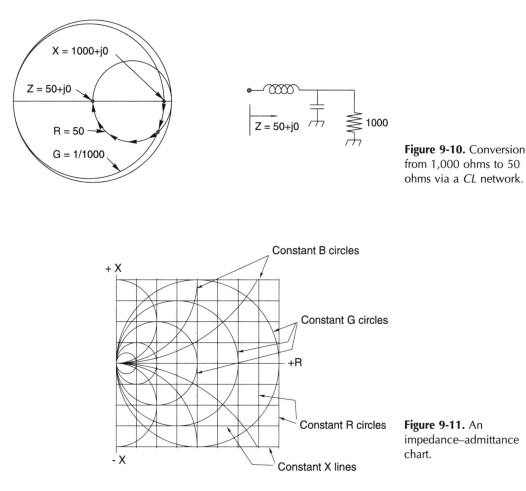

Figure 9-10. Conversion from 1,000 ohms to 50 ohms via a *CL* network.

Figure 9-11. An impedance–admittance chart.

The Smith chart is a favorite because it handles ladder networks that include transmission lines as well as passive components. If we did not care about transmission lines, then any chart that maps R, X into G, B would do. For example, take the R, X plane (half-plane, since we will exclude negative R). Draw in the curves for $G =$ constant and $B =$ constant. The resulting chart, shown in Figure 9-11, can be used to design lumped element L, C, R ladder networks, such as the networks of Figures 9-9 and 9-10.

PROBLEMS

1. Using a 50 ohm network analyzer, it is found that a certain device, when tested at 1 GHz, has a (complex) reflection coefficient of 0.6 at an angle of $-22°$ (standard polar coordinates: the positive x-axis is zero degrees, and angles increase in the counterclockwise direction).

(a) Calculate the impedance, $R + jX$.

(b) Find the component values for both the equivalent series $R_s C_s$ circuit and the equivalent parallel $R_p C_p$ circuit that, at 1 GHz, represent the device.

2. The circuit below matches a 1,000 ohm load to a 50 ohm source at a frequency of 10 MHz. The characteristic impedance of the cable is 50 ohms.

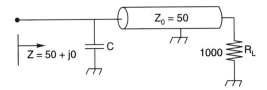

(a) Make a Smith chart sketch that shows the derivation of this circuit.

(b) Find the length of the (shortest) cable and the value of the capacitor. Specify the length in degrees and the capacitance in picofarads. Calculate these values rather than reading them from an accurately drawn Smith chart.

(c) Use your ladder network analysis program (see Chapter 1, Problem 3, and Chapter 8, Problem 1) to find the transmission from 9 to 11 MHz in steps of 0.1 MHz.

3. Find a transmission line element to replace the capacitor in the circuit of Problem 2.

4. Suppose that a transmission line has small shunt susceptance (capacitive or inductive) at a point x. By itself, this will cause a small reflection. If an identical shunt reactance is placed one quarter-wave from the first, its reflection will compensate the first, and the cable will have essentially perfect transmission. Show that this is the case (a) analytically using the "tan tan" formula for Z' and (b) using the Smith chart (the area around the center of the chart).

POWER SUPPLIES 10

The term "power supply" most often refers to a 60 Hz line-operated transformer–rectifier–filter combination. The power supplies discussed below are of this type, but we will see later that the same basic elements and methods of analysis are found in switching power supplies and in switching AM modulators.

FULL-WAVE RECTIFIER

Probably the most straightforward circuit combines a full-wave rectifier with a choke-input filter. As before, a "choke" is an inductor that ideally has infinite inductance and no dc resistance. Two examples are shown in Figure 10-1. For either circuit, the voltage at point "V" is shown in Figure 10-2. The rectifier circuit delivers power on both halves of the input ac waveform, hence the term *full wave*. The bridge rectifier (right-hand circuit) does not require a center-tapped transformer secondary, and makes better use of the copper in the secondary since it uses all of it on both

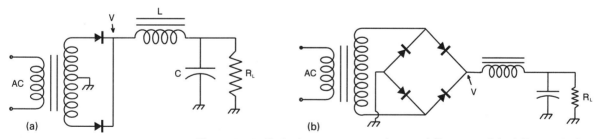

Figure 10-1. Choke-input power supplies: (a) full-wave and (b) full-wave bridge.

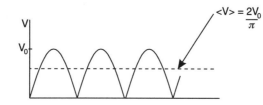

Figure 10-2. Voltage waveform at the filter input.

halves of the cycle. The two-rectifier circuit, on the other hand, has only half as much rectifier voltage drop and, for that reason, is usually chosen for low-voltage supplies (e.g. 5 V and less). This keeps the efficiency reasonably high by keeping the diode dissipation much lower than the power delivered to the load. In these two circuits the choke could just as well be put in the ground return; this is actually preferred because it will place less voltage stress on the choke.

INHERENT REGULATION OF THE CHOKE-INPUT POWER SUPPLY

Here regulation means voltage stability with respect to changes in the load, R_L. (A separate consideration is regulation with respect to variations in the input ac voltage.) It is important to see that the choke-input circuit is inherently regulated as long as the current through the choke is always positive. Here is why: If the current through L is always positive then either one rectifier or the other is forward biased and has nearly zero voltage drop. The left side of L is therefore always connected to the transformer secondary, and the voltage will be as shown in Figure 10-2, complete half-cycle cusps from zero to the peak secondary voltage. This waveform, in as much as it derives from a perfect voltage source (zero impedance), is independent of load current. Its average value is $2V_0/\pi$ where V_0 is the peak voltage. Since there can be no dc voltage across an (ideal) inductor, the dc component on R_L is the same, $2V_0/\pi$, and the regulation is perfect except for the small ohmic losses in the rectifiers, transformer windings, and choke, and the leakage inductance in the transformer.

Let us see what is necessary to maintain a net positive current in the choke. The L-section LC filter suppresses ripple, so the ac voltage at the right-hand side of the inductor is negligible compared to the ac voltage at the left-hand side. At the left-hand side, most of the ac voltage is the 120 Hz component of the rectifier output. If the peak rectifier voltage is V_0, then the peak voltage of the 120 Hz component is $4V_0/(3\pi) = 0.42V_0$. The peak ac current through the choke will be $4V_0/(3\pi\omega L)$. To keep the current from bottoming out, the dc current must be greater than this peak ac current. Therefore, we have $2V_0/(\pi R_L) > 4V_0/(3\pi\omega L)$ or $L > 2R_L/(3\omega)$. This minimum inductance is known as the critical inductance. In some applications the load may vary such that the current sometimes drops to very low values, even zero. A minimum current can be guaranteed by paralleling a resistor (known as a "bleeder") across the load. Unfortunately, the resistor will consume some power. High-voltage power supplies very often must include a bleeder for safety: it discharges the filter capacitor when the supply is turned off. Another method to maintain constant current is to use a "swinging" choke, which has high inductance when needed (i.e. at low current) but is allowed to suffer reduced but still sufficient inductance at high current (where the core saturates).

RIPPLE

The L and C form a simple L-section ripple filter. In the above single-phase full-wave power supply the peak value of the 120 Hz component at the input of the filter (left side of the choke) is $0.42V_0$. The capacitor will normally have a reactance much smaller than the load resistance, R_L, so we can find the output ripple voltage by considering just the L and C as a voltage divider:

$$V_{\text{ripple (peak)}} = 0.42V_0X_C/(X_L - X_C).$$

Sometimes resonant LC circuits are used, as in Figure 10-3, to suppress the primary ripple frequency even more. A conventional LC network can follow to filter the higher harmonics.

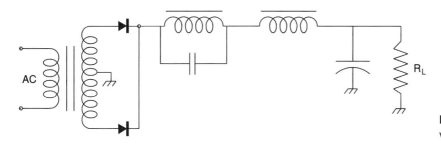

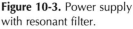

Figure 10-3. Power supply with resonant filter.

HALF-WAVE RECTIFIER

The simplest rectifier circuit uses a single diode (Figure 10-4). Obviously this arrangement does not make good use of the transformer secondary, and also has considerably larger ac component than the full-wave circuit. Moreover, the fundamental ac component is at 60 Hz instead of 120 Hz,

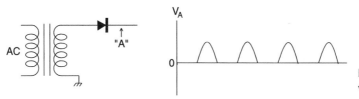

Figure 10-4. Simple half-wave rectifier circuit.

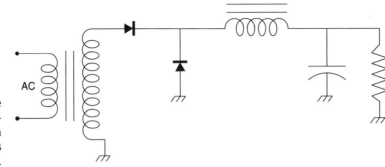

Figure 10-5. A half-wave circuit can also have choke-input regulation if a freewheeling diode is included.

and therefore requires more *LC* filtering for a given amount of output ripple. Good regulation requires a choke-input filter but we will not have positive current flowing constantly through the choke if it is connected to a single diode that is back-biased half the time. The solution is to add a "freewheeling" diode (common in switching power supply circuits). Figure 10-5 shows this two-diode arrangement. When the voltage at the top of the secondary becomes negative and "disconnects" the top diode, the freewheeling diode continues to supply current to the inductor. The voltage at the left-hand side of the choke is therefore always the positive secondary voltage or zero volts, and the circuit can have the inherent regulation of the full-wave choke-input power supply.

The simplest power supplies use an *RC* filter (no choke). Power is lost in the resistor and regulation is poor; the average voltage at the rectifier output depends on the load. The resistor, *R*, limits the current to protect the diode. Usually the transformer has enough ohmic loss and leakage inductance, so this resistor is omitted.

ELECTRONICALLY REGULATED POWER SUPPLIES

If a basic power supply, such as any of those shown above, is followed by a series "pass transistor," then feedback circuitry can control the pass transistor to maintain a constant output voltage. A typical circuit is shown in Figure 10-6. Note that the pass transistor must dissipate power and that such a regulator circuit is really just a class-A amplifier. A parallel transistor can be used as an active bleeder resistor to make a shunt-regulated power supply.

Transformers and chokes are expensive and heavy, so are seldom found in consumer electronics. In a television set, for example, the main power supply is often a simple half-wave or full-wave bridge rectifier operating off the line to produce about 140 V. (These so-called "hot chassis" sets require that a 1:1 isolation transformer be used during servicing.) The horizontal amplifier circuit, which is the biggest power consumer in the set, can use a

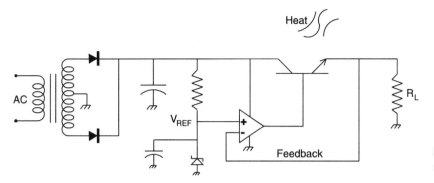

Figure 10-6. Electronically regulated power supply.

transistor that operates well at this voltage. If the circuit had to operate at lower voltage, power would be wasted in a dropping resistor, or the simple half-wave rectifier would have to be replaced by a switching rectifier (which we will consider later). The rest of the circuit blocks in a television set operate on lower voltages. It would be possible to put these blocks in a series string across the 140 V, but, instead, various lower supply voltages are derived from the horizontal output, which, as we will see later, does double duty as an amplifier and a switching power supply.

THREE-PHASE RECTIFIERS

Big power supplies, like big motors, benefit from three-phase power, which allows more efficient power transmission and lighter transformers. Let us examine a few common circuits.

Figure 10-7 shows a three-phase half-wave supply. This is already an improvement over the full-wave single-phase supply. The voltage at the input to the choke has a higher ripple frequency (180 versus 120 Hz) and much less ripple (it never goes to zero). Choke-input filters can be used without adding any freewheeling diode(s).

A three-phase full-wave rectifier is shown in Figure 10-8. This circuit uses center-tapped secondary windings. The voltage at the input to the

Figure 10-7. Three-phase half-wave power supply.

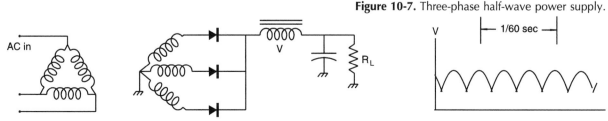

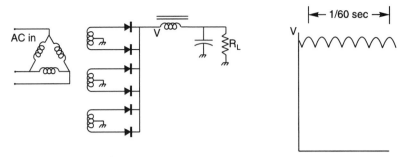

Figure 10-8. Three-phase full-wave power supply.

choke now has very low ripple, and the fundamental ripple frequency is 360 Hz (does not require as large L and C). Note that it is really a six-phase circuit; each rectifier connects to one of the six phases.

A three-phase full-wave bridge circuit is shown in Figure 10-9. This bridge rectifier has the same relation to the previous circuit as the four-diode full-wave bridge has to the two-diode ordinary full-wave circuit. Again there are six equivalent phases. You may recognize this circuit as the one used in automobile alternators.

Figure 10-9. Three-phase full-wave bridge power supply.

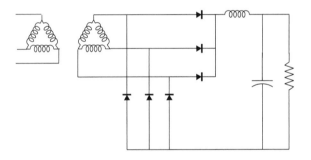

PROBLEMS

1. The half-wave power supply shown below supplies 25 V dc at 0.2 A to the load (shown as a resistor). Assume the components are ideal and that the capacitor is large enough so that the ac voltage at the output (ripple) is very small.

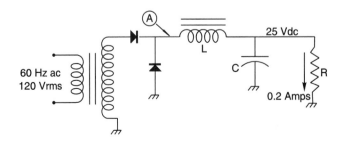

(a) Assuming the inductance of the choke is large enough to maintain positive current, sketch the voltage waveform at point "A."

(b) The value of the average voltage at "A" must be 25 V since there is no dc drop across the choke. Find the root mean square (rms) voltage, $V_0/\sqrt{2}$ across the transformer secondary. (*Answer*: 55.5 V rms.)

(c) Calculate the peak value of the dominant ac component (the 60 Hz component) of the voltage at "A," that is, find its Fourier amplitude. If you have trouble with this, estimate the magnitude of this component from your sketch. (*Answer*: 39.3 V peak.)

(d) Use the result from Problem 1(c) to find the critical inductance of the choke, that is, the inductance just large enough to keep the current in the choke always positive. (*Answer*: 0.52 H.)

(e) What value should the capacitor have to make the 60 Hz ripple at the output be less than 100 mV r.m.s.? (*Answer*: 3700 μF.)

2. Consider the common half-wave capacitor-input power supply shown below. The resistors r and R represent respectively the ohmic resistance and the load resistance of the diode. Assume that the transformer is perfect and that C is large enough to make any ac voltage at the output negligible. Suppose that the peak voltage on the transformer secondary is 10 V, and that the values of r and R are 1 and 20 ohms, respectively. Calculate the output voltage, V_{dc}.

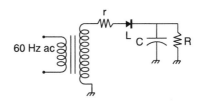

Hint: The average current through the diode is equal to the current in the load, $V_{dc}/20$. Current flows in the diode only when the voltage at the top of the transformer secondary is greater than V_{dc}; the instantaneous value of the current is $(10\cos\theta - V_{dc})/1$. Integrate this current over one cycle to find its average and set this expression equal to $V_{dc}/20$. You will need a calculator to solve for the angle at which the current stops flowing. (*Answer*: 7.51 V.)

3. A not so common choke-input half-wave power supply is shown below. This circuit, without the freewheeling diode, will not provide the inherent regulation of the circuit of Problem 1. If the load is constant, however, regulation is not important, and with this circuit the voltage can be lowered (with no power loss) by increasing the inductance of the choke. Assume as before that the value of the capacitor is large enough to make any ac voltage at the output negligible.

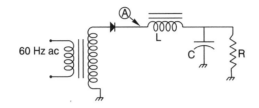

(a) Draw a sketch showing the voltage at point "A" and the current in the inductor. *Hint:* Remember that $V = L\, dI/dt$. It may help you to consider the cases where R is very large and where R is very small.

(b) If the 60 Hz voltage at the transformer secondary is $10\sin(\omega t)$ and the output voltage is 5 V, find the duty cycle of the diode, that is, the angle at which the diode goes into conduction (where the diode becomes forward biased) and the angle at which it goes out of conduction (where the choke goes dry, that is, the choke current goes to zero.)

(c) If the output current is 1 A, find the value of L.

AMPLITUDE MODULATION 11

Modulation means adding information to an otherwise pure sinusoidal carrier wave by varying the amplitude or the phase (or both). The simplest, amplitude modulation (AM) is on/off keying. This binary AM can be accomplished with just a switch (telegraph key) connected in series with the power source. The earliest voice transmissions used a carbon microphone as a variable resistance in series with the antenna. Amplitude modulation is used in the long-wave, middle-wave, and short-wave broadcast bands.* Let us examine AM, first in the time domain and then in the frequency domain.

AM IN THE TIME DOMAIN

Without modulation (when the music or speech is silent) the voltage and current at the antenna are pure sine waves at the carrier frequency. The rated power of a station is defined as the transmitter output power when the modulation is zero. The presence of an audio signal changes the amplitude of the carrier. Figure 11-1 shows a hypothetical AM transmitter and receiver. The audio signal (amplified microphone voltage) has positive and negative excursions, but its average value is zero. Suppose the audio voltage is bounded by $+V_m$ and $-V_m$. A dc bias voltage of V_m volts is added to the audio voltage. The sum, $V_m + V_{audio}$, is always positive, and is used to multiply the carrier wave, $\sin(\omega_c t)$. The resulting product is the AM signal; the amplitude of the RF sine wave is proportional to the biased audio signal. The simulation in Figure 11-2 shows the various waveforms in the transmitter and receiver of Figure 11-1 corresponding to a random segment of an audio waveform. The biased audio waveform is called the modulation envelope. Note that at full modulation where $V_{audio} = +V_m$,

*The long-wave (LW) band, extends from 153 to 179 kHz. The middle-wave (MW) band extends from 520 to 1,700 kHz. Short-wave bands are usually identified by wavelength: 75, 60, 49, 41, 31, 25, 19, 16, 13, and 11 m. The spacing between LW frequency assignments is 9 kHz. MW spacings are 10 kHz in the Western hemisphere and 9 kHz elsewhere. Short-wave frequency assignments are less coordinated, but almost all short-wave stations operate on frequencies that are an integral number of kilohertz.

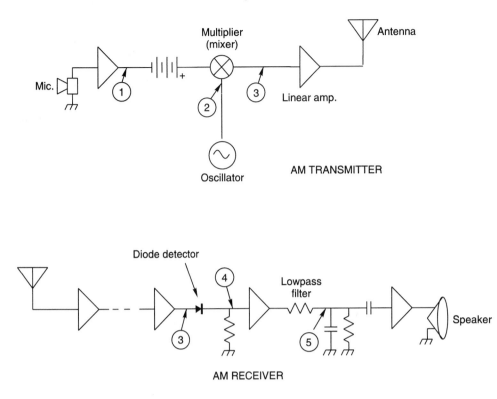

Figure 11-1. Hypothetical AM transmitter and receiver.

the carrier is multiplied by $2V_m$ whereas at zero modulation the carrier is multiplied by V_m (bias only). This factor of two in amplitude means the fully (100%) modulated signal has four times the peak power of the un-modulated signal (carrier wave alone). It follows that the antenna system for a 50,000 W AM transmitter must be capable of handling 200,000 W peaks without breakdown. The average power of the modulated signal is determined by the average square of the modulation envelope. For example, in the case of 100% modulation by a single audio tone, the average power of the modulated signal is greater than the carrier by a factor of $< (1 + \cos \theta)^2 > = 3/2$.

Figure 11-2 also shows how the receiver demodulates the signal, that is, how it detects the modulation envelope. The detector is just a rectifier diode that eliminates the negative cycles of the modulated RF signal. A simple RC low-pass then produces the average voltage of the positive loops. (The average voltage of these sinusoidal loops is just their peak voltage times $2/\pi$, so the average is proportional to the peak, that is, the envelope.) Finally, ac coupling removes the bias, leaving an audio signal identical to the signal from the microphone.

$t := 0, .01 .. 5$

$Vaud(t) := 3.6 \cdot \sin(t) - 5.3 \cdot \cos(3 \cdot t)$

$Carrier(t) := \sin(40 \cdot t)$

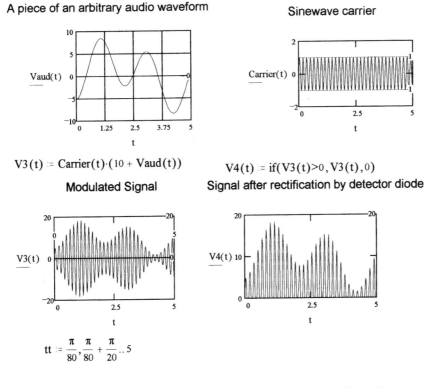

A piece of an arbitrary audio waveform

Sinewave carrier

$V3(t) := Carrier(t) \cdot (10 + Vaud(t))$

Modulated Signal

$V4(t) := if(V3(t) > 0, V3(t), 0)$

Signal after rectification by detector diode

$tt := \dfrac{\pi}{80}, \dfrac{\pi}{80} + \dfrac{\pi}{20} .. 5$

The envelope, shown below, is recovered by a low-pass filter filter.

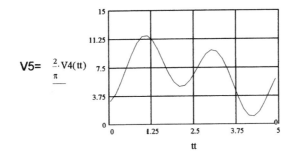

$V5 = \dfrac{2}{\pi} \cdot V4(tt)$

Figure 11-2. Waveforms in the AM system of Figure 11-1.

AM IN THE FREQUENCY DOMAIN

Let us look at the spectrum of the AM signal, that is, how the power is distributed in frequency. This is easy when the audio signal is a simple sine wave, say $V_a \, sin(\omega_a t)$. The voltage at point 4 in Figure 11-1 is then $\sin(\omega_c t)[V_m + V_a \sin(\omega_a t)] = V_m \sin(\omega_c t) + 1/2 V_a \cos[(\omega_c - \omega_a)t] - 1/2 V_a \cos[(\omega_c + \omega_a)t]$. These three terms represent the *carrier* at ω_c with power

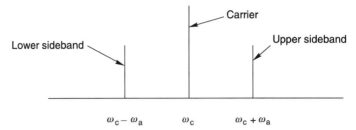

Figure 11-3. Spectrum of an AM signal with single-tone modulation.

$V_m^2/2$, a *lower sideband* at $\omega_c - \omega_a$ with power $V_a^2/8$, and an *upper sideband* at $\omega_c + \omega_a$, also with power $V_a^2/8$. This spectrum is shown in Figure 11-3. Note that the carrier, the component at ω_c, is always present and that its amplitude is independent of the modulation level. At 100% sine wave modulation $V_a = V_m$, and the average power in each sideband is one-quarter of the carrier power. The maximum average power is therefore 3/2 times the carrier power, as we saw earlier from the time domain description. The sidebands, which carry all the information, account here for only one-third of the total power transmitted. With speech waveforms the sidebands may contain even less power. Question: could the carrier be eliminated to save energy? A typical audio signal has components at frequencies up to at least 10 kHz. Each component produces a pair of sidebands that straddle the carrier. AM broadcast stations restrict their audio to not exceed 10 kHz, but this still produces a spectrum 20 kHz wide. The spacing between frequency assignments in the AM broadcast band is only 10 kHz. To prevent overlap, no two stations in a given listening area are assigned the same frequency or adjacent frequencies.

HIGH-LEVEL MODULATION

The simple transmitter shown in Figure 11-1 is entirely practical. (Of course we would fix it up so that a battery would not be needed to supply the bias voltage.) The linear RF power amplifier would be the major part of this transmitter. Because of the constant carrier, this power amplifier would be quite inefficient (see Problem 4). We have already seen that class-C and class-D amplifiers, which are extremely nonlinear but have high efficiency for all signal levels, are well suited for AM modulation because their RF output amplitude is equal to the applied dc supply voltage. These amplifiers are therefore equivalent to multipliers that form the product of the power supply voltage times a unit sine wave at the RF frequency. In *high-level modulation*, a high-power biased audio voltage powers a class-C or class-D final amplifier. Let us look at several circuits for high-level modulation.

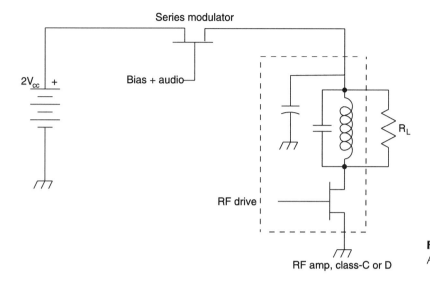

Figure 11-4. Class-A series AM modulator.

CLASS-A MODULATOR

A simple high-level modulator* is the class-A series modulator shown in Figure 11-4. Here the series "pass transistor" can operate as a follower, making this circuit inherently linear – a plus. But let us look at the efficiency. With no audio (carrier only) the pass transistor is already partially conducting, supplying V_{cc} to the amplifier. Assume, for simplicity, that the RF amplifier is 100% efficient (which is almost the case for a class-D amplifier). With no modulation (carrier only) the total supply voltage, $2V_{cc}$, divides equally across the series transistor and the amplifier. Since the currents are the same, the power dissipated as heat by the pass transistor is equal to the RF output power. The modulator is therefore only 50% efficient. The efficiency of this modulator increases to 75% for 100% modulation (see Problem 5). This modulator is simply too costly to operate in high-power broadcast stations.

CLASS-B MODULATOR

The traditional AM transmitter uses a class-B modulator in which audio voltage developed by a high-power class-B amplifier is added to the dc bias. In the circuit of Figure 11-5, the dc bias is fed through the secondary of the modulation transformer. Audio voltage produced by the modulator

---

*In standard AM practice the term *high-level modulator* refers to the high-power audio amplifier used to generate the varying supply voltage for the amplifier, but it is really the class-C or class-D amplifier that does the modulation, that is, effectively multiplies an RF sine wave by the (biased) audio voltage.

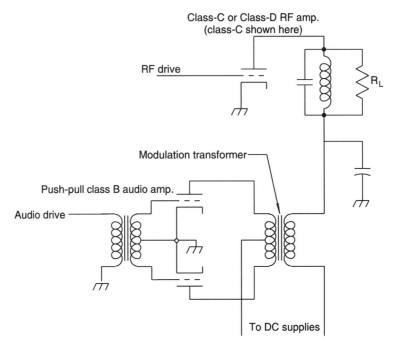

Class-C or Class-D RF amp.
(class-C shown here)

RF drive

R_L

Modulation transformer

Push-pull class B audio amp.

Audio drive

To DC supplies

Figure 11-5. Class-B plate-modulated AM transmitter.

appears across the secondary and adds to the bias voltage. With no audio present, the class-B audio amplifier consumes negligible power, and the bias voltage supply provides power for the carrier. At 100% sine wave modulation the modulator must supply audio power equal to half the bias supply power. A 50,000 W transmitter thus requires at least a 25,000 W audio amplifier for high-level modulation. (This result is for a single tone but is essentially the same for speech.) The carrier-only efficiency is just that of the amplifier alone. With 100% modulation, an AM transmitter using high modulation will have an overall efficiency given by

$$\eta = P_{\text{out}}/P_{\text{dc}} = 1.5\,\eta_{\text{RF}}/(1 + 0.5/\eta_{\text{mod}}) \tag{11-1}$$

where P_{out} is the RF output power, P_{dc} is the power furnished by the dc supplies, and η_{RF} and η_{mod} are the efficiencies of the RF amplifier and the modulator, respectively. For the transmitter shown in Figure 11-5 we might have $\eta_{\text{RF}} = 80\%$ and $\eta_{\text{mod}} = 0.78\%$ for an overall efficiency of 73%.

CLASS-S MODULATOR

There are several newer methods to produce AM with even higher efficiency. All use switching techniques. The pulse-width modulator shown in Figure 11-6 is a high-efficiency version of the series modulator of Figure 11-4. It uses a switching regulator circuit (sometimes called a class-S ampli-

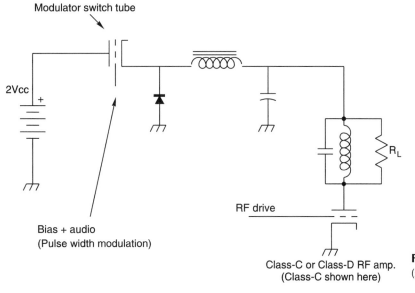

Modulator switch tube

2Vcc

Bias + audio
(Pulse width modulation)

R_L

RF drive

Class-C or Class-D RF amp.
(Class-C shown here)

Figure 11-6. Class-S (switching) AM modulator.

fier) to supply V_{cc} to a class-C or class-D amplifier. We will see this kind of circuit later in switching power supplies. Its operation is the same as that of a choke-input power supply: the voltage at the left side of the choke is always either $2V_{cc}$ (switch tube on) or zero (switch tube off). The duty cycle of the switch tube therefore determines the average voltage at the choke input. The choke and filter capacitor remove the chopping waveform of the switch tube. A low-level circuit produces a stream of pulses to drive the switch. The duty cycle of the pulse stream is proportional to the biased audio voltage.

DIGITAL-TO-ANALOG MODULATOR

Probably the most efficient high-level modulator is just a high-power digital-to-analog converter. It uses solid state switches to add the voltage of many separate low-voltage power supplies rather than tubes or transistors to drop the voltage of a single high-voltage supply. Modulators of this type have efficiencies above 95% and are built by Continental Electronics and Brown Boveri. Figure 11-7 is a simplified schematic diagram of one of these digital-to-analog modulators. An analog-to-digital converter is used at the audio input. This can be omitted if the audio signal is available in digital format. Still another design combines the RF square wave outputs of many class-D amplifiers. Only enough amplifiers are made active at any point on the audio waveform to achieve the desired RF output amplitude. In this scheme there is control circuitry but no actual modulator.

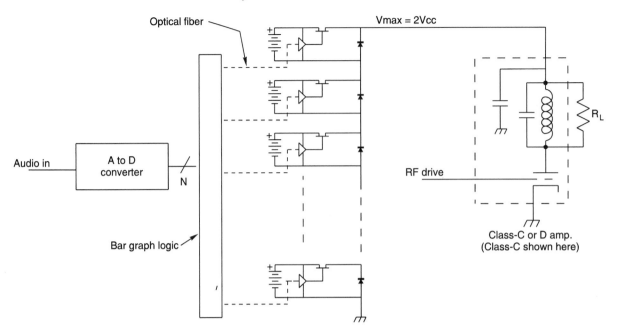

Figure 11-7. Digital-to-analog AM modulator.

CURRENT PRACTICE

Standard AM broadcast band transmitters in the USA are limited to 50 kW (Station WLW in Cincinnati ran a famous experimental 500 kW transmitter in the late 1930s). Many 1–2 MW AM stations operate in other parts of the world in the long-wave, medium-wave, and short-wave bands. These superpower transmitters all use vacuum tubes, usually as class-C amplifiers. AM stations are beginning to use all solid-state transmitters that combine power from a number of amplifiers in the 1 kW range. The advantages include lower (safer) voltages, indefinite transistor life rather than expensive tube changes every 1 or 2 years, and better "availability;" a defective module only lowers the power slightly and can be replaced while the transmitter remains on the air.

PROBLEMS

1. You may have observed someone listening to distorted sound from an AM radio whose tuning is not centered on the station. Often this mistuning is done deliberately when the listener has impaired high-frequency hearing and/or the radio has insufficient bandwidth. What is going on here? Why would a radio not have sufficient bandwidth and why would insufficient bandwidth cause some listeners to tune slightly off station?

2 We have seen that an AM transmitter, 100% modulated with an audio sine wave, has sidebands whose total power is equal to half the carrier power. Consider 100% modulation by an audio square wave. What is the ratio of sideband power to carrier power?

3. Suppose you are trying to listen to a distant AM station but another station on the same frequency is coming in at about the same strength. Will you hear both programs clearly or will they somehow interfere with each other?

4. In the AM transmitter of Figure 11-1 the modulated signal is produced at a low power level. A linear RF amplifier then produces the required high power. No high-power modulation equipment (high-voltage switchers, transformers, etc.) is needed. The disadvantage is low efficiency, even with a class-B amplifier, since the amplifier must be capable of supplying a peak power of four times the carrier power. Show that when the audio level is zero (carrier only), the efficiency of the class-B amplifier is 39% ($\pi/8$).

5. Show that the efficiency of the transmitter of Figure 11-4 is 75% under conditions of 100% modulation by a single audio tone. (Assume the RF amplifier is 100% efficient.)

6. During periods where the audio signal level is low, the amplitude of an AM signal varies only slightly from the carrier level. The modulation envelope, which carries all the information, rides on top of the high-power carrier. If the average amplitude could be decreased without decreasing the amplitude of the modulation, power could be saved. Discuss how this might be accomplished at the transmitter, and what consequences, if any, it might have at the receiver.

12 SUPPRESSED CARRIER AM

Ordinary AM is inefficient because the carrier wave, which carries no information, accounts for most of the power. Why not save power by transmitting only the sidebands and then add a locally generated carrier in the receiver? It is easy enough to modify the transmitter to eliminate the carrier. Look at the circuit diagram given previously for an AM transmitter (see Figure 11-1). Replace the battery by a zero-volt battery (a wire), and the carrier is gone. The resulting signal is known as *double-sideband suppressed carrier* (DSBSC). To restore the missing carrier we might try the receiver circuit shown in Figure 12-1. The locally generated carrier (beat frequency oscillator or BFO) is simply *added* to the intermediate frequency (IF) signal just ahead of the envelope detector. In a superheterodyne receiver we only have to generate this carrier at one frequency, the IF. As you would expect, the added carrier must have the right frequency. But it must also have the right phase. Suppose, for example, the modulation is a single audio tone at 400 Hz. If the replacement carrier is off by 45° in phase, the output of the envelope detector will be severely distorted. And if the phase is off by 90° the output of the detector will be a tone at 800 Hz (100% distortion!). These example waveforms are shown in Figure 12-2.

Generating a replacement carrier with the correct phase is fairly easy if the transmitter provides a low-amplitude "pilot" carrier – just enough for the receiver to use as a phase reference. In our model transmitter we would go back to using a bias battery, but it would have only a small voltage. The receiver can isolate the pilot by means of a narrow bandpass filter, amplify it, and finally add it back in just ahead of the envelope detector as in

Figure 12-1. Hypothetical DSBSC receiver: additive carrier reinsertion and envelope detection

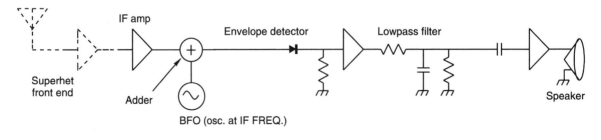

Demonstration that Detection of Double Sideband A.M. Requires the Correct Carrier Phase

$t := 0, .1 .. 150$

$carrier(t) := sin(t)$

$audio(t) := sin(.05 \cdot t)$ **Audio is sine wave at 1/20 carrier frequency**

$dssc(t) := audio(t) \cdot carrier(t)$ **Double sideband suppressed carrier (sidebands alone)**

Note that dssc(t) = 1/2(cos(.95t) − cos(1.05)t)

$sig(t) := carrier(t) + dssc(t)$ **Carrier plus sidebands = (1+audio)(carrier)= normal AM**

Sidebands plus correct carrier (note that the envelope is the sinewave audio)

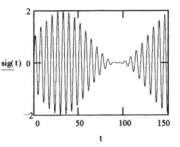

$alt(t) := dssc(t) + cos\left(t + \dfrac{\pi}{4}\right)$ **Give the re-injected carrier a 45 deg. phase error.**

Sidebands with carrier 45 deg. out of phase (Note envelope distortion)

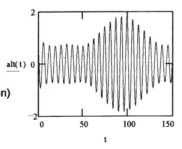

$alt2(t) := dssc(t) + cos(t)$ **Here the re-injected carrier has a 90 deg. phase error.**

Sidebands with carrier 90 deg. out of phase

(Note that envelope has twice the audio frequency - very severe distortion)

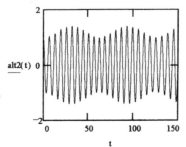

Figure 12-2. Detection of DSSC by *adding* a local carrier.

Figure 12-1. An equivalent circuit to generate the local carrier could use a phase-locked loop, for which the pilot signal would be the reference.

SINGLE SIDEBAND

The requirement of correct BFO phasing can be (almost) lifted if only one of the two sidebands is transmitted. The sidebands are mirror images of each other, so they carry the same information, and we do not really need both. We will discuss later three methods to eliminate one of the sidebands, but the first and most obvious method is to use a narrow bandpass filter centered on the desired sideband. The resulting signal is known as *single-sideband suppressed carrier* (SSBSC) or simply as *single-sideband* (SSB). Take the previous example of a single 400 Hz audio tone. The SSB transmitter will put out a single frequency, 400 Hz above the frequency of the (suppressed) carrier if we have selected the upper sideband. This signal will appear in the receiver 400 Hz away from the IF center frequency. Suppose we are still using the BFO and envelope detector. The envelope of the signal-plus-BFO will indeed be an undistorted sine wave at 400 Hz. So far, so good. But suppose the signal had been two tones, one at 400 Hz and one at 600 Hz. When the BFO carrier is added, the resulting envelope will have tones at 400 and 600 Hz, but it will also have a component at 200 Hz (600–400) that is surely distortion. Unwanted components are produced by the IF components beating with each other; their strength is proportional to the product of the two IF signals. The strength of each wanted component, on the other hand, is proportional to the product of its amplitude times the amplitude of the BFO. If the BFO amplitude is much greater than the IF signal the distortion can be reduced.

PRODUCT DETECTOR

The distortion described above can be eliminated by using a product detector instead of an envelope detector. In the receiver shown in Figure 12-3, the product detector is just the familiar multiplier (a.k.a. mixer). The output of this detector is the sum of the products of each IF signal component times the BFO. It does not produce cross-products of the various IF signal components. Product detectors are also called "baseband mixers" because they translate the sideband components all the way down to their original audio frequencies.

While the product detector does not produce cross-products (intermodulation distortion) of the IF signal components, the injected carrier should ideally have the correct phase. The wrong phase is of no consequence in our single-tone example. But when the signal has many compo-

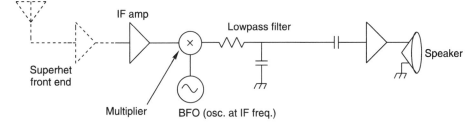

Figure 12-3. SSBSC receiver using a product detector.

nents, their relative phases are important. A waveform will be distorted if all its spectral components are given an identical phase shift (see Problem 3). It is only when every component is given a phase shift proportional to its frequency that a waveform is not distorted but only delayed in time. Therefore, an SSB transmitter, like the double-sideband transmitter, should really transmit a pilot carrier, and the receiver should lock its BFO phase to this pilot. Some SSB systems do just that. But for voice communications it is common to use no pilot. Speech remains intelligible and almost natural-sounding even when the BFO phase is wrong. The BFO can even be slightly off frequency. When the frequency is too high, a demodulated upper sideband (USB) voice signal has a lower than natural pitch. When it is too low, the pitch is higher than natural. Finally, what happens if we try to use a product detector with a free-running (not phase-locked) BFO to receive DSBSC? If the BFO phase is off by 90° the detector produces no output. If the BFO phase happens to be correct, the output will have maximum amplitude. For an intermediate phase error the amplitude will be reduced.

OTHER ADVANTAGES OF SSB

SSB, besides not wasting power on a carrier, uses only half as much bandwidth, so a given band can hold twice as many channels. Halving the receiver bandwidth also halves the background noise, so there is a 3dB improvement over AM in the signal-to-noise ratio. When a spectrum is crowded with AM signals, their carriers produce annoying beat notes in a receiver (a carrier anywhere within the receiver passband appears to be a sideband belonging to the spectrum of the desired signal). With SSB transmitters there are no carriers and no beat notes. Radio amateurs, the military, and aircraft flying over oceans use SSB for shortwave (1.8–30 MHz) voice communication.

GENERATION OF SSB

There are at least three well-known methods to generate SSB.

FILTER METHOD

In the filter method of Figure 12-4, a sharp bandpass filter removes the unwanted sideband. This filter usually uses crystal or other high-Q mechanical resonators, and has many sections. It is never tuneable, so the SSB signal is generated at a single frequency (a transmitter IF frequency) and then mixed up or down to the desired frequency. The filter, in as much as it does not have infinitely steep skirts, will cut off some of the low-frequency end of the voice channel.

PHASING METHOD

The phasing method uses two multipliers and two phase-shift networks to implement the trigonometric identity

$$\cos(\omega_c t)\cos(\omega_a t) \pm \sin(\omega_c t)\sin(\omega_a t) = \cos[(\omega_c \mp \omega_a)t]. \qquad (12\text{-}1)$$

The subscripts a and c denote the audio and (suppressed) carrier frequencies. If the audio signal is $\cos(\omega_a t)$, we have to generate the $\sin(\omega_a t)$ needed

Figure 12-4. SSB generator – filter method.

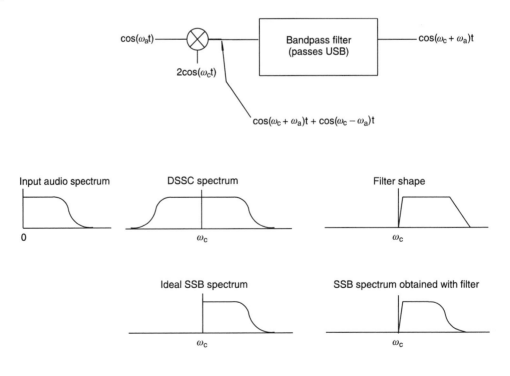

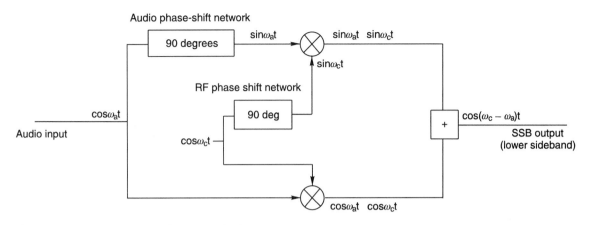

Figure 12-5. SSB generator – phasing method.

for the second term. A 90° audio phase shift network can be built with passive components, digital signal-processing techniques, and so on. The audio phase-shift network is usually implemented with two networks, one ahead of each mixer. Their phase *difference* is close to 90° throughout the audio band. The second term also requires that we supply $\cos(\omega_c t)$ but, since ω_c is fixed, anything that provides a 90° shift at this one frequency will suffice. Figure 12-5 illustrates the phasing method. If the adder is changed to a subtractor (by inverting the polarity of one input), the output will be the upper rather than the lower sideband.

WEAVER METHOD

A third method (due to Weaver, 1956 [2]) needs neither a sharp bandpass filter nor phase-shift networks. Weaver's method, shown in Figure 12-6, uses four multipliers and two low-pass filters. The trick is to mix the audio signal with a first oscillator whose frequency, ω_0, is in the center of the audio band. The outputs of the first set of mixers have the desired 90° phase difference. The second set of mixers then works the same way as the two mixers in the phasing method. Referring to Figure 12-6,

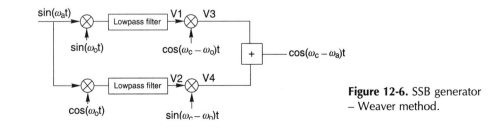

Figure 12-6. SSB generator – Weaver method.

$$V1 = \text{lowpass}[\sin(\omega_a t) \sin(\omega_0 t)] = \cos[(\omega_a - \omega_0)t]/2 \qquad (12\text{-}2)$$

$$V2 = \text{lowpass}[\sin(\omega_a t) \cos(\omega_0 t)] = \sin[(\omega_a - \omega_0)t]/2 \qquad (12\text{-}3)$$

$$V3 = \cos[(\omega_c - \omega_0)t] \cos[(\omega_a - \omega_0)t]/2 = \cos[(\omega_a - \omega_c)t]/4$$
$$+ \cos[(\omega_a + \omega_c - 2\omega_0)t]/4 \qquad (12\text{-}4)$$

$$V4 = \sin[(\omega_c - \omega_0)t] \sin[(\omega_a - \omega_0)t]/2 = \cos[(\omega_a - \omega_c)t]/4$$
$$- \cos[(\omega_a + \omega_c - 2\omega_0)t]/4 \qquad (12\text{-}5)$$

$$V_{\text{out}} = V3 + V4 = \cos[(\omega_c - \omega_a)t]/2. \qquad (12\text{-}6)$$

We see from Equation (12-6) that this particular arrangement generates the lower sideband.

SSB WITH CLASS-C OR CLASS-D AMPLIFIERS

The three methods described above for generating an SSB signal are done using low-level circuitry. Linear amplifiers then produce the required power for the antenna. There is, however, a way to use a class-C or class-D amplifier with simultaneous phase modulation and amplitude modulation to produce SSB. This follows from the fact that any narrow-band signal (continuous wave, AM, FM, phase modulation, SSB, DSBSC, narrow-band noise, etc.) is essentially a sinusoid at the center frequency, ω_0, whose phase and amplitude vary on a timescale much longer than $1/\omega_0$. Suppose we have generated SSB at low level. We can envelope detect it, amplify it, and use the amplified envelope to AM modulate the class-C or class-D amplifier. At the same time we can phase modulate the amplifier with the phase determined from the low-level SSB signal. The low-level signal is simply amplitude limited and then used to drive the modulated amplifier. What is the point of all this? The high-power SSB signal is produced with an amplifier that has close to 100% efficiency. Moreover, an existing AM transmitter can be converted to SSB by making only minor modifications.

BIBLIOGRAPHY

1. W. E. Sabin and E. O. Schoenike (1987), (eds) *Single-Sideband Systems and Circuits*. New York: McGraw-Hill.
2. D. K. Weaver, Jr (1956), "A third method of generation and detection of single-sideband signals", *Proc. IRE* 1703–6.

1. Some short-wave broadcasters have experimented with SSB. What are the possible advantages of SSB broadcasting over conventional AM broadcasting? What are some disadvantages? Do you think SSB will eventually replace AM for broadcasting?

2. For battery-operated transceivers (cellular phones, etc.) SSB AM could provide much greater battery life then AM (or FM). Are there disadvantages to using SSB for this application?

3. Demonstrate for yourself the kind of phase distortion that will occur when the BFO in a product detector does not have the same phase as the suppressed carrier. Use the Fourier decomposition of a square wave: $V(t) = \sin(\omega t) + \frac{1}{3}\sin(3\omega t) + \frac{1}{5}\sin(5\omega t) + \dots$. Have your computer plot $V(\theta) = \sin(\theta) + \frac{1}{3}\sin(3\theta) + \frac{1}{5}\sin(5\theta) + \dots + \frac{1}{9}\sin(9\theta)$ for θ from 0 to 2π. Then plot $V'(\theta) = \sin(\theta + 1) + \frac{1}{3}\sin(3\theta + 1) + \frac{1}{5}\sin(5\theta + 1) + \dots + \frac{1}{9}\sin(9\theta + 1)$, that is, the same function but with every Fourier component given an equal phase shift (here 1 radian). Would you expect these waveforms to sound the same? Consider the case in which the phase shift is 180°. This just inverts the waveform. Is this a special case or is an inverted audio waveform actually distorted?

4. The phase shift networks used in the phasing method of single sideband generation have flat amplitude versus frequency response – they are known as *all-pass* filters. All-pass filters have equal numbers of poles (all in the left-hand plane) and zeros (all in the right-hand plane). Find the amplitude and phase response of the network shown below, which is a first-order all-pass filter. (Note that the inverting op amp is used only to provide an inverted version of the input.)

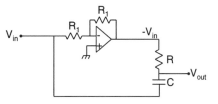

5. The most general all-pass filter can be obtained by cascading first-order all-pass sections (see Problem 4) with second-order all-pass sections. Find the amplitude and phase response of the network shown below, a second-order all-pass filter.

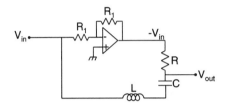

OSCILLATORS

The oscillator is a basic electronic building block used as the frequency-determining element of a transmitter or receiver (local oscillator (LO) or beat frequency oscillator (BFO)) or as a master clock in a computer, a wristwatch, or whatever. It has one function: to divide time into equal intervals. The invention of mechanical oscillators (clocks) made it possible to divide time into intervals much smaller than the earth's rotation period and much more regular than a human pulse rate. Mechanical clocks have an escapement mechanism that lets a weight fall or a spring unwind in a sequence of regular steps. Electronic oscillators are analogs of mechanical clocks.

RELAXATION OSCILLATORS

The earliest clocks used a "verge and folliot" escapement that resembled a torsional pendulum but was not a pendulum at all. Instead it operated as follows: torque derived from a weight or spring was applied to a pivoted mass. The mass accelerated according to torque $= I\,d^2\theta/dt^2$ (the angular version of $F = M\,d^2x/dt^2$) and moved accordingly. When θ reached a threshold, θ_0, the mechanism reversed and applied torque in the opposite direction. The mass then accelerated in the opposite direction. When it reached $-\theta_0$ the torque again reversed, and so on. The period depended on the moment of inertia, on the magnitude of the torque, and on how long the torque was applied. This clock is a "bang-bang" control system like most home-heating systems, where the temperature cycles between the turn-on and turn-off points of the thermostat.

Electronic bang-bang oscillators are common. One of the simplest is shown in Figure 13-1. When the voltage on the capacitor builds up to

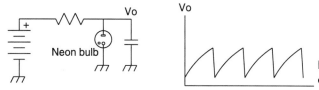

Figure 13-1. A neon bulb oscillator.

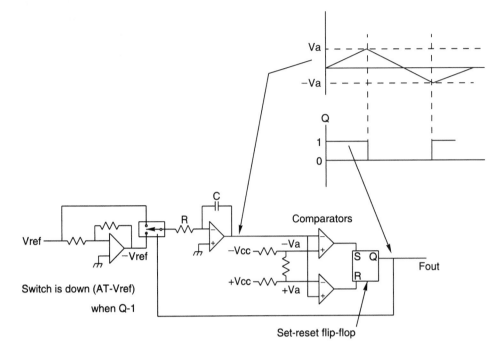

Figure 13-2. A bang-bang electronic oscilator.

about 60 V the neon bulb fires. The capacitor then discharges through the bulb until the voltage decays to about 40 V. The bulb extinguishes (for lack of sufficient voltage to maintain ionization), and the cycle begins anew.

A more precise circuit, shown in Figure 13-2, uses an operational amplifier (op amp), two comparators, a set-reset flip-flop and an electronic switch. You can see that the frequency is sensitive to the reference voltage and also to the values of R, C, and V_{cc} (since the comparators are referenced to V_{cc}). Still, this is a useful and popular circuit, and, in a slightly simplified form, is the circuit used in the 555 timer chip. Since the frequency depends linearly on V_{ref}, this oscillator is a natural voltage-to-frequency converter or, in RF terminology, a voltage-controlled oscillator (VCO).

Clock makers improved stability dramatically by using a true pendulum (a moving mass with a restoring force supplied by a hair spring or gravity). The pendulum has its own natural frequency, independent of amplitude. It moves sinusoidally – classic simple harmonic motion. The pendulum clock escapement uses positive feedback to push the pendulum in the direction of its motion, like pushing a swing. If you look carefully at how a pendulum clock works you will probably decide that it runs push–pull class C since there are two brief pushes on each cycle.

SINE WAVE ELECTRONIC OSCILLATORS

Electronic versions of the pendulum clock use LC resonant circuits, mechanical resonators (e.g. quartz crystals), or other resonators. A resonator with some initial energy (inductor current, capacitor charge, mechanical kinetic energy, etc.) will oscillate sinusoidally with an exponentially decaying amplitude. The decay is, of course, due to energy loss in the load, if any, and in the internal loss of the finite-Q resonator. This is shown in Figure 13-3, where the resonator is a parallel LCR circuit. We can prevent the exponential decay from occurring if we supply energy to make up for the energy dissipated in R. To do this we must pump current into the resonant circuit when its voltage is positive and/or pull current out when its voltage is negative. Figure 13-4 shows how a transistor and a dc supply are connected to provide this energy. The single transistor cannot supply negative current, but we can set it up with a dc bias as a class-A amplifier so that current values less than the bias current are equivalent to negative current. A transistor, by virtue of its high output impedance, is perfectly suited for use as a current source in this application. All that remains to complete this oscillator circuit is to provide the transistor drive, that is, the base-to-emitter voltage. We want to increase the conduction of the transistor when the output voltage (emitter voltage) increases, and we see that the emitter voltage has the correct polarity to be the drive signal.

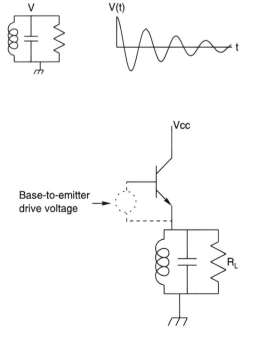

Figure 13-3. Damped oscillation in a parallel RLC circuit.

Figure 13-4. Transistor and dc supply to overcome damping.

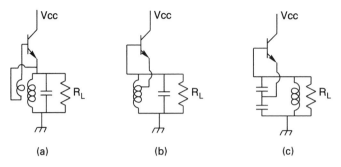

Figure 13-5. (a) Armstrong, (b) Hartley, and (c) Colpitts drive circuits.

(a)　　　　(b)　　　　(c)

The base-to-emitter voltage, which will have low amplitude (a fraction of a volt), must be a scaled down version of the sine wave on the emitter. Figure 13-5 shows three methods to come up with this drive signal. The Armstrong oscillator adds a secondary winding to the inductor to make a step-down transformer. The secondary voltage, which is the base-to-emitter voltage, is the desired small fraction of the emitter voltage. The Hartley oscillator accomplishes the same thing by connecting the emitter to a tap slightly below the top of the inductor. This is just an autotransformer version of the Armstrong oscillator if the magnetic flux links all the turns of the inductor. But the top and bottom portions of the inductor do not really have to be magnetically coupled at all; most of the current in the inductor(s) is from energy stored in the high-Q resonant circuit. This current is common to the two inductors, so they essentially form a voltage divider. (Note, though, that the ratio of voltages on the top and bottom portions of the inductor ranges from the turns ratio, when they are fully coupled, to the square of the turns ratio, when they have no coupling.) If we consider the totally decoupled Hartley oscillator – no mutual inductance – and then replace the inductors by capacitors of equal (but opposite) reactance and replace the capacitor by an inductor, we get the Colpitts oscillator.

Picking the Harltey circuit as an example, Figure 13-6 shows a finished circuit. It includes the standard biasing arrangement to determine the operating point and insure thermal stability. (A resistor voltage divider determines the base voltage, and an emitter resistor determines the emitter current.) A blocking capacitor allows the base to be dc biased with respect to the emitter. A bypass capacitor puts the bottom of the resonant circuit at RF ground.

Note that an oscillator is essentially an amplifier with a positive-feedback loop. But, because there is no input signal to have a terminal in common with the output signal, oscillators, unlike amplifiers, do not have common-emitter, common-collector, and common-base versions. Nevertheless, depending on the (arbitrary) point in the circuit that is an ac ground point, there are grounded-emitter, grounded-collector, and grounded-base versions of the three standard feedback oscillators. Figure

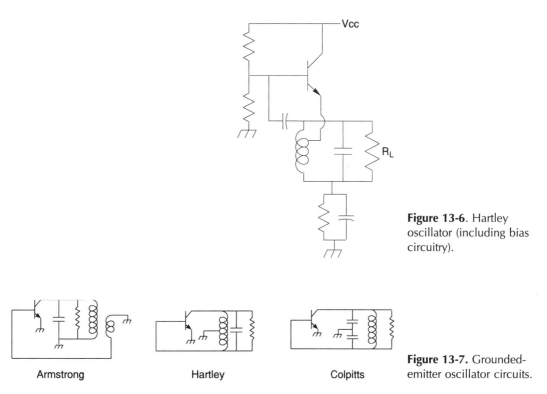

Figure 13-6. Hartley oscillator (including bias circuitry).

Armstrong   Hartley   Colpitts

Figure 13-7. Grounded-emitter oscillator circuits.

13-7 shows the basic RF circuitry of the grounded emitter oscillators. Most RF oscillators (below microwave frequencies) are Colpitts or Hartley oscillators, though, as drawn, they may be hard to recognize. With slight modifications, they may have other names.

AN UNINTENTIONAL OSCILLATOR

In RF work it is not unusual to have a newly built amplifier oscillate. One way this happens is shown in Figure 13-8. The circuit is a basic common-emitter amplifier with parallel resonant circuits on the input and output (as bandpass filters and/or to cancel the input and output capacitances of the transistor). When the parasitic base-to-collector capacitance of the transistor is included, the circuit has the topology of the decoupled Hartley oscillator. If the feedback is sufficient, it will oscillate. The frequency will be somewhat lower than that of the input and output circuits so that they look inductive as shown in the center figure. It is also called a TPTG oscillator, for "tuned-plate tuned-grid" (though with a transistor instead of a tube, it is a tuned-collector tuned-base oscillator). With luck, the loop gain will be less than unity at the frequency for which the total loop phase shift is 360°, and the amplifier will be stable. If not, it must be *neutralized*

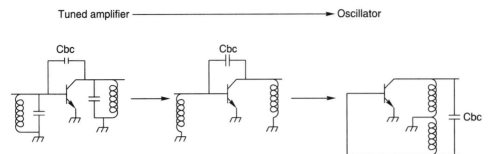

Figure 13-8. Tuned amplifier as an oscillator.

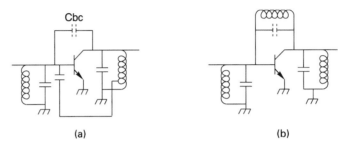

Figure 13-9. Amplifier neutralization.

to avoid oscillation. Two methods of neutralization are shown in Figure 13-9. In Figure 13-9a a secondary winding is added to provide an out-of-phase voltage which is capacitor coupled to the base to cancel the in-phase voltage coupled through C_{bc}. In Figure 13-9b an inductor from collector to base resonates C_{bc} to effectively remove it. (A dc blocking capacitor would be placed in series with this inductor.) In grounded-base transistor amplifiers and grounded-grid vacuum tube amplifiers the input circuit is shielded from the output circuit. These are stable without neutralization (but do not provide as much power gain).

SERIES RESONANT OSCILLATORS

The oscillators discussed above were all derived from the parallel resonant circuit shown in Figure 13-1. We could just as well have started with a series LCR circuit. Like the parallel circuit, a series LCR circuit executes an exponentially damped oscillation unless we can replenish the dissipated energy. In this case we need to put a voltage source in the loop that will be positive when the current is positive and negative when the current is negative, as shown in Figure 13-10. While a bare transistor with base-to-emitter voltage drive made a good current source for a parallel-mode oscillator, a low-impedance voltage source is needed for a series-mode

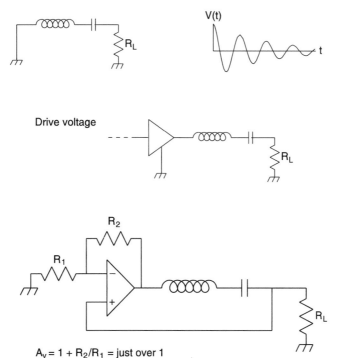

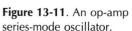

$A_v = 1 + R_2/R_1$ = just over 1

Figure 13-10. Series-mode oscillator operation.

Figure 13-11. An op-amp series-mode oscillator.

oscillator. In the series-mode oscillator shown in Figure 13-11, an op amp with feedback is such a voltage source. Since no phase inversion is provided by the tank circuit, the amplifier is connected to be noninverting. The op amp would limit this particular circuit to relatively low frequencies – no higher than a few megahertz. An emitter follower is an amplifier with low output impedance, and can be used in a series oscillator (see Problem 4). When the series *LC* circuit is replaced by a multisection *RC* ladder network, the resulting audio oscillator is commonly known as a phase-shift oscillator (even though all the feedback oscillators discussed here oscillate at the frequency at which the overall loop phase shift is zero).

NEGATIVE-RESISTANCE OSCILLATORS

In a parallel oscillator the transistor *sources* more current to the *RLC* circuit when the voltage increases. The transistor therefore functions as a negative resistance (an ordinary resistor *sinks* more (or sinks less) current when the voltage increases). But the transistor is a three-terminal device and, to function as a negative *R*, the third terminal is provided with a drive signal derived within the *LCR* tank. Figure 13-12 shows how two transistors can be used to make a negative resistance that is simply paralleled

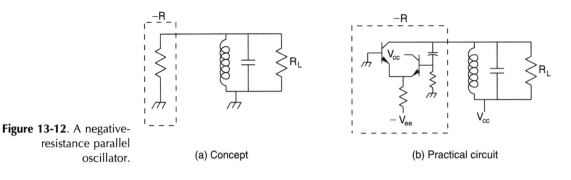

Figure 13-12. A negative-resistance parallel oscillator.

(a) Concept   (b) Practical circuit

with the *LCR* tank. The two transistors form an emitter-coupled differential amplifier in which the resistor to $-V_{ee}$ acts as a constant-current source, supplying a bias current, I_0. The input to the amplifier is the base voltage of the right-hand transistor. The output is the collector current of the left-hand transistor. The ratio of input to output is $-4V_T/I_0$, where V_T is the thermal voltage, $26\,\text{mV}$. This ratio is just the negative resistance, because the input and output are tied together. This negative resistance oscillator uses a parallel resonant circuit, but a series resonant version is certainly possible as well. You can see that the classic feedback oscillators are not very different from this negative-resistance oscillator. In those circuits the phase inversion is provided by passive components. Here it is provided by the phase inversion of an additional transistor.

Any circuit element or device that has a negative slope on at least some portion of its *I–V* curve can, in principle, be used as a negative resistance. Tunnel diodes can be used to build oscillators up into the microwave frequency range. At microwave frequencies, where the equivalent circuits for transistors contain many reactances, single-transistor negative-resistance oscillators are common. A plasma discharge exhibits negative resistance, and provided a pre–vacuum tube method to generate coherent sine waves. High-efficiency Poulsen arc transmitters, circa World War I, provided low-frequency sine waves at powers exceeding $100\,\text{kW}$.

OSCILLATOR DYNAMICS

These resonant oscillators are basically linear amplifiers with positive feedback. At turn-on they can get started by virtue of their own noise if they run as class A. A class C oscillator can get started from the initial turn-on transient. Once running, the signal level is ultimately limited by the power supply rails, a severe nonlinearity. The fact that the amplitude does not increase indefinitely shows that some nonlinearity is operative in every real oscillator. For waveform purity, the nonlinearity should be smooth and symmetric, for example $V_{\text{out}} = AV_{\text{in}} - BV_{\text{in}}^3$ where B/A is very small. (You can model an oscillator and experiment with the form of the nonlinearity and the amount of feedback – see Problem 3.)

STABILITY

Long-term (seconds to years) frequency fluctuations are due to component ageing and changes in ambient temperature, and are called drift. Short-term fluctuations, known as oscillator noise, are due to the combination of noise produced in the active device and the finite loaded Q of the resonant circuit. The higher the Q, the faster the loop phase shift changes with frequency. Any disturbances (transistor fluctuations, power supply variations changing the parasitic capacitances of the transistor, etc.) that tend to change the phase shift will cause the frequency to move slightly to reestablish the overall 360° shift. The higher the resonator Q, the smaller the frequency shift. Note that this is the *loaded* Q, so the most stable oscillators, besides having the highest Q resonators, are loaded as lightly as possible. The topic of oscillator noise will be taken up later in more detail to find just how narrow the best possible "lineshape" of an oscillator can be.

DESIGN EXAMPLE – COLPITTS OSCILLATOR

Let us design a complete grounded-emitter Colpitts oscillator beginning with the basic circuit of Figure 13-7. We again use the common biasing circuit, which is insensitive to device variations and provides protection from thermal runaway. The biased transistor is shown in Figure 13-13a. In

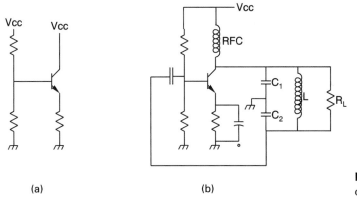

(a) (b)

Figure 13-13. Colpitts oscillator.

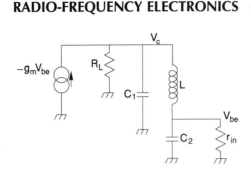

Figure 13-14. Colpitts oscillator equivalent circuit.

Figure 13-13b we have added the Colpitts resonant circuit. A large-value bypass capacitor puts the emitter at RF ground. The bottom of the inductor is connected to the base to provide the feedback connection. This connection is made through a blocking capacitor to avoid disturbing the bias on the base. Finally, the dc connection to the collector is made through an RF choke to avoid shorting the collector RF to ground.

We can design the bias circuit just as if we were building an amplifier. The collector waveform will have a peak-to-peak value of $2V_{cc}$, so the output power will be $P_0 = V_{cc}^2/R_L$. If we assume the oscillator will have class-A operation (conduction throughout the entire cycle) the efficiency will be 1/2. The input power will therefore be $2V_{cc}^2/(2R_L)$, and the average input current will be V_{cc}/R_L. The bias network is designed to obtain at least this value of quiescent current. An equivalent circuit for the oscillator of Figure 13-13b is shown in Figure 13-14. The load resistor is between the collector and ground, so the peak-to-peak output voltage will indeed be $2V_{cc}$. The transistor is represented by a simplified equivalent circuit consisting of a current generator and an input resistor. Normally C_2 will be much larger than C_1, so the collector sees a parallel resonant circuit with a resonant (angular) frequency given by $\omega = 1/\sqrt{LC_1}$ (which will be the oscillation frequency). Note that the necessary phase inversion comes directly from the voltage divider formed by L and C_2. If we assume that $r_{in} \gg X_{C_2}$ then $V_{be} = V_c \mid X_{C_2} \mid /(\mid X_{C_2} \mid - \mid X_L \mid)$, which inverts the phase for all frequencies above the resonant frequency of C_2, and L. Assuming $X_{C_2} \ll X_L$, the oscillation frequency will be the resonant frequency of L and C_1. The inductor is chosen to produce a high loaded Q, that is, $R_L/\omega L$ is made as large as possible. Note, however, that the equivalent parallel resistance of the inductor itself is part of the load, and it is the total loaded Q that is to be maximized. This consideration normally excludes very low values of inductance. From inspection of the equivalent circuit we can directly write down the condition for oscillation:

$$V_{be} \leqq \frac{V_c \mid X_{C_2} \mid}{\mid X_{C_2} \mid - \mid X_L \mid} = \frac{-V_{be} g_m R_L \mid X_{C_2} \mid}{\mid X_{C_2} \mid - \mid X_L \mid} \tag{13-1}$$

which becomes

$$\frac{g_m R_L \, |X_{C_2}|}{|X_L| - |X_{C_2}|} \geq 1. \qquad (13\text{-}2)$$

In practice, one usually makes C_2 somewhat smaller than this unity gain value to compensate for device variation and stray losses.

NUMERICAL EXAMPLE

Suppose we need an oscillator to supply 5 mW (+7 dBm) at 5 MHz. This oscillator is to operate from a +5 V power supply. We will use the circuit arrangement of Figure 13-13b with a common small-signal transistor, a 2N3904. We will assume there is no need for high efficiency in this low-power circuit, so we will bias the transistor to 10 mA (50 mW from the supply). This higher than necessary bias level provides more transconductance, because $g_m = I_{cc}/V_T$. Here $g_m = 10\,\text{mA}/26\,\text{mV} = 0.4$. This bias current will produce a drop of 1 V across the emitter bias resistor, leaving us with a maximum possible output of 4 V peak. Let us be pessimistic and assume we will have an output of only 3 V peak. This determines the value of the load resistance, $R_L = V_{peak}^2/2P = 3^2/(2 \times 0.005) = 900\,\text{ohms}$. (A matching network can be made to convert the load impedance, whatever it is, to 900 ohms.) We will use an inductor of 10 μH, which has a reactance of 314 ohms at 5 MHz. The loaded Q of the oscillator tank circuit will therefore be $R_L/X_L = 900/314$ or about 3. (In a crystal oscillator, the loaded Q could be over 1,000.) An amplifier with this tank circuit would have a fractional bandwidth of 1/3, but in an oscillator the positive feedback multiplies the Q to provide a much smaller fractional linewidth. (This will be explained in a later discussion of oscillator noise.) We already anticipate that $C_2 \gg C_1$, so C_1 resonates with L at 5 MHz. This gives $C_1 = 101\,\text{pF}$. From the above analysis, the condition for oscillation is

$$X_{C_2} = X_L/g_m R_L \sim X_{C_1}/g_m R_L$$

so

$$C_2 = C_1 g_m R_L = 101 \times 0.4 \times 900\,pF = 0.036\,\mu F.$$

To insure enough feedback, we will let C be about half this value, or 0.02 μF. The final circuit is shown in Figure 13.15. We should go back to justify the simplification of neglecting r_{in}, the input resistance of the transistor. This equivalent resistance is given approximately by $r_{in} = \beta V_T/I_0$. For this ordinary small-signal transistor, β is about 100 so $r_{in} = 100 \times 026/0.01 = 260\,\text{ohms}$. In many circuits this would be

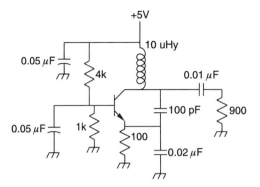

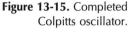

Figure 13-15. Completed Colpitts oscillator.

considered a low impedance, maybe even a short, but here, in comparison with the 1.6 ohm reactance of C_2, we can consider it an open circuit.

PROBLEMS

1. Draw a schematic diagram (without component values) for a bipolar transistor Colpitts oscillator with the collector at ground for both dc RF. Include the biasing circuit. The oscillator is to run from a positive dc supply.

2. A simple computer simulation can illustrate how an oscillator builds up to an amplitude determined by the nonlinearity of its active element. The program in the box models the negative resistance oscillator of Figure

```
'QBasic simulation of negative resistance oscillator of
'Figure 13-12a.
SCREEN 2
R = 1: L = 1/6.2832: C = L 'the parallel RLC circuit: 1 ohms,
1/2pi henries, 1/2pi farads.
RN = .9 'run program also with RN = 0.2 to see nonsinusoidal
'waveform.
E = 0.1    'negative resistance: I = (1/RN)*(V-EV^3).
V = 1: U = 0    'initial conditions, V is voltage, U is dV/dt.
DT = .005    'step size in seconds.
FOR I = 1 TO 3000
T = T + DT    'increment the time.
VNEW = V + U * DT
U = U + (DT / C) * ((1 / RN) * (U - 3 * E * V * V * U) - V / L - U / R)
V = VNEW
PSET (40 * T, 100 + 5 * V) 'plot the point.
NEXT I
```

13-12a. The LC resonant frequency is 1 Hz. This network is in parallel with a negative-resistance element whose voltage versus current relation is given by $I = -(1/R_n) \times (V - \varepsilon V^3)$, to model the circuit of Figure 13-12. The small-signal (negative) resistance is just $-R_n$. The term $-\varepsilon V^3$ makes the resistance become less negative for large signals. The program integrates the second-order differential equation for $V(t)$ and plots the voltage versus time from an arbitrary initial condition, $V = 1\,V$.

Run this program. Change the value of the load resistor R. Find the minimum value of R for sustained oscillation. Experiment with the values of R and R_n. You will find that when the loaded Q of the RLC circuit is high, the oscillation will be sinusoidal even when the value of the negative resistor is only a fraction of R. When Q is low (as it is for $R = 1$), a low value of R_N such as $R_n = 0.2$ ohm will produce a distinctly distorted waveform.

3. Design (without specifying component values) a single-transistor series mode oscillator based on the emitter follower circuit.

14 PHASE LOCK LOOPS

A phase lock loop (PLL) circuit forces the phase of a voltage-controlled oscillator (VCO) to follow the phase of a reference signal. Once lock is achieved, the average frequency of the VCO will be exactly equal to the average frequency of the reference. In one class of applications the PLL is used to generate a stable signal whose frequency is determined by an unstable (noisy) reference signal. Here the PLL is, in effect, a narrow bandpass filter that passes a carrier while rejecting its noise sidebands. Examples include telemetry receivers that lock onto weak telemetry signals from spacecraft, and various "clock smoother" circuits. In another class of applications the PLL is designed to follow all the phase fluctuations of the reference. An example is the PLL-based FM detector, where the VCO reproduces the input signal (normally in the intermediate frequency (IF) band). The voltage applied by the loop to the VCO is proportional to the instantaneous frequency (in as much as the VCO has a linear voltage-to-frequency characteristic), and this voltage is the audio output of the detector.

PHASE ADJUSTMENT BY MEANS OF FREQUENCY CONTROL

In a PLL, the VCO frequency, rather than phase, is determined by the control voltage, but it is easy to see how frequency control can set phase. Suppose you have two mechanical clocks. You want to make clock B agree with clock A. You notice that clock B is consistently 5 min behind clock A. You could exert direct phase control, using the time adjustment knob to set clock B ahead 5 min to agree with clock A. Or you could use frequency control, regulating the speed of clock B to run somewhat faster. Once clock B catches up with clock A you reset the frequency control to its original value. You may have done an electronic version of this in the laboratory (Figure 14-1). Suppose you have a two-channel oscilloscope. The first channel displays a sine wave from a fixed-frequency reference oscillator. The scope is synchronized to this reference oscillator so channel A displays a motionless sine wave. Channel B is connected to a variable-

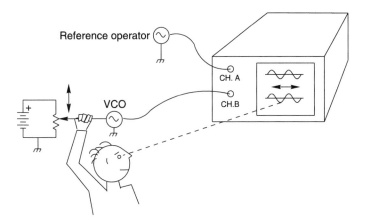

Figure 14-1. PLL – principle of operation.

frequency oscillator, which might be a laboratory instrument with a frequency knob or a VCO fitted with a potentiometer. Suppose the frequencies are the same. Then the sine wave seen on channel B is also motionless. If you lower the VCO frequency slightly, the channel B trace will drift to the right. And if you raise the VCO frequency, the trace will drift to the left. To align the two traces, you can shift the frequency slightly, to let the channel B trace drift into position, and then return the VCO frequency to the reference frequency value to stop the drift. The operator keeps the traces aligned, and is therefore an element of this PLL. To automate the loop we use an electronic phase detector, that is, an element that produces a voltage proportional to the phase difference between the reference and the VCO. The PLL circuit then looks like that shown in Figure 14-2. If the VCO phase gets behind (lags) the reference phase, the phase detector will produce a positive "error voltage" that will speed up the VCO. As the phase error then decreases, the error voltage also decreases, and the VCO slows back down. The exact way in which equilibrium is established, that is, the mathematical form of $\theta(t)$, depends on the yet-unspecified loop filter or "compensation network", shown as dashed box in Figure 14-2. We will examine phase detectors and loop filters further below.

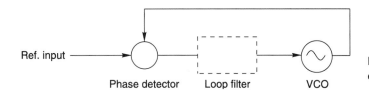

Figure 14-2. PLL block diagram.

MECHANICAL ANALOG OF A PLL

The PLL is a feedback control system – a servo. The electromechanical system shown in Figure 14-3 is a positioning system that makes the output shaft angle, θ, track the input shaft angle, θ_{ref}. The gear teeth are in constant mesh but the top gear can slide axially along the middle (idler) gear. A dc motor provides the power to turn the output shaft. The input shaft might be connected to the steering wheel of a ship, while the output shaft drives the rudder.

Study the operation of this servo system. (Assume for now that the voltage, V_{bias}, is set to zero.) When the input and output angles are equal, the slider on the potentiometer is centered, and produces zero error voltage. But suppose that the reference phase gets slightly ahead of the output phase – maybe the input crank has been abruptly moved clockwise. During this motion, the threads on the input shaft have caused the top gear to slide to the left, producing a positive voltage at the output of the potentiometer. This voltage is passed on by the loop filter to the power amplifier, and drives the motor clockwise (increasing θ). This rotates the top gear clockwise. The input shaft is now stationary, so the top gear moves on the threads to the right, reducing the potentiometer output voltage back to zero. The output phase is again in agreement with the

Figure 14-3. Positioning system with feedback control, a mechanical PLL.

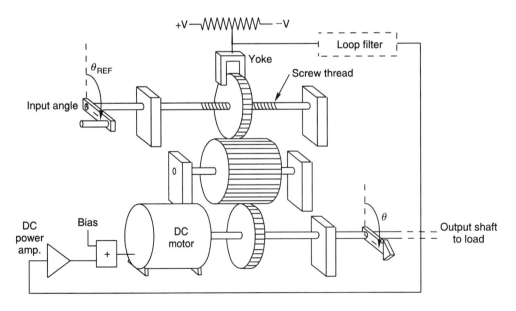

input or "reference" phase. In this system the gear–screw thread mechanism is the phase detector. (Note that the range of this phase detector can be many turns since it is limited only by how far the top gear can slide along the middle gear.) The negative-feedback error signal provides the drive to force the (high-power) output to agree with the (low-power) input command.

We see how this mechanical system operates as a power-steering servo. To see that it is also equivalent to a PLL, suppose that an operator is cranking the input shaft at a constant rate. The negative feedback will bring the output shaft up to the same rate. At equilibrium, the top gear is being turned by the middle gear, but the threaded shaft is turning just as fast. The top gear therefore does not move along the threads, but stays at an equilibrium position. The output frequency is perfectly locked to the input frequency. How about the phases; will they also agree exactly? They did in the nonrotating (power-steering) mode of operation, but now, with continuous rotation, they may not. The key is the loop filter. Suppose the loop filter is just a piece of wire (no filter at all). Because the motor is running continuously there will have to be a phase error in order to move the potentiometer off zero and produce a drive voltage. We can avoid this error by introducing the necessary voltage as V_{bias} at the motor input. Then the phase error can be zero – but only for operation at one particular speed. If the cranking rate (reference frequency) changes, the bias voltage will not be quite right, and some phase error must be established to adjust the motor voltage (analogous to VCO control voltage). How much phase error (*tracking* error) is necessary depends on the gain of the power amplifier. Increasing the gain of the amplifier/motor and/or the gain (sensitivity) of the phase detector will reduce the error. Set up this way, with no loop filter, this system is a *type I servo*; there is no phase error if the input is a constant (the power-steering mode) but there will be a phase error when the input is a ramp (PLL operation: the input phase is a ramp, $\theta_{\text{ref}}(t) = \omega_{\text{ref}}t$). If the loop filter contains an integrator, the system becomes a *type II servo*. The motor is driven by the integral of the error voltage history. The steady-state error voltage can be zero. In fact, it *must* be zero or the integrator output would be changing and the loop would not be in steady-state equilibrium.

NOTE   The type of servo is given by the number of integrators contained in its loop. Even without a loop filter, a PLL contains one integrator – its VCO (the motor in the mechanical system). The VCO frequency is, by definition, the time derivative of the phase. The phase is therefore proportional to the integral of the control voltage.

LOOP DYNAMICS

We have seen how the loop operates under static conditions when the input phase is either constant (power-steering mode) or a constant ramp (PLL mode). Let us now look at its response to disturbances in the input phase. We have already described qualitatively the response to a step function; the loop catches up with the reference. The way in which it catches up, that is, quickly, slowly, with an overshoot, or with no over-shoot, depends on the loop filter and the characteristics of the phase detector, amplifier, and motor. An exact analysis is straightforward if the entire system is *linear*, that is, if the system can be described with linear differential equations. In this case it can be analyzed with the standard techniques applied to linear electronic circuits – complex numbers, Fourier and Laplace transforms, and so on. In the PLL, the loop filter, at least, is linear because it consists of passive components (resistors and capacitors and, in a type II PLL, an operational amplifier (op amp). The VCO is linear if its frequency is a linear, that is, a straight line function of the control voltage. Of course, over a small operating region, any smooth voltage–frequency characteristic is approximately linear. The most commonly used phase detector is a simple multiplier (mixer). We will see that it produces an error voltage proportional to the sine of the phase difference. For small arguments, $\sin x = x$, so it is also linear over some restricted region. (The phase detector in our mechanical analog is linear over a range limited only by the length of the screw thread.) The motor speed, however, will generally depend on the inertia and friction of the load as well as the back electromotive force generated in the armature. Nevertheless, the motor in such a system can be linearized by wrapping an interior "tachometer feedback loop" around it.)

A linear system is described by its transfer function, that is, the complex amplitude of the sinusoidal output response to a unit amplitude input sine wave at frequency ω. For the PLL this input is a sine wave phase modulation at frequency ω superposed on top of the otherwise steady phase ramp, that is, $\theta_{\mathrm{in}} = \omega_{\mathrm{ref}}t + e^{j\omega t}$, and the output phase is $\theta_{\mathrm{out}} = \omega_{\mathrm{ref}}t + A\,e^{j\omega t}$. The principle of superposition allows us to disregard the continuous phase ramp while we find the response to the phase modulation. But before we can perform the analysis, we must specify the loop filter.

LOOP FILTER

A standard filter for type II loops uses an op amp to produce a weighted sum of the phase detector output plus the integral of the phase detector

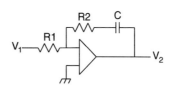

$\tau_1 = R_1 C$ $\tau_2 = R_2 C$

Figure 14-4. Loop filter circuit.

output. Figure 14-4 shows this circuit. Note that if we let C go to infinity and set $R_2 = R_1$, the response of this filter is just unity (except for a minus sign), so this circuit will serve to analyze the type I PLL as well as the type II PLL. Remembering that the negative input of the op amp is a virtual ground, we can write the transfer function of this filter:

$$-\frac{V_2}{V_1} = \frac{R_2 + 1/j\omega C}{R_1} = \frac{R_2}{R_1} + \frac{1}{j\omega C R_1} = \frac{\tau_2}{\tau_1} - \frac{j}{\omega \tau_1}. \qquad (14\text{-}1)$$

In the time domain, the relation between V_2 and V_1 is given by

$$\frac{dV_2}{dt} = -\frac{R_2}{R_1}\frac{dV_1}{dt} - \frac{V_1}{R_1 C}. \qquad (14\text{-}2)$$

LINEAR ANALYSIS OF THE PLL

Figure 14-5 is a block diagram of the loop, including the loop filter of Figure 14-4. Let us find the frequency response with respect to the reference phase, that is, $\theta(\omega)$ when the reference phase is $e^{j\omega t}$. (Remember that this excitation and response are in addition to the constant phase ramp $\omega_{\text{ref}} t$.) From inspection of Figure 14-5 we can write

$$-K_D(\theta - \theta_{\text{ref}})\left(\frac{\tau_2}{\tau_1} - \frac{j}{\omega \tau_1}\right)K_O = j\omega\theta. \qquad (14\text{-}3)$$

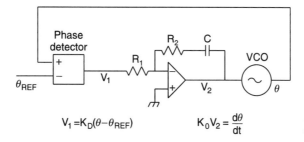

$V_1 = K_D(\theta - \theta_{\text{REF}})$ $K_O V_2 = \dfrac{d\theta}{dt}$

Figure 14-5. PLL block diagram.

Solving for θ, we have the frequency response

$$\frac{\theta(\omega)}{\theta_{\text{ref}}(\omega)} = \frac{1}{1 - 1/(K_D K_O/(\omega^2 \tau_1) + jK_D K_O \tau_2/(\omega \tau_1))} .$$ (14-4)

FREQUENCY RESPONSE OF THE TYPE I LOOP

Let us look at the frequency response for the type I loop by letting C go to infinity and letting $R_1 = R_2$, effectively eliminating the loop filter. The frequency response, Equation (14-4), becomes

$$\frac{\theta(\omega)}{\theta)\text{ref}(\omega)} \rightarrow \frac{1}{1 + j\omega/K}$$ (14-5)

where $K = K_D K_O$, the *loop gain*. The frequency response is identical to that of a simple RC low-pass filter with a time constant $RC = 1/K$. We saw earlier that the type I PLL needs high loop gain to have good tracking accuracy (ability to follow a changing reference frequency without incurring a large phase error). Here we see that high gain implies high bandwidth; the type I PLL cannot have both high gain (good tracking) and narrow bandwidth (ability to ignore noise – fast phase fluctuations – of the reference).

FREQUENCY RESPONSE OF THE TYPE II LOOP

To deal with the type II loop (the most common PLL) it is standard to define two constants, the *natural frequency*, ω_n, and the *damping coefficient*, ζ, as follows:

$$\omega_n = \sqrt{\frac{K_O K_D}{\tau_1}} \quad \zeta = \frac{\tau_2 \omega_n}{2} .$$ (14-6)

We will see soon that these names relate to the transient response of the loop but, for now, we will substitute them into Equation (14-4), the expression for the frequency response, to get

$$\frac{\theta(\omega)}{\theta_{\text{ref}}(\omega)} = \frac{\omega_n^2 + 2j\omega\zeta\omega_n}{\omega_n^2 + 2j\omega\zeta\omega_n - \omega^2} .$$ (14-7)

This transfer function is plotted against ω/ω_n for several different damping coefficients (Figure 14-6).

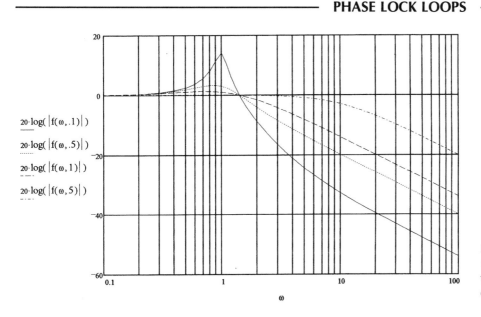

$20 \cdot \log(|f(\omega, .1)|)$

$20 \cdot \log(|f(\omega, .5)|)$

$20 \cdot \log(|f(\omega, 1)|)$

$20 \cdot \log(|f(\omega, 5)|)$

Figure 14-6. Frequency response versus ω/ω_n for a type II loop with various damping coefficients.

TRANSIENT RESPONSE

The transient response can be determined from the frequency response by using Fourier or Laplace transforms but this loop is simple enough that we can work directly in the time domain. Equation (14-2), gives the time response for the loop filter. We assume the loop has been disturbed but the reference is now constant. By inspection of Figure (14-5) we can write

$$\frac{dV_1}{dt} = K_D K_O V_2 = K V_2. \tag{14-8}$$

Combining this with Equation (14-2) we get

$$\frac{d^2 V_1}{dt^2} + K \frac{\tau_2}{\tau_1} \frac{dV_1}{dt} + \frac{K}{\tau_1} V_1 = 0 \quad \text{or} \quad \frac{d^2 V_1}{dt^2} + 2\omega_n \zeta \frac{dV_1}{dt} + \omega_n^2 V_1 = 0. \tag{14-9}$$

Assuming a solution of the form $e^{j\omega t}$, we find an equation for ω:

$$-\omega^2 + 2j\omega_n \zeta \omega + \omega_n^2 = 0. \tag{14-10}$$

The roots of this equation are

$$\omega = -\omega_n \zeta \pm \omega_n \sqrt{\zeta^2 - 1}. \tag{14-11}$$

When ζ is less than unity, the two solutions for ω are complex conjugates, giving sine and cosine oscillations with damped exponential envelopes. This is the underdamped situation. When ζ is greater than unity the two solutions for ω are both damped exponentials, but with different time constants. If ζ is exactly unity, the two solutions are $e^{-\omega t}$ and $te^{-\omega t}$. In

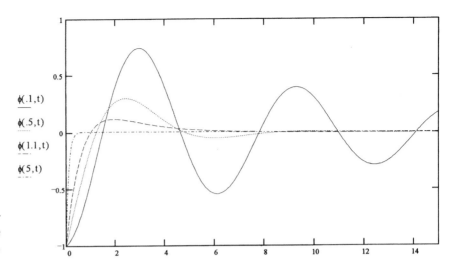

$\phi(.1,t)$

$\phi(.5,t)$

$\phi(1.1,t)$

$\phi(5,t)$

Figure 14-7. Phase error for a −1 radian step in the reference phase at $t = 0$.

every case, a linear combination of the two appropriate solutions can match any given set of initial conditions, that is, the $t = 0$ values of V_1 and $d/dt\, V_1$. Figure 14-7 shows the transient phase error for a step function in the reference phase of 1 radian at $t = 0$. This initial condition, $V_1(0) = -1$, determines the other initial condition, $dV_1/dt = 2\zeta\omega_n$ through the action of the loop filter circuit. Curves are shown in Figure 14-7 for the recovery from this transient for damping coefficients of 0.1, 0.5, 1.1, and 5. In each case, $\omega_n = 1$. The underdamped cases show the trademark oscillatory behavior. When the damping is 5, the fast exponential recovery is essentially that of a wideband type I loop.

NOTE Equation (14-10) (which is the same as the denominator of the transfer function) is known as the *characteristic equation* of the system. Here it is a second-order equation, and this type II loop is therefore also a second-order loop. The order of given system is the order of the differential equation describing its transient response.

MULTIPLIER AS A PHASE DETECTOR

A multiplier (mixer) is the most commonly used phase detector, especially at higher frequencies. When the reference sine wave is multiplied by the VCO sine wave, the usual sum and difference frequencies are generated. The desired baseband output is a difference component. If the VCO and reference frequencies are equal, this baseband component will be a dc voltage, zero when the VCO and reference phases differ by 90° and linearly proportional to the phase difference in the vicinity of 90°. Therefore, when a multiplier is used as a phase detector, the output phase will differ from the reference phase by 90° rather than by zero degrees. If this is a problem (in most applications it is not), a 90° phase shift network can be put at one of the multiplier inputs or in the output line. Note that the multiplier phase detector will also put out zero volts when the phase shifts are different by 90° in the other direction. The loop will not stay long in this position of metastable equilibrium. As soon as it "falls off", the feedback will be positive, rather than negative, and the loop will rush to the stable equilibrium point.

RANGE AND STABILITY

The type II loop can operate over the full range of the VCO because its integrator can apply whatever bias is needed. The type I loop, if $K_D K_O$, is small and if the phase detector has a limited range, may not be able to track over the full range of the VCO. Both the loops discussed above are unconditionally stable since the transient responses are decaying exponentials for any combination of the parameters K_D, K_O, R_1, R_2, and C. (When you build one and it oscillates, it is usually because you have high-frequency poles you did not consider (maybe in your op amp or circuit parasitics), and have really built a higher-order loop).

ACQUISITION TIME

A high-gain type I loop can achieve lock very quickly. A type II loop with a small bandwidth can be very slow. Acquisition depends on some of the beat note from the phase detector getting through the filter to FM modulate the VCO. That beat note puts a pair of small sidebands on the VCO output. One of these sidebands will be at the reference frequency and will mix with the reference to produce dc with the correct polarity to push to the VCO toward the reference frequency. The integrator gradually builds up this dc until the beat frequency comes within the loop bandwidth. From that point the acquisition is very fast. This reasoning, applied to the type II loop, predicts an acquisition time of about $4f^2/B^3$ seconds, where f is the initial frequency error and B is the bandwidth of the loop (see Problem 6). If the bandwidth is 10 Hz and the initial frequency error is 1 kHz, the predicted time is about 1 h. (In practice, unavoidable offsets in the op amp would make the integrator drift to one of the power supply rails and never come loose). Obviously in such a case some assistance is needed for lock-up. One common method is to add search capability, that is, circuitry that causes the VCO voltage to sweep up and down until some significant dc component comes out of the phase detector. At that point the search circuit turns off and the loop locks itself up. The integrator can form part of the sweep circuit, which can be similar to the sawtooth oscillator described earlier. Another way to aid acquisition is to use a frequency/phase detector instead of a simple phase detector. This is a digital phase detector which, when the input frequencies are different, puts out dc whose polarity indicates whether the VCO is above or below the reference. This dc quickly pumps the integrator up or down to the correct voltage for lock to occur. You will see these circuits described in Motorola literature. In digitally controlled PLLs, it is common for a microprocessor with a look-up table in ROM (read-only memory) to pre-tune the VCO to the commanded frequency. This pretuning allows the loop to acquire lock quickly.

PLL RECEIVER

There are many inventive circuits and applications in the literature on PLLs. Here is an example. Deep space probes usually provide very weak telemetry signals. The receiver bandwidth must be made very small to reject noise that is outside the narrow band containing the modulated signal. A PLL can be used to do the narrow-band filtering and to do the detection as well. One simple modulation scheme uses slow frequency-shift keying (FSK), where the signal is at one frequency for a "0" and then slides over to a second frequency for a "1". The loop bandwidth of the PLL is made just wide enough to follow the keying. The control voltage on the VCO is used as the detected signal output. (With a wide loop bandwidth, such a circuit can demodulate ordinary FM audio broadcast signals.) In this narrow-band example, the PLL circuit has the advantage that it will track the signal automatically when the transmitter frequency drifts or is Doppler shifted. Another modulation scheme uses phase-shift keying (PSK). Suppose the transmitter phase is shifted 180° to distinguish a "1" from a "0". This would confuse the PLL receiver; for random data the average output from the phase detector would be zero, and the PLL would be unable to lock. A simple cure is to use the "doubling loop" shown in Figure 14-8. The incoming signal is first sent through a doubler to produce an output that looks like the output of a full-wave rectifier. (The waveform loops are positive no matter what the modulation does.) The VCO output frequency is then divided by two to produce a phase-locked reference. This reference is used as one input of a mixer. The other input of the mixer is the raw signal, and the output is the recovered modulation.

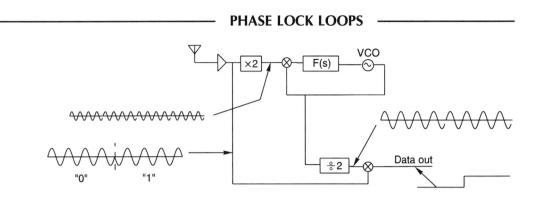

Figure 14-8. Doubling loop for detection of binary phase-shift modulation.

BIBLIOGRAPHY
1. F. M. Gardner (1979), *Phaselock Techniques*, 2nd Edition, New York: John Wiley.
2. B. Kuo (1987), *Automatic Control Systems*," 5th Edition, Englewood Cliffs: Prentice Hall.

PROBLEMS

1. How would you modify the gear train in Figure 14-3 so that the output shaft would turn N/M times faster than the input shaft?

2. Show that Equation (14-2) describes the time domain input/output relation for the loop filter circuit of Figure (14-4).

3. Suppose the VCO in the block diagram of Figure (14-5) is noisy and that its noise can be represented as an equivalent noise voltage added at the control input of the VCO. Redraw the block diagram showing a summing block just in front of the VCO. Find the frequency transfer function for this noise input. Show that the loop is able to "clean up" the low-frequency noise of the VCO.

4*. Write a computer program to numerically simulate the PLL shown in Figure (14-5). Use a multiplier-type phase detector, and investigate the process of lock-up. Let the phase detector output be proportional to $\cos(\theta - \theta_{\text{ref}})$. Use numerical integration on the simultaneous first-order differential equations for $V_1(t)$ and $V_2(t)$.

5*. Invent a phase detector circuit that would have a range of many multiples of 2π rather than the restricted range of the multiplier phase detector.

6*. Derive the formula given for lock-up time, $\tau_{\text{lock-up}} \approx 4(\Delta\omega)^2/B^3$, where $\Delta\omega$ is the initial frequency error and B is the loop bandwidth (in

*These more difficult problems could be used as projects.

radians per second). Consider the type II loop of Figure 14-5. Find the ac component on the VCO control input. (Assume that $\Delta\omega$ is high enough that the gain of the loop filter for this ac voltage is just $R_2 R_1$.) Find the amplitude of the sidebands caused at the output of the VCO from this ac control voltage component. Use this amplitude to find the dc component at the output of the phase detector caused by the product of the reference signal and the VCO sideband at the reference frequency. This dc component will be integrated by the loop filter. Find dV/dt at the output of the loop filter. Find an expression for $\tau_{\text{lock-up}}$ by estimating the time for the dc control voltage to change by the amount necessary to eliminate the initial error.

FREQUENCY SYNTHESIZERS 15

By *frequency synthesizer* we usually mean a signal generator which can be switched to put out any one of a discrete set of frequencies and whose frequency stability is derived from a standard oscillator, either a built-in crystal oscillator or an external "station standard." Most laboratory synthesizers generate sine waves, but some low-frequency synthesized function generators also generate square and triangular waves. General-purpose synthesizers have high resolution; the step between frequencies is usually better than 1 Hz and may be millihertz or even microhertz. Many television receivers and communications receivers have synthesized local oscillators. Special-purpose synthesizers may generate only a single frequency.

At least three general techniques are used for frequency synthesis. *Direct synthesizers* use frequency multipliers, frequency dividers, and mixers. *Indirect synthesizers* use phase lock loops. *Direct digital synthesizers* use a digital accumulator to produce a staircase sawtooth, a look-up table to change the sawtooth to a staircase sinusoid, and a digital-to-analog converter to provide the analog output. Some designs combine these three techniques.

DIRECT SYNTHESIS

The building blocks for direct synthesis are already familiar. Frequency multiplication can be done with almost any nonlinear element. We have already seen with power supplies that a full-wave rectifier is an efficient frequency doubler. A limiting amplifier (limiter) or diode clipper circuit will convert sine waves into square waves, which include all the odd harmonics of its fundamental frequency. A delta function pulse train (in practice, a train of very narrow pulses) contains all the harmonics. When frequencies in only a narrow range are to be multiplied, class-C amplifiers can be used. (A swing does not have to be given a push on each and every cycle.) Frequency division, used in all three types of synthesizers, is almost always done digitally; the input frequency is used as the clock for a digital counter made of flip-flops and logic gates. Frequency translation can be done with any of the standard mixer circuits.

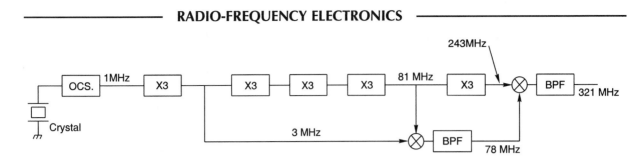

Figure 15-1. A direct synthesizer to produce 321 MHz from a 1 MHz reference.

Single-frequency synthesizers are usually ad hoc designs; the arrangement of mixers, multipliers, and dividers depends on the ratio of the desired frequency and the reference frequency. As an example of direct synthesis, the circuit of Figure 15-1 generates 321 MHz from a 1 MHz standard. One prime factor of 321 is 107. It would be difficult to build a ×107 multiplier. This design uses only triplers and mixers. Laboratory instruments that cover an entire range of frequencies must, of course, use some general scheme of operation.

MIX AND DIVIDE DIRECT SYNTHESIS

Some laboratory synthesizers use an interesting mix and divide module. An n-digit synthesizer would use n identical modules. An example is shown in Figure 15-2. Each module has access to a 16 MHz source and ten other

Figure 15-2. A mix and divide direct synthesizer.

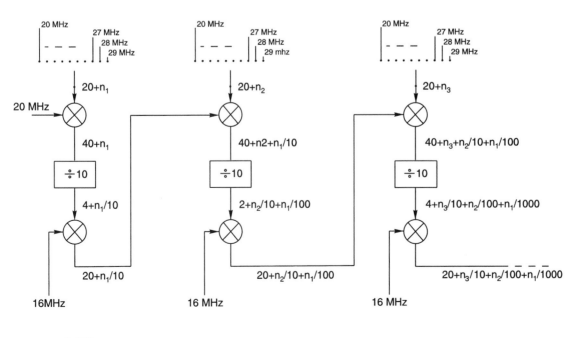

reference sources from 20 to 29 MHz. (These references are derived from an internal or external standard, usually at 5 MHz.) In this kind of design the internal frequencies must be chosen carefully so that after each mixer the undesired sideband can be filtered out easily.

INDIRECT SYNTHESIS

The phase-lock loop circuit of Figure 15-3 is an indirect synthesizer to generate the frequency $N f_{std}/M$ where N and M are integers. If the $\div N$ and/or the $\div M$ blocks are variable modulus counters, the synthesizer frequency is adjustable. The 321 MHz synthesizer of Figure 15-1 could built as the indirect synthesizer of Figure 15-3 by using a divide-by-321 counter (most likely a divide-by-3 counter followed by a divide-by-107 counter). Circuits like that of Figure 15-3 are used as local oscillators for digitally tuned radio and television receivers. For an AM broadcast band receiver, the frequency steps would be 10 kHz (the spacing between assigned frequencies). This requires that the reference frequency of the phase detector, f_{std}/M, be 10 kHz, so only the modulus N would be adjustable. What about a synthesized local oscillator for a short-wave radio? We would probably want to be able to tune with a resolution of, say, 10 Hz. The reference frequency in this simple circuit of Figure 15-3 would have to be 10 Hz and the loop bandwidth would have to be about 2 Hz. This low bandwidth would make fast switching difficult. Moreover, with such a small loop bandwidth, the close-in noise of the voltage controlled oscillator (VCO) would not be cleaned up by the loop, and the performance of the radio would suffer. For this application a more sophisticated circuit is needed. One method is to synthesize a VHF or UHF frequency with steps of 1 or 10 kHz, divide it to produce the necessary smaller steps, and then mix it to a higher frequency. Other circuits use complicated multiple loops. The newest receivers use LO synthesizers based on the principle of direct digital synthesis.

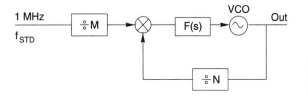

Figure 15-3. Basic indirect (phase lock loop) synthesizer.

DIRECT DIGITAL SYNTHESIS

This technique, illustrated in Figure 15-4, uses a digital register and a two-input adder as a phase accumulator. The output of the adder is latched into the register on every cycle of a high-frequency clock. The inputs to the adder are the current register contents (the phase) and an addend (the phase increment), which is adjustable. At every clock cycle, the total accumulated phase increases by an amount equal to the addend. Since the accumulator has a finite length, it rolls over and its (digital) output has a sawtooth shape. The output of the accumulator is used to address a ROM (read-only memory) that converts the stairstep digital sawtooth to a stairstep digital sine wave. Finally, a digital-to-analog converter provides an analog output, and high-frequency switching noise is removed by a low-pass filter whose cutoff frequency can be just below the clock frequency. The frequency can be changed in very fine increments if the number of bits in the accumulator is large. With a 32-bit accumulator, for example, the frequency resolution will be $f_{clock}/2^{32}$. For a clock frequency of 10 Mhz the resolution is $10^7/2^{32} = 0.0023\,\text{Hz}$. The most significant bit (MSB) of the accumulator toggles at the desired output frequency so one might think that an analog bandpass filter to isolate fundamental frequency would complete the synthesizer. The problem is that while the MSB has the right average frequency it usually is quite irregular. If the addend is a power of two the MSB will toggle at uniform time intervals, but otherwise it will have jitter that depends on the value of the addend. The solution is to use more than just the top bit. If we take the top eight bits and regard their total value as an analog number, then the phase error is never greater than $360/2^8$ degrees. We could feed the eight-bit digital sawtooth into a digital-to-analog converter and then use the low-pass filter to produce a

Figure 15-4. Direct digital synthesizer (DDS).

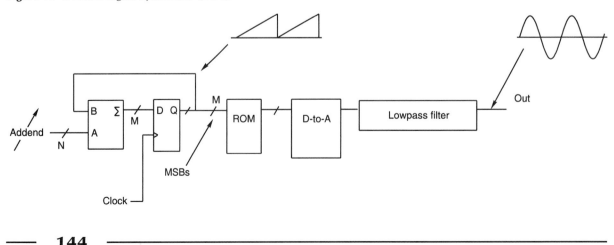

sine wave (i.e. isolate the fundamental component), but a neater method is first to feed the digital sawtooth into a ROM look-up table that converts it into a digital *sine wave* to drive the digital-to-analog converter, as shown in Figure 15-4. The sine wave has no harmonics, so the bandpass filter is not needed.*

The direct digital synthesizer (DDS) has been around for at least 20 years, but speeds have been increasing as they are cast into silicon. Present-day monolithic DDS chips use clock frequencies up into the UHF range (hundreds of megahertz). Note that the DDS can use the technique of digital pipelining in the adder/accumulator to simultaneously achieve high-output frequencies and high-frequency resolution. The pipe delay is generally of no consequence.

NOISE SPECTRUM OF THE DDS

Let us look more closely at the DDS to see that it does produce a small residual phase noise, that is, the frequency spectrum is not just a spike at the desired frequency. First, consider the simple case of a 4-bit accumulator. If the addend is 4 and the accumulator starting value is 0, the subsequent states are 4, 8, 12, 0, 4, 8, 12, ... The output frequency is 4/16 times the clock frequency. This is shown in Figure 15-5, where the desired sine wave output is plotted together with the periodically sampled-and-held† output voltage of the analog-to-digital converter. Now let the addend be 5. The sine wave frequency will be 5/16 times the clock frequency, as shown in Figure 15-6. In this case the accumulator cycle is 0, 5, 10, 15, 4, 9, 14, 3, 8, 13, 2, 7, 12, 1, 6, 11, 0, 5, 10, ... Since the output repeats in sixteen clock cycles, the spectrum would, in general, have com-

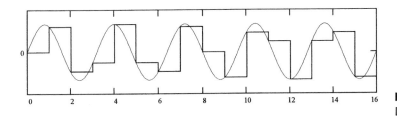

Figure 15-5. Sixteen-bit DDS with addend = 4.

*This bandpass filter could actually be just a low-pass filter with a cutoff frequency less than one octave above the desired output frequency. Note, however, that this filter would have to be tuneable if the frequency range of the synthesizer is to be more than one octave.
†In general, a special filter is needed to convert a sampled-and-held version of a waveform back to its original form. But here, with only a single spectral component, this operation can be omitted.

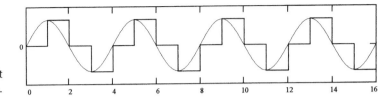

Figure 15-6. Sixteen-bit DDS with addend = 5.

ponents beginning at $f_{\mathrm{clk}}/16$ and spaced by $f_{\mathrm{clk}}/16$. But if the values from the analog-to-digital converter are perfectly accurate (fall exactly on the sine curve as shown in Figure 15-6) the only nonzero component is the desired one at $5f_{\mathrm{clk}}/16$. When the addend was 4 (see Figure 15-5) the only nonzero component is at $4f_{\mathrm{clk}}/16$. But inevitably, the voltage from the ROM and digital-to-analog converter combination will have some systematic error. To have no error, the digital-to-analog converter would have to produce 2^N different values, that is, a value for every state of the accumulator. These synthesizers often use 32-bit accumulators. No digital-to-analog converters are available with that dynamic range, so the digital-to-analog converter is connected to only the top 12 or 16 bits of the accumulator, and there is a round-off error. How does this error affect the frequency spectrum of the synthesizer? We can consider the output of the digital-to-analog converter to consist of the ideal sampled and held sine wave plus an additive voltage caused by round-off. In as much as these errors are random, their contribution to the output spectrum will be a comb of frequency components whose spacing is given by the master clock frequency divided by the length of the accumulator cycle. In the examples of Figures 15-5 and 15-6, the lengths of the accumulator cycle were 4 and 16, respectively. To find the (smallest) cycle length L in the general case of an N-bit accumulator and an addend A, we note that $LA \bmod 2^N = 0$. Therefore, $LA = M \times 2^N$, where M is an integer. LA is a multiple of both A and 2^N, so we find their least common multiple (LCM). L is then just LCM/A. As an example, suppose we have a 32-bit accumulator and $A = 2^8$ (certainly not a randomly chosen number). Here the least common multiple is 2^{32}, so $L = 2^{32}$, so $L = 2^{32}/2^8 = 2^{24}$. In this example, since A is a power of 2, the accumulator cycle length is the same as the cycle length of the output sine wave (as in Figure 15-5). The only noise consists of small components at the harmonics of the output frequency – about the same as the spectrum of an analog sine wave oscillator with an almost purely sinusoidal output waveform. As a second example, suppose A is any odd number. This case covers half of the frequencies produced by the DDS. Here the least common multiple is $A \times 2^N$, and the cycle length is 2^N. This completely changes the character of the noise. Now the noise will be an almost continuous "grass" of weak spectral components separated by $f_{clk}/2^N$. You can see that the noise, though it may be extremely small, can completely change character from one frequency setting to the next.

SWITCHING SPEED AND PHASE CONTINUITY

Indirect synthesizers cannot change frequency faster than the time needed for their phase lock loops to capture and settle. Direct synthesizers and DDSs can switch almost instantly. Sometimes it is necessary to switch frequencies without losing phase continuity. The DDS is perfect for this since the addend is changed and the phase rate changes but there is no sudden phase jump. In other applications it is necessary that, when the synthesizer is retuned to a previously selected frequency, the phase takes on the value it would have had if the frequency had never been changed. This second kind of continuity might be called phase memory. A frequency synthesizer that provides the first kind of phase continuity clearly will not provide the second, and vice versa. Continuity of the second kind can be obtained with a direct synthesizer that uses only mixers and multipliers (no dividers).

PHASE NOISE FROM MULTIPLIERS AND DIVIDERS

It is important to see how any noise on the reference signal of a synthesizer determines the noise on its output signal. Let us examine how noise is affected by the operations of frequency multiplication and division. We will assume the input noise sidebands are much weaker than the carrier. Suppose the input signal has a discrete sideband at 60 Hz. (Such a sideband would normally be one of a pair, but for this argument we can consider them one at a time.) Let this sideband have a level of $-40\,\text{dBc}$, i.e. its power is 40 dB below the carrier power. If this signal drives a $\times N$ frequency multiplier, it turns out that the output signal will also have a sideband at 60 Hz but its level will be $-40 + 20\log(N)\,\text{dBc}$. The relative sideband power increases by the *square* of the multiplication factor.

Let us verify this, at least for a specific case – a particular frequency tripler. Let the input signal be $\cos(\omega t) + \alpha \cos[(\omega + \delta\omega)t]$, that is, a carrier at ω having a sideband at $\omega + \delta\omega$ with a relative power of α^2 or $20\log(\alpha)\,\text{dBc}$. We assume that $\alpha \ll 1$. Here the tripler will be a circuit whose output voltage is the cube of the input voltage. Expanding the output and keeping only terms of order α or higher whose frequencies are at or near $3f$ we have

$$\{\cos(\omega t) + \alpha \cos[(\omega + \delta\omega)t]\}^3 \to \cos^3(\omega t) + 3\alpha \cos^2(\omega t) \cos[(\omega + \delta\omega)t]$$
$$\to 1/2 \cos(\omega t) \cos(2\omega t) + 3/2\alpha \cos(2\omega t) \cos[(\omega + \delta)t]$$
$$\to 1/4 \cos(3\omega t) + 3/4\alpha \cos[(3\omega + \delta\omega)t] =$$
$$1/4\{\cos(3\omega t) + 3\alpha \cos[(3\omega + \delta\omega)t]\}.$$

Note that the carrier-to-sideband spacing is still $\delta\omega$ but that the relative amplitude of the sideband has gone up by 3, the multiplication factor. The relative power of the sideband has therefore increased by 3^2, the square of the multiplication factor. A continuous distribution of phase noise $S(\delta\omega)$ is like a continuous set of discrete sidebands so, if the noise spectrum of a multiplier input is $S(\delta\omega)$, the noise spectrum of the output will be $n^2 S(\delta\omega)$. Sideband noise enhancement is a direct consequence of multiplication. If the multiplier circuit itself is noisy, the output phase noise will increase by more than n^2. Fortunately, most multipliers contribute negligible additive noise.

Division, the inverse of multiplication, reduces the phase noise power by the square of the division factor. Mixers just translate the spectrum of a signal, they do not have a fundamental effect on noise. Additive noise from mixers is usually negligible. If a direct synthesizer is built with ideal components, the relation between the output phase noise and the phase noise of the standard will be as if the synthesizer were just a multiplier or divider (no matter what internal operations are used). The phase noise produced by indirect synthesizers depends on the quality of the internal VCOs and the bandwidths of the loop filters.

BIBLIOGRAPHY

1. J. A. Crawford (1994), *Frequency Synthesizer Design Handbook*, Boston: Artech House.
2. V. Manassewitsch (1980), *Frequency Synthesizers Theory and Design* 2nd Edition, New York: John Wiley.

PROBLEMS

1. Design a direct synthesizer (ad hoc combination of mixers, dividers, and multipliers) to produce a frequency of 105.3 MHz from a 10 MHz reference. Avoid multipliers higher than ×5 and do not let the two inputs of any mixer have a frequency ratio higher than 5:1 (or lower than 1:5).

2. Design a DDS with 10 kHz steps for use as the tuneable local oscillator in a middle-wave broadcast band (530–1700 kHz) AM receiver with a 455 kHz intermediate frequency. Assume a high-side local oscillator, that is, the synthesizer frequencies range from $530 + 455$ to $1,700 + 455$. An accurate reference frequency is available at 10 MHz.

3. Design a synthesizer with the range of 1–2 MHz that has phase memory, that is, when the synthesizer is reset to an earlier frequency its phase will be the same as if it had been left set to the earlier frequency. The required step size is 50 kHz and the available frequency reference is 5 Mhz. *Hint*: One approach is to generate a frequency comb with a spacing of 50 kHz and then phase lock a tuneable oscillator to the desired tooth of the comb.

4. Explain why a direct synthesizer that includes one or more dividers will not have phase memory.

16

SWITCHING CONVERTERS

Switching converters are (dc to dc) power converter circuits based on switching elements (often one transistor and one diode) and an inductor (sometimes in the form of a transformer). They are often used as local "board level" power converters in systems with a single power bus. For example, a bus voltage of 20 V dc can be converted wherever needed to power 5 V logic circuitry, to furnish various positive and negative voltages for analog circuitry, and in high-voltage display systems. By varying the duty cycle of the switching element, the dc output can be precisely regulated. If the duty cycle is deliberately modulated, the switching converter becomes a high-efficiency "class-S" power amplifier used, for example, as an AM modulator for a class-C or class-D RF amplifier. The virtues of switching converters are their high efficiency and light weight.

BASIC SWITCHER TOPOLOGIES

Three common converter circuits are shown in Figure 16-1. All use a switch, an inductor, a diode, and a filter capacitor. (The dc input supply (V_b) is shown as a battery.) In all these circuits, an inductor (rather than a resistor) provides the means by which current is transferred at a controlled rate and without energy loss from the supply side to the load side.

BUCK CIRCUIT

CONTINUOUS MODE

The buck circuit is the same as a choke-input power supply fed by a half-wave rectifier with a freewheeling diode except that here the filter sees rectangular voltage pulses rather than the positive loops of a sine wave.

Figure 16-1. Three common dc-to-dc converters.

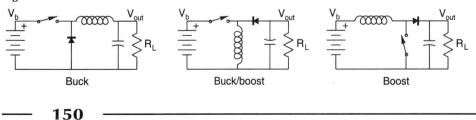

Buck   Buck/boost   Boost

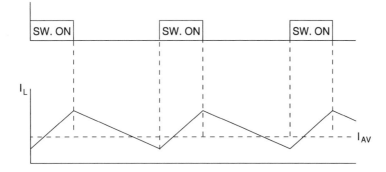

Figure 16-2. Buck supply operation cycle (continuous mode).

When the switch is open, the diode provides the left side of the inductor with a current path from ground. If the choke inductance is large enough to maintain nonzero positive current in the inductor, the average voltage at the input of the choke will be the supply voltage, V_b, times the duty factor of the switch (the fraction of the time the switch is on). The output voltage is equal to this average voltage, so it will remain constant with respect to load changes. (Perfect inherent regulation with respect to load changes requires ideal components, for example no dc resistance in the choke.) The duty factor may, of course, be changed to control the output voltage. Feedback control of the duty factor can be used to obtain regulation with respect to variations of the input voltage. Feedback can also provide very precise regulation with respect to load changes but, due to the inherent regulation, it will not have to make more than slight adjustments.

The current through the choke ramps up while the switch is closed and ramps down when it is open. Since $V = L\,dI/dt$, the slope of the up-ramp is a constant, equal to the supply voltage minus the output voltage divided by the inductance. The slope of the down-ramp is also constant, equal to the output voltage divided by the inductance. This operation is shown in Figure 16-2. When $V_{out} < V_b/2$ (i.e. the duty factor is $< 1/2$) the charging slope is steeper than the discharging slope, and vice versa.

DISCONTINUOUS MODE

If the inductor value is too low, the down-ramp will drain the inductor dry (zero current). The freewheeling diode will stop conducting and the left-hand side of the choke will float. The inherent regulation will be lost because we will find that the output voltage will now be a function of the load. Figure 16-3 shows the operation cycle in this discontinuous mode. We assume the capacitor is large enough to allow us to treat the output voltage as a constant. The length of the up-ramp is τ_1, the on time of the switch. In the discontinuous mode, the length of the down-ramp is the time needed for the inductor current to reach zero. To find the output

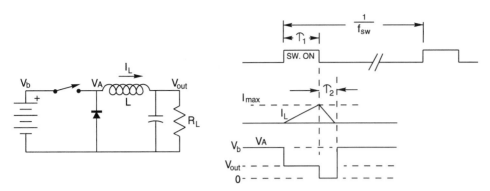

Figure 16-3. Buck supply operation cycle (discontinuous mode).

voltage we first note that the average current through the choke must be equal to the current in the load, V_{out}/R. From Figure 16-3, the average current through the choke is the area under the triangle divided by the length of the cycle, $1/f_{sw}$, where f_{sw} is the switching frequency. The duration of the down-ramp, τ_2, is given by LI_{max}/V_{out}. Finally, the up-ramp shows that I_{max} is equal to $(V_b - V_{out})\tau_1/L$. Putting all this together yields a quadratic equation for the output voltage, V_{out}:

$$\left(\frac{V_{out}}{V_b}\right)^2 + \frac{f_{sw}\tau_1^2 R}{2L}\left(\frac{V_{out}}{V_b}\right) - \frac{f_{sw}\tau_1^2 R}{2L} = 0. \tag{16-1}$$

But while this lets us find V_{out} in the discontinuous mode, we would prefer to avoid this mode altogether. To find the necessary critical inductance, we note that the discontinuous mode becomes continuous when there is no space between the triangular current pulses in Figure 16-3. The average current is then just half the height of the triangle, $I_{max}/2$ or $1/2[\tau_1/L(V_b - V_{out})]$. Since this average current has to be equal to V_{out}/R, we have

$$L_{critical} = \frac{\tau_1 R}{2}\left(\frac{V_b}{V_{out}}\right). \tag{16-2}$$

BUCK/BOOST CIRCUIT

CONTINUOUS MODE

The buck/boost circuit works quite differently than the buck converter. In this circuit, energy is pumped into the inductor while the switch is closed and is then transferred to the load when the switch is open. The inductor current versus time plot of Figure 16-2 also applies for the buck/boost circuit in the continuous mode. While the switch is closed, the inductor current ramps up linearly. The diode is back-biased during this time, so the load side is disconnected. When the switch opens, the inductor current remains continuous, supplied now by the load via the diode. Since current is drawn from the load, its voltage is negative. This mode is easy to analyze. For the up-ramp we have $\Delta I = V_b \tau_1 / L$. For the down-ramp we have $\Delta I = -V_{out} \tau_2 / L$. The up and down ramps have equal current excursions, ΔI, so we have

$$V_{out} = -V_b \frac{\tau_1}{\tau_2}. \tag{16-3}$$

Note that, in the continuous mode, the output voltage is also inherently regulated, that is, independent of the load.

DISCONTINUOUS MODE

Figure 16-4 shows the voltage and current waveforms for this circuit in the discontinuous mode. The analysis is similar to that of the buck circuit. The maximum inductor current is given by $I_{max} = \tau_1 V_b / L$ and also by $I_{max} = -\tau_2 V_{out} / L$, so $\tau_2 = \tau_1 V_b / |V_{out}|$. Again, the load current is given by $|V_{out}| / R$, and must equal the average current furnished by the inductor. The latter is the area under the τ_2 triangle divided by the cycle length, $1/f_{sw}$, so we have

$$\frac{|V_{out}|}{R} = \frac{1}{2} \tau_2 I_{max} f_{sw} = \frac{1}{2} \frac{V_b \tau_1}{|V_{out}|} \frac{V_b \tau_1}{L} f_{sw}. \tag{16-4}$$

Solving for V_{out} gives

$$V_{out} = V_b \tau_1 \sqrt{\frac{f_{sw} R}{2L}}. \tag{16-5}$$

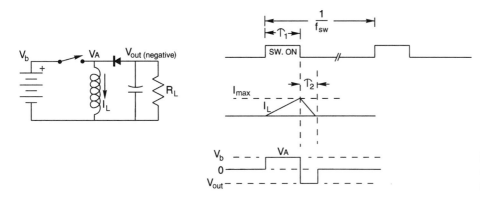

Figure 16-4. Buck/boost operation (discontinuous mode).

As before, the output voltage in this mode is a function of the load – the inherent regulation is lost. The critical inductance needed to stay out of this discontinuous mode is the subject of Problem 3.

BOOST CIRCUIT

CONTINUOUS MODE

The boost circuit, like the buck/boost circuit, uses the inductor for energy storage. For the boost circuit, Figure 16-2 again represents the current in the inductor for the continuous mode. The charge and discharge of the inductor become $\Delta I/\tau_1 = V_b/L$ and $\Delta I/\tau_2 = (V_{out} - V_b)/L$, respectively. The boosted voltage is therefore

$$V_{out} = V_b\left(1 + \frac{\tau_1}{\tau_2}\right). \tag{16-6}$$

Once again the continuous mode provides inherent regulation.

DISCONTINUOUS MODE

Finally, we have the discontinuous mode for the boost circuit. Figure 16-5 shows I_L and V_A versus time. Again the current ramps up from zero when the switch is closed and ramps back down to zero when it is opened. As before, we assume the filter capacitor, C, is large so that V_{out} is constant.

In this case the maximum inductor current is given by $I_{max} = \tau_1 V_b/L$ and also by $I_{max} = \tau_2(V_{out} - V_b)/L$, so $\tau_2 = \tau_1 V_b/(V_{out} - V_b)$. Again, the current into the load is given by V_{out}/R, and must equal the average current furnished by the inductor. The latter is the area under the τ_2 triangle divided by the cycle length, $1/f_{sw}$, so we have

Figure 16-5. Boost supply operation cycle (discontinuous mode).

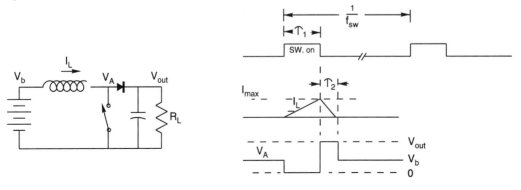

$$\frac{V_{\text{out}}}{R} = \frac{1}{2} \tau_2 I_{\text{max}} f_{\text{sw}} = \frac{1}{2} \frac{V_{\text{b}}\tau_1}{V_{\text{out}} - V_{\text{b}}} \frac{V_{\text{b}}\tau_1}{L} f_{\text{sw}}. \qquad (16\text{-}7)$$

Rearranging, we have a quadratic equation for V_{out}:

$$V_{\text{out}}(V_{\text{out}} - V_{\text{b}}) = V_{\text{b}}^2 \tau_1^2 \frac{f_{\text{sw}} R}{2L}. \qquad (16\text{-}8)$$

OTHER CONVERTER TOPOLOGIES

The three converter circuits described above are not the only possible topologies (see the Bibliography). For example, in "complementary" designs such as the Cùk converter, the dc voltage supply is replaced by a current supply (a voltage supply with a series choke), the energy storage inductor is replaced by a capacitor, and the low-impedance load is replaced by a high-impedance load (by adding a series input inductor). In "resonant converters," sinusoidal transients are used to reduce losses in the switching element by causing either the voltage or the current to be at zero when the switching takes place.

TRANSFORMER-COUPLED CONVERTERS

When the dc input is derived from a rectifier directly on the ac power line the three basic converters are modified by incorporating an isolation transformer. The transformer provides isolation from the power line, required for safety whenever external connections will be made to the circuitry powered by the converter. This circuit, a *switching power supply*, is found, for example, in every desktop personal computer. Operating at a high frequency, this transformer can be much smaller and lighter than a 60 Hz transformer of equivalent power. Transformer versions of the buck/boost and boost converters are known as *flyback* converters, and can provide voltage of either polarity. Flyback converters require no inductor apart from the transformer. The inductor, needed for energy storage, is provided by the magnetizing inductance of the transformer. Positive and negative output flyback circuits are shown in Figure 16-6. When the switch is closed, the diode disconnects the secondary, and a ramp current flows in the primary, storing magnetic energy. When the switch is opened the transformer discharges, just as in the simple buck/boost circuit except that now the discharge is through the secondary winding. (The magnetic energy does not care which winding carries the discharge current.)

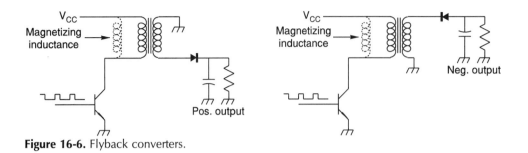

Figure 16-6. Flyback converters.

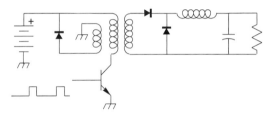

Figure 16-7. Forward converter.

The transformer version of the buck converter is the *forward converter* shown in Figure 16-7. Here the original choke remains; the transformer is an additional component. While the switch is pulsed on, the transformer secondary supplies power to the load. Here the transformer is used as a real transformer rather than as an energy storage device. The turns ratio of the transformer, as well as the duty factor of the switch and the primary supply voltage, determine the output voltage. There is a problem, however, with opening the switch because there will be current flowing in the magnetizing inductance. (Remember that the magnetizing inductance is an equivalent shunt inductance on the primary side equal to the inductance of the primary winding or, just as well, a shunt inductance on the secondary side equal to the inductance of the secondary winding.) Even if the transformer has a high magnetizing inductance, there will be some stored energy when the switch is opened. A third winding can be provided, as shown in Figure 16-7, to discharge this energy back into the primary supply (so that it is not wasted). This winding is sometimes called a reset winding since it resets the transformer to zero, readying it for the next pulse. In the forward converter, the transformer secondary delivers a stiff voltage pulse (i.e. a pulse from an equivalent low-impedance voltage generator). This filter therefore needs a filter choke and a freewheeling diode to provide inherent voltage regulation.

THE HORIZONTAL OUTPUT CIRCUIT IN CATHODE RAY TUBE TERMINALS AND TELEVISION SETS

Cathode ray tube (CRT) terminals (including television (TV) sets) use a flyback transformer in their horizontal output circuit. This is a more complicated circuit than the flyback converter discussed above because the horizontal output circuit has several functions. The first of these functions is to generate a linear current ramp in the horizontal deflection coil (which is in the magnetic deflection yoke on the neck of the CRT). Current in the deflection coil produces a magnetic field in order to exert a force ($F = qv \times B$) on the moving electrons. The deflection function is a double function because the deflection current must be negative while the beam is on the left-hand side of the screen and positive while the beam is on the right-hand side of the screen. The deflection circuit handles high power as it cycles energy in and out of the deflection coil. It is important that this energy be recycled. If it were dumped after every scan line, considerable power would be dissipated. (Since the vertical scanning is done at such a lower rate, the magnetic energy in the vertical deflection can be dumped.) The second function of this circuit is to generate the high voltage, some 10–20 kV, for the electron gun in the neck of the CRT. This function does not involve much power because the beam current is very low.

A typical circuit is shown in Figure 16-8. Q_1 is a power transistor, mounted on a heat sink. Transistors made for this application often have the diode, D_1, included on the silicon chip, forming a simple integrated circuit. (Circuit diagrams sometimes do not indicate this built-in diode). The inductor L is the horizontal deflection coil. C_2 is a large capacitor used for energy storage; its voltage will remain essentially constant at V_{cc}. (Its voltage must be V_{cc} because there is nominally no dc resistance between C_2 and V_{cc}.) To follow the operation of the circuit (Figure 16-9)

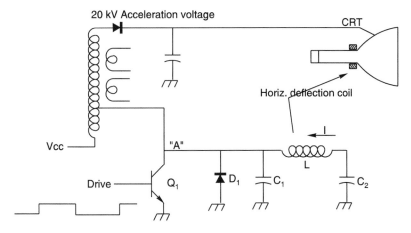

Figure 16-8. Horizontal deflection and high-voltage supply circuit for a CRT.

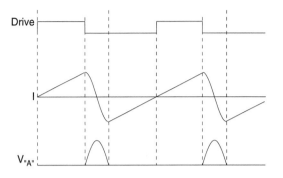

Figure 16-9. Deflection circuit operation cycle.

it is convenient to begin when the deflection current is zero, that is, when the electron beam is at the middle of the screen. The transistor turns on and current is pulled from C_2 through L. Since C_2 is very large, the voltage across L is fixed at V_{cc}, and the current ramps up linearly according to $dI/dt = V_{cc}/L$. The transistor turns off when the beam reaches the right-hand edge of the screen. The current in L, which has reached a maximum, makes a left turn into C_1, which, at this point, is discharged. The voltage on C_1 then begins a sinusoidal build-up at the resonant frequency of L and C_1. When the voltage on C_1 reaches its peak, the current is zero; all the energy is stored in C_1. The charge on C_1 then turns around and flows back into L. When the voltage reaches zero we again have full current in the coil, but it is now in the opposite direction. During this brief retrace interval the beam has moved to the left-hand edge of the screen. In the third and final phase, the diode can supply current from ground. The voltage across the coil is again V_{cc}, but the polarity has switched. The negative current ramps up to zero, and the whole cycle begins again. Phase 2, in which C_1 is charged and discharged is called the retrace or flyback phase. The voltage on C_1, V_A, reaches maybe 1,000 V at the peak (which the transistor must withstand). This voltage is determined by C_1; since the energy in the inductor is transferred to the capacitor, $C_1 V^2/2 = LI^2/2$. The pulse of voltage on C_1 is multiplied by the autotransformer some ten or twenty times, where it is rectified and filtered to provide the accelerating voltage on the second anode of the CRT.

Usually the flyback transformer has several auxiliary secondary windings. One supplies filament voltage to the CRT. Sometimes one of these voltages is added to V_{cc} to provide a boost or "bootstrap" voltage. Voltages from other secondary windings are rectified to provide supply voltages for other portions of the receiver or monitor. If the supply voltage for the horizontal oscillator is itself derived from one of these windings, some provision must be made to get the circuit started. (See Horowitz [1] for a discussion of a switching power supply with such a "kick-start" circuit.) It might seem as if the seamless joining of the deflection ramps at the middle of screen would require very accurate control of the width of the pulse driving the transistor. To see that this is not the case, refer to the timing diagram shown in Figure 16-9 and suppose that the drive pulse is longer than indicated. The retrace always begins when the transistor is turned off, so we can imagine that the leading edge of the drive pulse arrives early. But the early turn-on of the transistor happens while D_1 is conducting, so V_A is already zero. No current flows in the transistor until D_1 has drained the indicator dry. The transistor then begins to conduct just as if it had been turned on precisely at the zero crossing of the inductor current. The width of the drive pulse can therefore be considerably longer than indicated – up to as long as the horizontal period minus the time for the half-sinusoid of the retrace. Too short a drive pulse, however, will let the inductor run dry, stopping the deflection at the middle of the screen until the drive pulse arrives.

BIBLIOGRAPHY
1. P. Horowitz and W. Hill (1989), *The Art of Electronics*, 2nd Edition, New York: Cambridge University Press.
2. G. Kassakian, M. F. Schlect and G. C. Verghese. (1991), *Principles of Power Electronics*. Reading, MA: Addison Wesley.
3. N. Mohan, M. Undeland and W. P. Robbins (1995), *Power Electronics*, 2nd Edition, New York: John Wiley.

PROBLEMS

1. Show that when an uncharged capacitor is brought to potential V by connecting it through a resistor to a battery of voltage V, the energy supplied by battery is twice the energy deposited in the capacitor (CV^2 versus $CV^2/2$), that is, the transfer is only 50% efficient even if the value of the resistor is made very small.

2. Draw the voltage waveform at point "A" for *continuous* mode operation of the buck, buck/boost, and boost converters. (See Figures 16-3, 16-4, and 16-5.)

3. Find the critical inductance needed to stay out of the discontinuous mode for the buck/boost converter of Figure 16-4 and for the boost converter of Figure 16-5.

4. The diagram below shows a circuit for a class-S amplifier (switching amplifier). This could be an audio amplifier, in which case the load resistor represents a loudspeaker. The transistors operate as switches, that is, fully on or fully off. The voltage on the left-hand side of the inductor is always

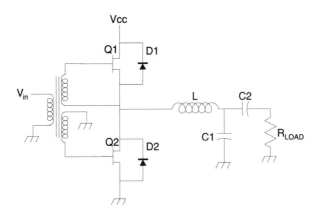

either zero or V_{cc}. The average value of this voltage appears at the right-hand side of the inductor and also on the load (except that C_2 blocks dc).

(a) When the signal is zero, the duty cycle (fraction of time that V_{cc} is connected to L) must be 50%. For this case, draw the waveforms of the currents in the inductor, Q_1, D_1, Q_2, and D_2.

(b) For this zero-signal case, explain why there is no net energy flow from the power supply.

5. The circuit shown below, a *blocking oscillator*, is essentially a self-excited switching converter. The transformer has three windings. Assume they are perfectly coupled and have equal inductance, L. The current from the dc supply is a positive ramp during the first half of the cycle while the transistor is on. If the magnetic core does not saturate, this ramp ends when the transistor goes out of saturation, that is when $I_c = \beta I_b$. The transistor quickly turns off and the stored energy in the transformer is returned to the power supply via the diode.

(a) Show that the frequency is given by $f = R/(4\beta L)$. (Assume the base-to-emitter junction is a short circuit.) Sketch the waveforms of the collector voltage and of the collector and diode currents.

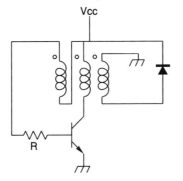

(b) Show that, if the magnetic core saturates before the transistor goes out of saturation, this circuit is a VCO or voltage-to-frequency converter.

DIRECTIONAL POWER METERS AND STANDING WAVES

We have seen that independent waves can travel in the two directions along a transmission line. As you might imagine, it is possible to build an instrument that measures separately the power flowing in each direction – a directional wattmeter. Such an instrument normally taps off only a small fraction of the power, so it can be installed permanently in a transmission line such as the line between a transmitter and antenna. A high forward reading indicates the transmitter is working properly and a low reverse reading indicates that the antenna is not shorted or open. The basis of a directional wattmeter is a *directional coupler*. (We will later look at an important class of directional couplers known as *hybrids*, where the power tapped off is half the incident power rather than just a small sample.) The instrument described below uses lumped components and is suitable for frequencies up to about 50 MHz.

AN IN-LINE DIRECTIONAL WATTMETER

If, at even only one point on a transmission line, we know both the voltage *and* the current we can distinguish between a forward traveling wave and a reverse traveling wave. Recall the general equations for the voltage and current on the line:

$$V = V_\mathrm{f} e^{-jkx} + V_\mathrm{r} e^{jkx} \tag{17-1a}$$

$$IZ_0 = V_\mathrm{f} e^{-jkx} - V_\mathrm{r} e^{jkx}. \tag{17-1b}$$

(Of course everything is multiplied by $e^{j\omega t}$, and the actual voltage and currents are then the real parts of the expressions in Equations (17-1).) Solving for the forward and reverse amplitudes, we have

$$V_\mathrm{f} = \frac{e^{jkx}}{2}(V + IZ_0) \tag{17-2a}$$

$$V_\mathrm{r} = \frac{e^{-jkx}}{2}(V - IZ_0). \tag{17-2b}$$

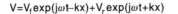
$V=V_f\exp(j\omega t-kx)+V_r\exp(j\omega t+kx)$

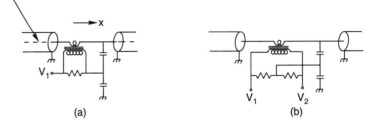

(a) (b)

Figure 17-1. Directional power sensors for (a) forward power and (b) forward and reverse power.

Knowing both I and V at any given point x along the line, we can use Equations (17-2a) and (17-2b) to solve for V_f and V_r. The circuit shown in Figure 17-1 can be thought of as an analog computer that carries out the calculations of Equations (17-2). In Figure 17-1a, the current through the resistor will be proportional to the line current (we assume the transformer is ideal). The resistor R, transformed by the turns ratio into R/N^2, becomes a very low resistance (current-sensing shunt) in series with the line. Denote the voltage across the resistor as βI. This voltage is added to a sample, αV, of the line voltage to produce a sum V_1. The addition operation is simply the result of V_1 being the voltage of the resistor in series with the sample of the line voltage. If $\beta/\alpha = Z_0$, the sum, by Equations (17-2), will be

$$V_1 = \alpha V + \beta 1 = \alpha(V + Z_0 I) = (2\alpha e^{-jkx})V_f + 0V_r \qquad (17\text{-}3)$$

and V_1 is proportional to the voltage in the forward wave. To sense reverse power, we can subtract rather than add, forming V_2 as follows:

$$V_2 = \alpha V - \beta I = \alpha(V - Z_0 I) = 0V_f + (2\alpha e^{jkx})V_r. \qquad (17\text{-}4)$$

The circuit in Figure 17-1b produces both V_1 and V_2. Remember that the time dependence is $e^{j\omega t}$, so V_1 and V_2 are sinusoidal voltages with magnitudes $2\alpha V_f$ and $2\alpha V_r$. Figure 17-2 shows how this circuit is often built. The one-turn primary winding is just the inner conductor of the transmission line passed through the toroidal transformer core. If the phases, as well as the magnitudes, of V_1 and V_2 are measured, the complex reflection coefficient can be determined. But often simple diode detectors are used to measure only the magnitudes of V_1 and V_2, making a "scalar" rather than a "vector" instrument. It is important to remember that this instrument is designed for a particular characteristic impedance, so its indications will be correct only when used with transmission lines of that impedance. All directional couplers have this property; they are built for a specific characteristic impedance (common standards are 50,

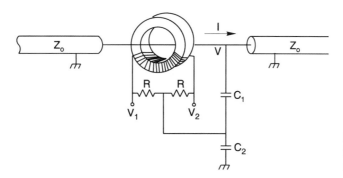

Figure 17-2.
Implementation of the
circuit of Figure 17-1b.

75, 300, and 600 ohms). This particular circuit introduces a slight reflection
in the line (see Problem 4), but there are many designs that do not.

RESISTIVE IMPEDANCE BRIDGE

The circuit described above is difficult to build for microwave frequencies,
where lumped components are far from ideal – they are too big. A surprisingly simple all-resistor bridge, however, can give the reflection coefficient
directly. Of course a resistive bridge is lossy, so this is a laboratory instrument rather than an in-line monitor. The circuit is shown in Figure 17-3.
By inspection we can write an expression of V_{out} as the difference between
the outputs of two voltage dividers:

$$V_{out} = \frac{V_{in}Z}{Z + 50} - \frac{V_{in}}{2}. \tag{17-5}$$

Solving for V_{out}/V_{in}, we find it proportional to the reflection coefficient, ρ:

$$2\frac{V_{out}}{V_{in}} = \frac{Z - 50}{Z + 50} = \rho. \tag{17-6}$$

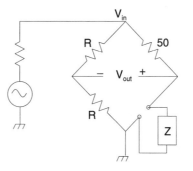

Figure 17-3. Resistive
impedance bridge.

If we can measure the relative phase between V_{out} and V_{in} we can calculate the complex impedance, Z, using Equation (9-1)

$$Z = Z_0 \frac{(1+\rho)}{(1-\rho)}. \tag{17-7}$$

The resistive bridge, however, like the high-frequency directional coupler, is often built as a scalar instrument; it measures just the magnitude of ρ. Diode detectors are built into the resistive network, and the whole structure can be very small. The small size helps to avoid parasitic reactances and makes this design suitable for use to quite high frequencies, at least 10 GHz. Combined with a sweep generator, a logarithmic amplifier, and a cathode ray tube display unit, this bridge forms a scalar network analyzer. You will see them in the Hewlett Packard catalog alongside the much more expensive vector network analyzers.

STANDING WAVES

We have seen that the most general single-frequency solution for voltage and current on a transmission line consists of two independent waves, one traveling left to right and the other traveling right to left. Simultaneous forward and reverse waves produce a *standing wave*; the magnitude of the voltage will not be constant along the line but will have maxima and minima (loops and nodes). A pure standing wave is set up when the forward and reverse waves have the same magnitude. The voltage at the nodes will be zero. When the magnitudes are not the same there will be a residual traveling wave, left by the larger of the two traveling waves. A transmission line terminated in a short circuit will have a node at the short and at every half-wave ahead of the short.

The analysis of standing waves is straightforward. Let the forward wave be $V_f e^{-jkx}$ and the reverse wave be $V_r e^{jkx}$. The voltage on the line is their sum:

$$V = V_f e^{-jkx} + V_r e^{jkx}. \tag{17-8}$$

Let $V_r = |\rho| e^{j\theta} V_f$, where $|\rho|$ is the magnitude of the reflection coefficient and θ is the phase. Equation (17-8) becomes

$$V = V_f e^{-jkx}(1 + |\rho| e^{j(2kx+\theta)}). \tag{17-9}$$

From inspection of the right-hand term we see that the amplitude of the voltage will be a maximum at values of x that make $e^{j(2kx+\theta)} = 1$, and will be a minimum at values of x that make the same expression equal to -1. At the maxima, the voltage amplitude will be $V_f(1 + |\rho|)$. At the minima it will be $V_f(I - |\rho|)$. The voltage standing wave ratio (VSWR) is defined as the ratio of maximum peak voltage to the minimum peak voltage (which are found at different positions along the line), so

$$VSWR = (1+ |\rho|)/(1- |\rho|). \qquad (17\text{-}10)$$

EFFECT OF STANDING WAVES ON AN ANTENNA TRANSMISSION LINE

If the impedance of an antenna is not equal to the characteristic impedance, Z_0, of the transmission line, there will necessarily be a standing wave on the line. The impedance seen by the transmitter will therefore depend on the length of the line. (The line rotates the antenna impedance clockwise around the Smith chart.) No matter what impedance appears at the transmitter, a matching network can make it equal to the impedance the transmitter was designed to feed. Often transmitters have a built-in adjustable matching network. The transmitter works happily into what it sees as a matched load. Is there any disadvantage to operating this way? The standing wave is not fundamentally a problem (it does not radiate power, for example) but if it is large, there will be increased ohmic line losses and possibly voltage breakdown. Suppose the mismatch results in a VSWR of 2:1. From Equation (17-10) the magnitude of the reflection coefficient, $|\rho|$, will be 1/3, so the reflected power will be 1/9 of the forward power. The forward wave will need slightly more power (12.5% more) to get the desired net radiated power, so the dissipation in the line will be increased by $1.125(1 + 1/9)$ or about 40% (see Problem 6). If the dissipation in the line is acceptable, say below 10% with a matched load, then with the 2:1 VSWR it will still be below 14% and we would conclude that this VSWR is of no great consequence. On the other hand, if the power is high enough that the voltage approaches the breakdown voltage of the cable, even a 2:1 VSWR might be intolerable.

PROBLEMS

1. A 50 ohm transmission line is used to connect a transmitter to an antenna whose impedance is $60 + j10$.

(a) What fraction of the incident power is reflected back into the transmission line?

(b) What is the VSWR on the line?

2. What is the shape of a constant VSWR curve on the Smith chart?

3. Consider the circuit shown below for a directional coupler to be used in a 50 ohm coaxial transmission line. Assume the toroidal transformer has a turns ratio of 1:40 and can be considered an ideal transformer. Find the value of C_2. Assume no current is drawn from the test points, V_1 and V_2.

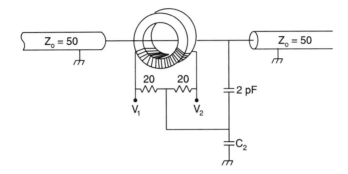

Hint: The current through the 20 ohm resistors will be the line current times 1/40 (the transformer ratio). The two capacitors form a simple voltage divider to produce a fraction of the voltage on the line. If power is flowing only in one direction on the line, the voltage at one of the test points must be zero. This voltage is the difference between the drop across one 20 ohm resistor and the output of the voltage divider.

4. The circuit of Problem 3 will create a small reflection. Calculate the reflection coefficient at 1 MHz and at 30 MHz. *Hint*: The circuit is equivalent to a small series resistor and a small parallel capacitor.

5. Design a circuit that measures the forward and reverse powers on a transmission line by sensing the voltages (but not the currents) at *two* points on the line separated by a distance d. What is a good choice of d?

6. A transmission line is mismatched, that is, terminated in an impedance other than Z_0, so a reflected wave is produced at the mismatch. The line is lossy; it has ohmic resistance, so any element of length, dx, will dissipate an amount of power proportional to $|I(x)|^2 dx$. Show that the power dissipated in the line is equal to the sum of the powers that would be dissipated if the forward and reverse waves are taken one at a time. (Assume the length of the line is an integral number of half-wave lengths or that it is very long.)

SMALL-SIGNAL RF AMPLIFIERS

18

Radio-frequency and intermediate-frequency amplifiers in radio receivers are obvious examples of small-signal RF amplifiers. The maximum output power of these amplifiers is typically from 1/100 to 1/10 W (10 to 20 dBm). Their use at extremely low power levels is ultimately limited by their own internally generated noise. Key characteristics of these class-A amplifiers are gain, bandwidth, input and output impedances, linearity, and noise.

LINEAR TWO-PORT NETWORKS

Small-signal amplifiers are nominally linear amplifiers; the output signal should be a faithful reproduction of the input signal.* A general definition of small-signal amplifiers could be that they are amplifiers built entirely of nominally linear elements (from which it follows that the overall circuit will also be linear). An amplifier, being an example of a two-port network ("two-port"), has a nominal input terminal, a nominal output terminal, and a common terminal (ground). The operation of any linear two-port can be described by four variables: its input and output voltages and currents. Any two of these variables can be considered independent variables ("input" or "cause"). The remaining two variables are then dependent variables ("output" or "effect"). If, for example, I_1 and I_2 are chosen as the input variables, the two-port is described by the equations

$$V_1 = Z_{11}I_1 + Z_{12}I_2 \tag{18-1}$$

$$V_2 = Z_{21}I_1 + Z_{22}I_2. \tag{18-2}$$

Note that the output is a linear function of the input variables; they appear raised only to the first power. By convention, the current at either end is

---

*While small-signal amplifiers are linear almost by definition, an important exception is the limiting amplifier or *limiter*. In these amplifiers, the gain decreases for increasing signal levels. A cascade of limiters can have an output level almost independent of input level. A limiter is used ahead of an FM detector if the particular FM detector is sensitive to amplitude variations as well as frequency variations. The most spectrally pure oscillators are based on amplifiers having a symmetric and gentle limiting characteristic such as the emitter-coupled differential amplifier.

positive when it flows into the network. For this choice of dependent variables, the four coefficients are known as the "Z-parameters". (We are implicitly dealing with ac circuit analysis, so these four parameters generally are complex and are functions of frequency.) By inspection of Equation (18-1) we see that Z_{11} is the input impedance of the network when the output current is zero, that is, when the output is open circuited. Z_{21}, the *forward transfer impedance*, is the open-circuit output voltage divided by the input current – a "transresistance". If we are given the load impedance we can use Equations (18-1) and (18-2) to calculate the power gain (see Problem 1). The *reverse transfer impedance*, Z_{12}, is a measure of reverse feed-through. If the RF amplifier preceding an unbalanced mixer in a superheterodyne receiver has reverse feed-through, some power from the local oscillator will get to the antenna. This two-port formalism provides more than just a top-level description of an amplifier. It is the basis for amplifier circuit analysis because the active devices (transistors) inside the amplifier can themselves be represented as two-port networks.

AMPLIFIER SPECIFICATIONS – GAIN, BANDWIDTH, AND IMPEDANCES

Gain (forward and reverse), bandwidth, input impedance, and output impedances could be called "linear specifications" because they can all be derived from the Z-parameters (or equivalent set of parameters) of an amplifier. The gain and bandwidth of an amplifier are ultimately limited by the characteristics of the transistor(s). Transistors have unavoidable built-in reactances, maybe just simple parallel input and output capacitances in a simple model for low frequencies or more than a dozen capacitors and inductors in a model for microwave frequencies. Amplifiers designed for narrow-band use (fractional bandwidths of 20% or less) use input and output matching networks to absorb or "cancel" these reactances. At higher frequencies, the shunt capacitive reactances become lower. The matching networks must then necessarily have higher loaded Qs, which means that bandwidth decreases. This limitation is fundamental; no matter how complicated the matching network, gain must be traded for bandwidth. Negative feedback around a transistor will lessen the effect of its reactances. But feedback decreases the gain, so again there is a trade-off between gain and bandwidth. In some applications the input and output impedances of an amplifier are critical. If a narrow-band filter is placed between two amplifiers, the amplifiers must present the proper impedances to the filter if the intended pass band shape is to be realized. The frequency dependence of the input and output impedances of an amplifier is, of course, related to its bandwidth, because the frequency response is normally determined by mismatch (i.e. reflection).

AMPLIFIER STABILITY

An amplifier is required to be stable (not oscillate) in its working environment. A 100 MHz amplifier, for example, will not be satisfactory if it simultaneously oscillates, even at a very different frequency, say 1 GHz. The oscillation invariably takes the circuit into large-amplitude excursions, and the superposition of amplification and oscillation is highly nonlinear. An amplifier that remains stable when presented with any combination of (passive) source and load impedances (but no external feedback paths) is said to be unconditionally stable. Unconditional stability is not always necessary. An IF amplifier in a receiver needs only to be stable in its never-changing working environment. The input RF amplifier in a short-wave radio, however, might be connected to any random arbitrary antenna so it should be unconditionally stable, at least with respect to input impedance. General-purpose "building block' amplifiers are usually designed to be unconditionally stable. A system designer can then realize a needed transfer function by cascading amplifiers, filters, and so on, and know that the combination will be stable. Stability, like gain and impedances, is entirely predictable from the two-port parameters. To find out whether the amplifier will be unconditionally stable it is necessary to show that the real part of the input (or output) impedance for any frequency is positive for any passive load (or source impedance).* Suppose that the analysis shows that for some combinations of load and frequency the real part of the input impedance is negative, but never more negative than 5 ohms. Then adding a series resistor of 5 ohms to the input of the amplifier would make it unconditionally stable. Such resistive remedies, however, always decrease gain and increase internally generated noise. The value of the reverse transfer parameter plays a key role in stability. For example, a sufficient (but not necessary) condition for an amplifier to be unconditionally stable is that its reverse transfer characteristic, such as Z_{12}

*A general analysis shows that both the input and output impedances will always have positive real parts if a certain stability factor K (a function of the two-port parameters) is greater than unity for all frequencies. Note, however, that all of this only assures that a two-port cannot be provoked into oscillation by varying its terminations. The two-port might already contain an embedded oscillator. A sufficient (but overly demanding) criterion for unconditional stability is that every internal two-port in a cascade of two-ports be unconditionally stable.

or Y_{12}, be zero. The addition of external circuitry to cancel the reverse transfer is called *neutralization*. In a narrow-band amplifier, for example, the troublesome base-to-collector capacitance can be cancelled by a parallel inductor.

OVERLOAD CHARACTERISTICS

Any amplifier will become nonlinear at high-enough signal levels, if only because the output runs up against the "rail" of the dc power supply. This behavior is beyond the realm of the small-signal parameters. A straightforward specification of the upper power limit of an amplifier is the *1 dB compression point*. This is the value of the output power at which the gain has dropped by 1 dB, that is, the point at which the output power is 79.4% of what would be predicted on the basis of low-power gain measurements.

INTERMODULATION

Small departures from linearity, even when the amplifier is far below compression, become a concern in a receiver when the pass band of an RF amplifier contains two or more signals at frequencies $f_a, f_b, f_c, \ldots$ that are much stronger than the desired signal (the signal that will be isolated downstream by a narrow band-pass filter). Nonlinearity can produce mixing products at frequencies of $n_a f_a + n_b f_b + n_c f_c, \ldots$, where $n_i = 0, \pm 1, \pm 2, \ldots$. In receivers the most troublesome of these products are the third-order products $2f_a - f_b$ and $2f_b - f_a$. (Third-order products will be inevitably produced if the output voltage of the amplifier contains even a small term proportional to the cube of the input voltage.) The special problem with these particular products is that they can fall within the IF pass band. To see this, suppose f_2 and f_3 are the frequencies of signals close enough to a desired frequency f_1 that they will pass through the broad-band front-end of a receiver. The local oscillator is tuned to convert f_1 to the center of the IF pass band. But the third-order products $2f_3 - f_2$ and/or $2f_2 - f_3$, being very close to f_1 can also fall within the narrow IF pass band and interfere with the desired signal. Second-order intermodulation is not so troublesome because the products can be filtered away.

A standard measurement of intermodulation is the *two-tone test*, which uses two closely spaced signals of equal amplitude A. On a log–log plot of output power versus input power each of these fundamental signals will fall on a 45° line, that is, slope = 1. The third-order products, however, will fall on a line with slope = 3 because the power in the third-order products

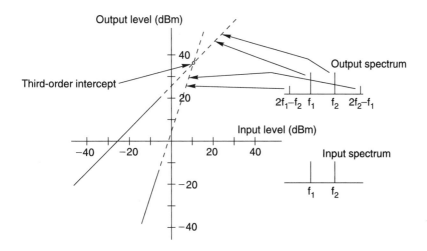

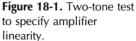

Figure 18-1. Two-tone test to specify amplifier linearity.

is proportional to the cube of the power in the pair of input signals. The *third-order intercept* is the point at which the third-order product would have as much power as each of the fundamental signals. Usually the number given for the intercept point is the output signal strength. The second-order intercept is defined the same way. Figure 18-1 shows a third-order intercept point of about +37 dBm. Generally an amplifier cannot be driven all the way to the intercept points; they are extrapolations from measurements made at much lower input levels. (The output strengths of the fundamental and the second- or third-order product need only be measured at one input level. Lines with slope = 1 and slope = 2 or 3 are then drawn through them to locate the intercepts.)

DYNAMIC RANGE

Every amplifier adds some noise to the signal. (Later we will discuss amplifier noise in some detail.) Very weak signals will be buried in this noise and lost. The *dynamic range* of an amplifier is therefore determined at the low end by the added noise and at the high end by non-linearity. In order to handle strong signals, a receiver should keep mixing products small by having as little amplification as possible prior to the narrowest bandpass filter. We will see, on the other hand, that if a receiver begins with a mixer or with a narrow-band filter, the loss in these elements adds noise and will render the receiver less sensitive than if the first element after the antenna had been a low-noise amplifier. A trade-off must often be made between sensitivity and dynamic range. Power dissipation is obviously important for battery-operated equipment where milliwatts may count. But to achieve the best possible dynamic range, a small-signal amplifier may have a fairly high-power quiescent point and have to dissipate as much as several watts of power even though it is handling extremely low signal levels.

NARROW-BAND AMPLIFIER CIRCUITS

Amplifiers for frequencies below about 30 MHz look very much like resistance-coupled audio amplifiers. The load resistors are replaced by shunt inductors, which cancel the transistor capacitances, which would otherwise tend to be short circuits at RF. These resonant circuits make a narrow-band amplifier. Often an even narrower bandpass is desired; the inductors are given smaller values and are shunted with external capacitors (effectively increasing the transistor capacitances). Focusing on one stage of an amplifier (or an amplifier of one stage), the fundamental design decisions are transistor selection and circuit configuration, that is, common emitter, common base, or common collector. (Here, and usually elsewhere, *emitter*, *base*, and *collector* can also mean *source*, *gate*, and *drain*.) The choice of transistor will be based on the ability to provide gain at the desired frequency, noise, and, perhaps, linearity. The orientation of the transistor might be as a common emitter when maximum gain is required, common base when the device is being pushed near its upper frequency limit or when the isolation between input and output is critical, or common collector when very low output impedance is needed. As far as noise goes, it turns out that the three orientations are equivalent when it comes to building a multistage (i.e. high-gain) amplifier.

WIDE-BAND AMPLIFIER CIRCUITS

Most wide-band amplifiers use feedback. An unbypassed emitter impedance provides series feedback. An impedance between the collector and the base provides shunt feedback. Modular general-purpose amplifiers from, for example, Avantek and Minicircuits, use resistive series and shunt feedback. These amplifiers are quite flat up to 1 or 2 GHz and have input and output impedances close to 50 ohms over the whole range. Resistive feedback is simple but degrades the noise performance of an amplifier. Wide-band low-noise amplifiers often use feedback networks made only of lossless elements, that is, reactors. The Miller* effect multiplies the effective input capacitance in a common-emitter amplifier. This capacitance can be neutralized, at least in narrow-band amplifiers. Wide-band amplifiers often use the *cascode* circuit, in which a common-emitter input stage drives a (low-impedance) common-base stage. Another good high-frequency circuit, the differential pair, uses an emitter follower stage (high-input impedance and low-output impedance) to drive a common-base stage.

TRANSISTOR EQUIVALENT CIRCUITS

An amplifier designer must have a precise electrical description of the transistor(s). For analysis, it is sufficient to have tables of the small-signal parameters of the transistor. (The transistor itself is a two-port "network.") These tables are usually given in the manufacturer's data sheets, but can also be measured in the laboratory if a vector network analyzer is available. A table of numbers, however, is a clumsy tool for design (synthesis), and the usual tactic is to represent the transistor by an equivalent circuit of resistors, capacitors, inductors, voltage-controlled voltage generators, voltage-controlled current generators, and so on. An equivalent circuit for a single frequency can be constructed directly from the small-signal parameters corresponding to that frequency (see Problem 4). This might be an adequate model for the design of a narrow-band amplifier, but remember that even an amplifier intended for only a narrow frequency range must be stable at *all* frequencies. For this reason, and for the design of wide-band amplifiers, equivalent circuits are concocted to represent the transistor over a wide frequency range. Normally the topology of an equivalent circuit is based on the construction and physics of the transistor; the element values are determined by least-square fitting programs that make the small-signal parameters of the model agree as closely as possible with the measured small-signal parameters of the actual transistor over the desired frequency range. Agreement can always be improved by adding more elements to the model, but an overly complicated model will block the intuition of the designer. Equivalent circuits have from one to

---

*The collector signal in a common-emitter amplifier has a larger magnitude (due to amplification) than the base signal. It also has the opposite sign. The voltage across the inherent base-to-collector capacitance of the transistor is therefore larger than the base voltage. As a result, more current flows in this capacitance than if the collector were grounded. The value of the capacitance is, in effect (Miller effect), multiplied.

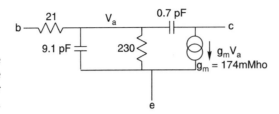

Figure 18-2. Simple equivalent circuit for the MRF901 bipolar transistor ($I_c = 5\,\text{mA}$, $V_{CE} = 5\,\text{V}$).

perhaps twenty parameters. Figure 18-2 shows a simple equivalent circuit for a common high-frequency transistor. This equivalent circuit was made by least-square fitting to the manufacturer's data from 100 to 2,000 MHz.

AMPLIFIER DESIGN

Designing a small-signal amplifier for microwave frequencies can be difficult, even when the only design goals are frequency response, gain, and stability. Computer-aided design is often used, especially when the amplifier must also have low noise and high dynamic range. The basic considerations have been discussed here, but amplifier design is a field in itself, and the reader is encouraged to consult the literature (see the Bibliography).

SIMPLE LOW-FREQUENCY AMPLIFIERS

At frequencies lower than a few megahertz we can strip the transistor equivalent circuits down to the barest essentials. Let us look at the simplest "back-of-an-envelope designs" for a common-emitter RF amplifier and then a common-base amplifier using bipolar transistors. This begins with a transistor properly biased into the active region at some collector current, I_o. The bias circuit can be the standard voltage divider for the base together with an emitter resistor to prevent thermal runaway. The emitter resistor, which might drop 20% of V_{cc}, is bypassed to ground the emitter for RF. A very simple model of a bipolar transistor is just a resistor r_b from the base to the emitter and a current generator from the collector to the emitter. The output resistance of the transistor is around 100 K ohms, but we can neglect it if we use a much smaller load resistance. This mismatch will prevent us from getting the maximum possible gain, but at high frequencies will still have lots of useful gain, and the mismatch helps achieve stability. (If necessary, we can neutralize the amplifier as discussed earlier.) The input circuit can provide a match to r_b. The common-emitter amplifier is shown in Figure 18-3. Recalling some transistor fundamentals, the mutual conductance of a transistor,

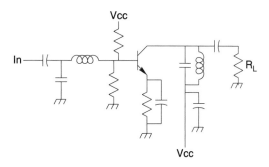

Figure 18-3. A common-emitter RF amplifier.

defined by i_c/v_b, is given by $g_m = I_o/V_{th}$ where $V_{th} = 0.026\,V$ and I_o is the collector bias current. The base resistance is $r_b = \beta/g_m$, where β is the transistor's current gain. The amplifier's output voltage is then given by $V_{out} = v_b g_m R_L$. Using these relations, power gain $= g_m R_L \beta = I_o R_L \beta/V_{th}$. Let us put in some typical numbers: $I_o = 5\,mA$, $\beta = 50$, and $R_L = 500$ ohms. This gives a power gain of about 36 dB, which is quite high despite the deliberate mismatch.

COMMON-BASE AMPLIFIER

A common-base circuit tends to be stable because the base isolates the input and output sides of the amplifier. The analysis is even easier with this circuit; we do not even have to know β of the transistor. Again we bias the transistor to some reasonable I_o. The bias circuitry is the same as above but we now need an RF choke to get dc to the emitter and another bypass capacitor to ground the base, as shown in Figure 18-4. In this circuit the input signal current is essentially equal to the output signal current because the base current is very low. The emitter resistance, r_e, is approxi-mately V_{th}/I_o, and, with $I_o = 5\,mA$ again, $r_c = 26\,mV/5\,mA = 5$ ohms. Suppose we want an input impedance of 50 ohms. We could build a matching network to convert the 5 ohms to 50 ohms, but let us use something

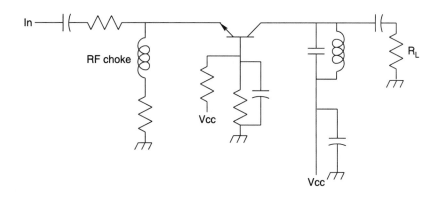

Figure 18-4. A common-base RF amplifier.

simpler, a 45 ohm series resistor, even if it will cost us some gain. The power gain will now be given by $G = R_L/50$. If we make $R_L = 500$ ohms, the amplifier will provide 10 dB of gain. If we had used an input-matching network, the gain would have been $500/5 = 100 = 20$ dB. (We could also keep the 45 ohm resistor and increase the load resistor to get more gain; R_L could be at least several thousand ohms.)

BIBLIOGRAPHY

1. R. S. Carson (1975), *High Frequency Amplifiers*, New York: John Wiley.
2. G. Gonzalez (1984), *Microwave Transistor Amplifier Analysis and Design*. Englewood Cliffs: Prentice Hall.
3. H. L. Krauss, C. W. Bostian and F. H. Raab (1980), *Solid State Radio Engineering*, New York: John Wiley.
4. G. D. Vendelin, A. M. Pavio and U. L. Rohde (1990), *Microwave Circuit Design Using Linear and Nonlinear Techniques*. New York: John Wiley.

PROBLEMS

1. Calculate the power gain (output power/input power) for the two-port network described by Equations (18-1) and (18-2) when the load impedance is Z_L.

2. A certain two-port component or network is described by its four Z-parameters according to the usual relations:

$$V_1 = Z_{11}I_1 + Z_{12}I_2$$
$$V_2 = Z_{21}I_1 + Z_{22}I_2.$$

One of many equivalent parameter sets is the Y-parameters:

$$I_1 = Y_{11}V_1 + Y_{12}V_2$$
$$I_2 = Y_{21}V_1 + Y_{22}V_2.$$

Derive expressions for the Y parameters in terms of the Z parameters.

3. (a) A certain amplifier with 20 dB of gain has a third-order intercept of 30 dBm (1 W at the output). If the input consists of a 0 dBm (0.001 W) signal at 100 MHz and another 0 dBm signal at 101 MHz, what will be the output power of the third-order products at 102 and 99 MHz?

(b) The same as Problem 3(a) except that input signal at 100 MHz increases in power to 10 dBm (0.01 W) while the input signal at 101 MHz remains at 0 dBm.

4. The Z-parameter description of a two-port corresponds in a one-to-one fashion to the equivalent circuit shown below, circuit (a). Find expressions for Z_A, Z_B, Z_C, and V in terms of the Z-parameters to make the circuits (a) and (b) equivalent.

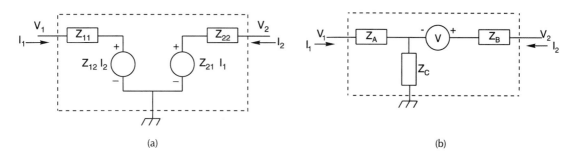

(a)   (b)

19

FILTERS II —
COUPLED-RESONATOR FILTERS

We have seen that the straightforward transformation from a prototype low-pass filter to a bandpass filter yields a circuit with alternating parallel resonant circuits and series resonant circuits, as shown in Figure 19-1. These simple bandpass filters work perfectly – when run on a network analysis program. But usually they call for component values that are not practical. The inductors in the shunt branches must be smaller than the inductors in the series branches by a factor of the order of the square of the fractional bandwidth. For a 5% bandwidth filter, the ratio of the inductor values would be of the order of 1:400. For a given center frequency we might be lucky to find a high-Q inductor of any value, let alone high-Q inductors with such different values. (Low-Q inductors make a filter lossy and change its pass band shape.) The series and shunt capacitor values would have the same ratio. Generally, Q is not a problem with capacitors, but very small values are impractical when they become comparable to the stray wiring capacitances.

This component value problem can be solved by transforming the prototype low-pass filters into *coupled-resonator bandpass filters*, which can be built with identical or almost identical *LC* resonant circuits. Figure 19-2 shows typical circuits.

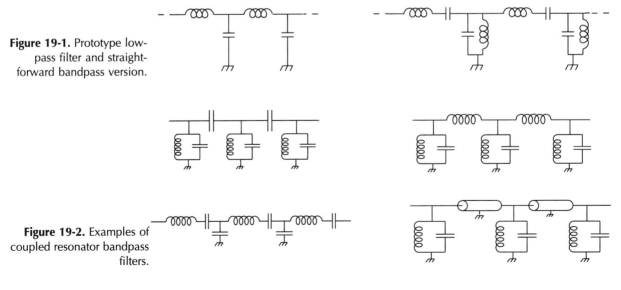

Figure 19-1. Prototype low-pass filter and straight-forward bandpass version.

Figure 19-2. Examples of coupled resonator bandpass filters.

IMPEDANCE INVERTERS

Coupled resonator filters are based on impedance inverters. Three examples of impedance inverters are shown in Figure 19-3, a 90° length of transmission line and two lumped LC circuits. In every case, an impedance Z, when seen through the inverter, becomes Z_0/Z, where Z_0 can be called the characteristic impedance of the inverter. For the transmission line inverter, Z_0 is the impedance of the line. For the LC inverters, both the reactance of the inductor, X_L, and the reactance of the capacitor, X_C, must be equal to the desired Z_0. Like the 90° cable, the lumped element circuits are perfect inverters only at one frequency, but in practice are adequate over a considerable range. An inverted capacitor is an inductor. An inverted inductor is a capacitor. Figure 19-4 shows an inverter (in this example the 90° transmission line) used to invert a parallel circuit, making an equivalent series circuit. The mathematics of this inversion is just

$$Z_{\text{in}} = Z_0^2 Y = Z_0^2 \left(\frac{1}{j\omega L_p} + j\omega C_p + \frac{1}{R_p} \right) = \frac{1}{j\omega (L_p/Z_0^2)} + j\omega (Z_0^2 C_p) + \frac{Z_0^2}{R_p}.$$

$$(19\text{-}1)$$

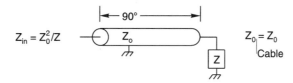

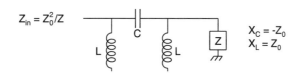

Figure 19-3. Three impedance inverter circuits.

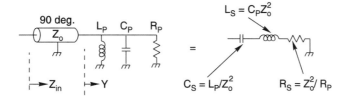

Figure 19-4. Impedance inverter makes a parallel circuit appear as a series circuit.

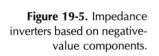

Figure 19-5. Impedance inverters based on negative-value components.

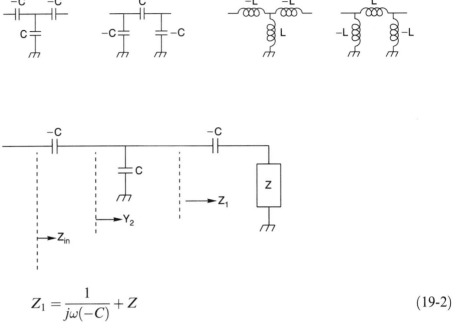

$$Z_1 = \frac{1}{j\omega(-C)} + Z \tag{19-2}$$

Figure 19-6. Operation of the T-form negative-capacitor inverter.

$$Y_2 = j\omega C + \frac{1}{Z_1} = \frac{j\omega C Z}{Z - (1/j\omega C)} \tag{19-3}$$

$$Z_{in} = \frac{1}{j\omega(-C)} + \frac{1}{Y_2} = \frac{1}{\omega^2 C^2 Z} = \frac{Z_0^2}{Z} \tag{19-4}$$

Let us look at four inverters that include inductors or capacitors with negative values. For these inverters, shown in Figure 19-5, $X_C = Z_0$ or $X_L = Z_0$. Figure 19-6 verifies the inverter action of the all-capacitor T-section inverter. You can use this kind of reasoning to verify the inverter action of the other circuits.

Because they contain negative capacitances or negative inductances, the four inverters in Figure 19-5 might seem to be only mathematical curiosities. Not at all; the negative elements can be absorbed by positive elements in the adjacent circuitry, as shown in Figure 19-7 with a pi section capacitor inverter between two parallel LC "tanks."

Ladder network filters have alternating series and shunt branches. Let us see how inverter *pairs* are used in ladder filters. Suppose we embed a series capacitor between a pair of inverters at some point along a ladder network. The combination of the capacitor and the inverter pair is equivalent to a shunt inductor, as shown in Figure 19-8. You can show just as

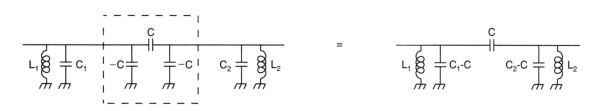

Figure 19-7. Negative capacitors absorbed into adjacent capacitors.

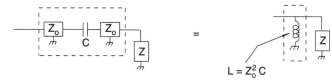

Figure 19-8. A series capacitor between inverters is equivalent to a shunt inductor.

easily that any series impedance, Z_s, together with a pair of bracketing inverters of characteristic impedance Z_0 is equivalent to a shunt admittance $Y_p = Z_0^{-2} Z_s$. Likewise, the combination of any shunt admittance Y and a pair of bracketing inverters is equivalent to a series impedance $Z = Z_0^2 Y$. Figure 19-9 illustrates this, showing how a series resonant series branch in an ordinary bandpass filter can be replaced by a parallel resonant shunt branch embedded between a pair of inverters. Likewise a parallel resonant shunt branch can be realized as a series resonant series branch embedded between a pair of inverters (Figure 19-10).

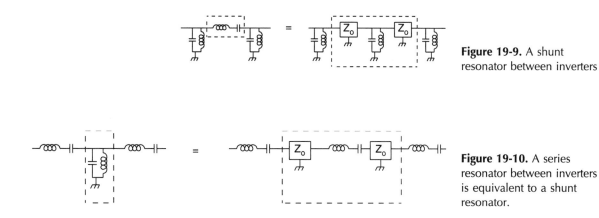

Figure 19-9. A shunt resonator between inverters

Figure 19-10. A series resonator between inverters is equivalent to a shunt resonator.

WORKED EXAMPLE – A BANDPASS FILTER WITH 1% FRACTIONAL BANDWIDTH

Consider a 50 ohm, 1 dB Chebyshev filter with a 10 MHz center frequency and a bandwidth of 100 KHz between the 1 dB points. The filter that results from the straightforward low-pass to bandpass transformation (see Chapter 4) is shown in Figure 19-11. The calculated response of this filter is shown in Figure 19-12. We might find 86 µH inductors with high Q at 10 MHz, but the 3.728 nH and 2.645 nH inductors would be tiny single turns of wire with very poor Q at 10 MHz. To get around these component limitations, we will convert this filter into a coupled-resonator filter. Suppose we have in hand some adjustable 0.3–0.5 µH inductors that, at 10 MHz, have very high Q. (We will see later just how much Q is required.) Let us first change the working impedance of the filter so that the parallel resonators at the end will use 0.5 µH, which is 134.1 times the original end inductors, and implies that the filter will be scaled to $50 \times 134.1 = 6,705$ ohms. We multiply the other inductors by 134.1 and

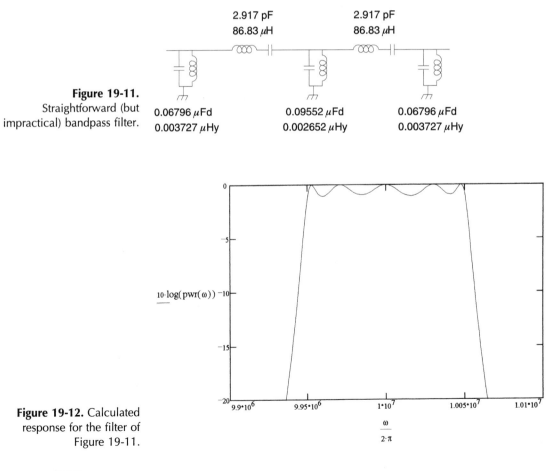

Figure 19-11. Straightforward (but impractical) bandpass filter.

Figure 19-12. Calculated response for the filter of Figure 19-11.

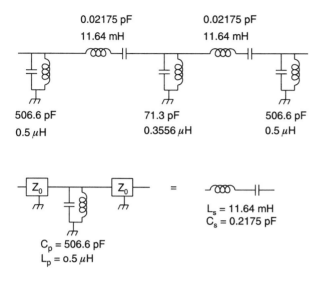

Figure 19-13. Filter of
Figure 19-11 scaled from 50
to 6,705 ohms.

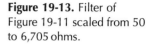

Figure 19-14. Inverters
transform a 0.5 μH shunt
inductor into an 11.644 mH
series inductor.

divide the capacitors by 134.1 to get the circuit of Figure 19-13. The
parallel resonators now use the desired inductors but the series resonators
call for inductors of 11.6 mH, a very large value for which we surely will
not find high-Q components. Moreover, the series capacitors are only
0.02 pF, a value far too small to be practical. We can solve this problem
by converting the series resonators into parallel resonators by using impedance inverters. Let us use the all-capacitor pi section inverters and the
same parallel resonators we used for the end sections. Figure 19-14 shows
how two inverters and the parallel resonator replace each series resonator.
We can calculate the characteristic impedance of the inverter, Z_0, as follows:

$$Z_0^2 Y = Z; \quad Z_0^2(j\omega C_\mathrm{p} + 1/j\omega L_\mathrm{p}) = j\omega L_\mathrm{s} + 1/j\omega C_\mathrm{s}$$

$$Z_0^2 = L_\mathrm{p}/C_\mathrm{s} = 0.5 \times 10^{-6}/0.02175 \times 10^{-12} = 4,795^2.$$

For this type of inverter we had seen that $Z_0 = X_\mathrm{c}$, so $C = 3.32$ pF. We
now have our coupled-resonator filter, but because it works at 6,705 ohms
we will add L-section matching networks at each end to convert it back to
50 ohms. The filter, at this point, is shown in Figure 19-15. All the reso-

Figure 19-15. Coupled-resonator version of the previous bandpass filter.

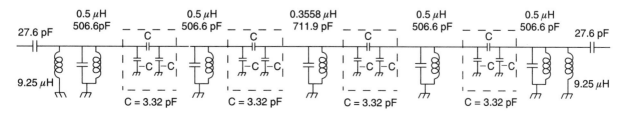

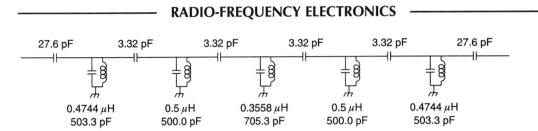

27.6 pF   3.32 pF   3.32 pF   3.32 pF   3.32 pF   27.6 pF

0.4744 μH   0.5 μH   0.3558 μH   0.5 μH   0.4744 μH
503.3 pF   500.0 pF   705.3 pF   500.0 pF   503.3 pF

Figure 19-16. Finished coupled-resonator filter.

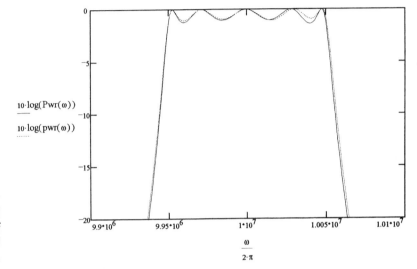

$$10 \cdot \log(\text{Pwr}(\omega))$$
$$10 \cdot \log(\text{pwr}(\omega))$$

$$\frac{\omega}{2 \cdot \pi}$$

Figure 19-17. Calculated response of the filters of Figure 19-15 (Pwr) and Figure 19-10 (pwr).

nators are now parallel resonators. (In other situations we might use inverters to convert parallel resonators into equivalent series resonators to make an all-series resonator filter – see Figure 19-2.)

The final clean-up step is to absorb the −3.32 pF capacitors into the resonator capacitors and combine the matching section inductors with the end-section resonator inductors, as shown in Figure 19-16.

The response of the finished filter is shown in Figure 19-17, and is almost identical to the response of the prototype filter of Figure 19-11. The difference, a fraction of a decibel, occurs because the inverters work perfectly only at the center frequency.

EFFECTS OF FINITE Q

These calculated filter responses assume components of infinitely high Q. We can calculate the effects of finite Q by paralleling the (lossless) inductors in our model with resistors equal to Q times the inductor reactances at the center frequency. If, for example, the Q is 500 (quite a high value for a coil), we would parallel the inductors in the filter of Figure 19-15 with resistors of about 15,000 ohms. Reanalyzing the circuit response, we would find that the filter will have a midband insertion loss of 7 dB and that the flat (within 1 dB) pass band response becomes rounded. The effect will be somewhat less for a filter with less steep skirts, such as a 0.01 dB Chebyshev or a Butterworth filter. But the real problem is still the small fractional bandwidth. For a filter with small fractional bandwidth to have the ideal shape of Figure 19-17, the resonators must be quartz or ceramic or other resonators with Qs in the thousands. An approximate analysis predicts that the midband loss *per section* in a bandpass filter will be of the order of

$$\frac{\text{power transmitted}}{\text{power incident}} = \left(1 - \frac{L_0/2}{Q \times \text{fractional bandwidth}}\right) \qquad (19\text{-}5)$$

where L_0 represents the inductor value in the normalized low-pass prototype filter. For our five-section filter we can take L_0 to be about 1.5 H. If the inductor Q is 500, the predicted transmission of the five-section filter is $5 \times 10 \log\{1 - (1.5/2)/[500 \times (1/100)]\} = -10$ dB, which is roughly equal to the actual value of -7 dB.

TUNING PROCEDURES

Filters with small fractional bandwidths and sharp skirts are extremely sensitive to component values. In the filter of Figure 19-16, for example, the resonators must be tuned very precisely or the shape will be distorted and the overall transmission will be lowered. (The values of the small coupling capacitors – all that remains of the impedance inverters – are not as critical.) Usually each resonator is adjustable by means of a variable capacitor or variable inductor. All the adjustments interact and, if the filter is totally out of tune, it may be hard to detect any transmission at all. A standard tuning procedure is to monitor the input impedance of the filter while tuning the resonators, one-by-one, beginning at the input end. While resonator N is being adjusted, resonator $N + 1$ is short circuited. The tuning of one resonator is done to produce a maximum input impedance while the tuning of the next is done to produce a minimum input impedance. The procedure must sometimes be customized, for example to account for matching sections at the ends.

OTHER FILTERS

The coupled-resonator technique is used from high frequencies through to microwaves. Not all RF filters, however, use the coupled-resonator technique. The intermediate frequency (IF) bandpass shape in television receivers is usually determined by a *SAW* (surface acoustic wave) bandpass filter. SAW filters are *FIR* (finite impulse response) filters, whereas all the *LC* filters we have discussed are *IIR* (infinite impulse response) networks. This classification is made according to the behavior of the output voltage following a delta function (infinitely sharp impulse) excitation. Digital filters can be designed to be either FIR or IIR filters.

BIBLIOGRAPHY

1. G. L. Matthaei, L. Young and E. M. T. Jones (1964), *Microwave Filters, Impedance-Matching Networks, and Coupling Structures.* New York: McGraw Hill (reprinted by Artech House, Boston, 1980).
2. D. G. Fink (1975), *Electronic Engineers' Handbook.* New York: McGraw-Hill.

PROBLEMS

1. Use your network analysis program to verify that the filter of Figure 19-15 does give the response shown in Figure 19–16.

2. Verify that the two *LC* circuits in Figure 19-3 are impedance inverters.

3. The filter shown below was developed in Chapter 4 as an example of the straightforward conversion from a prototype low-pass filter to a bandpass filter. This Butterworth (maximally flat) filter has a bandwidth of 10 kHz and a center frequency of 500 kHz.

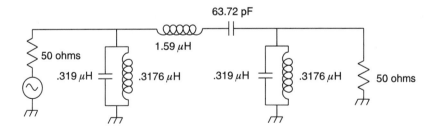

Suppose you have available some 30 µH inductors with a *Q* of 100 at 500 kHz. Convert the filter into a coupled resonator filter that uses these inductors. Use your ladder network analysis program to verify the performance of your filter.

4. A bandpass filter is to have the following specifications:

Center frequency: 10 MHz
Shape: three-section 1 dB Chebyshev filter
Bandwidth: 3 KHz (between outermost 1 dB points)
Source and load impedances: 50 ohms

Since the loaded Q of this filter is very high, $10^6/3,000 = 333$, it is important to use very high Q resonators. Suppose you have located some resonators (cavities, crystals, or whatever) with adequate Q. These resonators are all identical. At 10 MHz they exhibit a parallel resonance. As we move up from 10 MHz, they have a susceptance slope of 10^{-6} (1 mho/MHz).

(a) Find the LC equivalent circuit for these resonators (in the vicinity of 10 MHz).

(b) Design the specified filter around these resonators.

(c) Use your ladder network analysis program (see Chapter 1, Problem 3) to verify the frequency response of your design.

20 HYBRID COUPLERS

Hybrid couplers (also known as hybrid junctions or simply "hybrids") are four-port passive devices used as power splitters and power combiners. These nominally lossless devices are also used in mixers, modulators, and TR (transmit – receive) switches. An especially useful symbol for a hybrid is shown in Figure 20-1.

Most hybrids are unbalanced devices, that is, they have coaxial ports, so all four ports share a common ground, indicated in Figure 20-1 by a dotted ground symbol (but usually not shown). Each port has a characteristic impedance. Most packaged RF hybrids with coaxial ports are made so that the characteristic impedance of all four ports is 50 or 75 ohms. The symbol in Figure 20-1 is actually a signal flow diagram. Power incident on port 1 splits and exits through ports 2 and 3. If both of these ports are properly terminated there will be no reflections, and the impedance seen looking into port 1 will be equal to its characteristic impedance. In this case no power will reach port 4. (Opposite ports are said to be isolated.) But if port 2 and/or port 3 are not terminated in their own characteristic impedances, the power exiting these ports will be partially or completely reflected (the reflection coefficient, which depends on the mismatch, is calculated exactly as if the power had exited from a transmission line whose impedance is equal to that of the hybrid.) Any power reflected back into the hybrid splits and follows the signal paths as if it had come from an external source. You can see that with arbitary terminations the situation can become complicated. Usually, however, we are treating continuous-wave (cw) signals rather than pulses, so, rather than analyze multiple reflections, we only have to solve for the forward and reverse wave amplitudes on each of the four internal paths. This general problem sel-

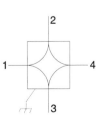

Figure 20-1. Schematic symbol for a hybrid coupler.

dom arises; hybrids usually have proper terminations, and the signal flows are simple and can be determined by inspection of the signal flow diagram.

DIRECTIONAL COUPLING

From inspection of the signal paths in Figure 20-1, we see that with the ports matched and with power flowing from port 1 to port 2 there will also be power flowing into port 3 but none into port 4. If the power is now reversed, to flow from port 2 to port 1, there will be power flowing into port 4 but none into port 3. Port 3 is therefore coupled to power flowing from port 1 to port 2. Likewise, port 4 is coupled to power flowing from port 2 to port 1. We see that, a hybrid, like the reflectometer discussed earlier, is a *directional coupler*. Here we will use the term *hybrid* only for 3 dB directional couplers, that is, couplers which split the incident power in half.

TRANSFORMER HYBRID

The name *hybrid transformer* was first applied around 1920 to the simple center-tapped transformer shown in Figure 20-2. The primary winding has N turns while each half of the secondary winding has $N/\sqrt{2}$ turns. For this simple hybrid the characteristic impedances of ports 1, 2, and 3 are equal and are twice the impedance of port 4. (Here these impedances, R, R, R, and $R/2$, are arbitrary, as long as the transformer behaves like a transformer, that is, its magnetizing inductance has a reactance substantially larger than R.) Let us confirm that this circuit has the power splitting and isolation characteristics of a hybrid. First consider a signal connected to port 1. If ports 2 and 3 have identical terminations they will have equal and opposite voltages. The voltage at port 4, because it is midway between

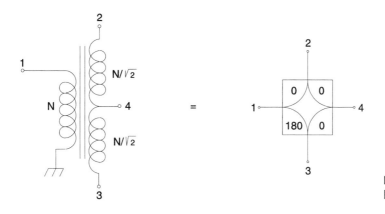

Figure 20-2. Transformer hybrid.

the voltages at ports 2 and 3, must be zero, and port 4 is indeed isolated from port 1. Next note that a signal applied to port 4 will appear unchanged at ports 2 and 3 but will not appear at port 1 (the currents to ports 2 and 3 are in opposite directions so there is no net flux in the transformer to provide a voltage at port 1 or produce an IX_L drop). You can verify that ports 2 and 3 are also isolated from each other (see Problem 1). Figure 20-2 also shows the symbol appropriate for this hybrid. The labels 0, 0, 0, and 180 indicate the phase shifts through the respective branches. A signal incident on port 1, for example, appears at port 2 with its phase unchanged (0° shift), and at port 3 with its polarity inverted (180° shift). Any hybrid with these four phase shifts is called a *180 degree hybrid*.

APPLICATIONS OF THE TRANSFORMER HYBRID

In telephony, hybrids allow a transmission line to carry independent signals in each direction. Figure 20-3 shows two telephone circuits. In the simple series circuit of Figure 20-3a, each user hears his or her own voice. In the circuit of Figure 20-3b, hybrids isolate each receiver from its own microphone. If we are using the hybrid of Figure 20-2, port 4 must be terminated in half the impedance used at ports 1, 2, and 3 or the cancellation will be incomplete. (The microphones and receivers must also have impedances corresponding to their ports.) In telephones the circuit is deliberately unbalanced to get a small amount of "side-tone," which enables the user to sense that the circuit is alive. Of course the cancellation should be as great as possible when this kind of full duplex circuit carries two-way digital data.

Figure 20-3. Hybrids allow full-duplex telephony over a single line.

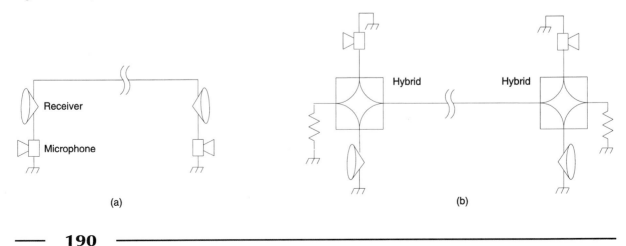

(a)   (b)

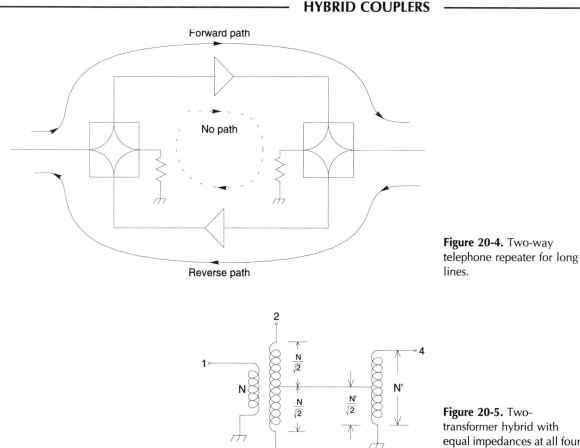

Figure 20-4. Two-way telephone repeater for long lines.

Figure 20-5. Two-transformer hybrid with equal impedances at all four ports.

The circuit of Figure 20-4 uses two hybrids and two amplifiers to make a two-way repeater for a long line. Here the hybrids let the signals in each direction be independently amplified without feedback – and oscillation.

It is convenient to have the characteristic impedances the same for all four ports of a hybrid. The transformer hybrid, fixed up to have equal impedances, is shown in Figure 20-5. This is the circuit you might find inside an off-the-shelf wideband 3 dB hybrid. Transformer hybrids made with toroidal cores work over large bandwidths, for example 10 KHz to 20 MHz and 1–500 MHz.

Sometimes, however, a hybrid has very different port types; a microwave hybrid might mix waveguide ports and coaxial ports.

QUADRATURE HYBRIDS

The transformer hybrid is naturally a 180° hybrid. Other circuits are natural 90° hybrids, the symbol for which is shown in Figure 20-6.

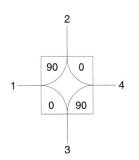

Figure 20-6. Symbol for a 90° (quadrature) hybrid.

We will see later that by adding lengths of transmission line we can convert a 180° hybrid to other hybrids, but first let us look at an interesting application of the 90° hybrid (often called a *quadrature hybrid*). Here the internal phase paths are 0 and 90°. A 90° path means a phase shift equal to that produced by a quarter-wave length of cable. At this point we can better describe the operation of a hybrid in terms of voltages. We will consider the hybrid to be connected to transmission lines (of the same impedance as the hybrid) in order to describe the signals in terms of incident and reflected waves. A signal incident at port 1 will be split equally into signals exiting ports 2 and 3. Since the power division is equal, the magnitudes of the voltages of the signals exiting ports 2 and 3 will be $1/\sqrt{2}$ times the magnitude of the incident voltage. The phases of the exiting signals will be delayed, as indicated on the symbol for the hybrid. For the hybrid of Figure 20-6, the signal exiting port 3 has no additional phase shift, but the signal exiting port 2 is multiplied by $e^{-j\pi/2}$. Suppose a signal is also incident at port 4. It will also split into signals exiting from ports 2 and 3. The total voltage of the waves exiting ports 2 and 3 is just the super-position of the waves originating from ports 1 and 4.

BALANCED AMPLIFIER

A common application for 90° hybrids is the balanced amplifier circuit shown in Figure 20-7. As long as the two amplifiers are identical they can have arbitrary input and output impedances, but the overall circuit will have input and output impedances of Z_0. To see how this happens, suppose that the hybrids are 50 ohm devices but that the input impedance of the amplifiers is not 50 ohms. Imagine that the interconnections are made using 50 ohm transmission line. The input lines have equal lengths and the output lines have equal lengths. The amplifiers are identical, so the two signals have equal phase changes upon reflection. An input signal is split

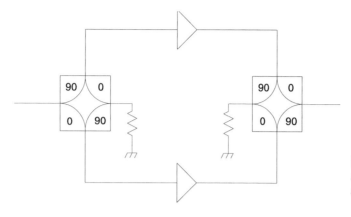

Figure 20-7. Balanced amplifier has constant input and output impedances.

by the input hybrid; half the power will be incident on the top amplifier and half on the bottom amplifier. But reflections from the amplifiers will be out of phase by 180° when they arrive back at the input of the hybrid because the signal on the upper path will have made a round trip through the 90° arm of the hybrid. The two reflections therefore cancel and there is no net reflection. The input impedance of the overall amplifier will be the characteristic impedance of the hybrid. The output side works the same way, and this combination of two arbitrary but identical amplifiers produces an amplifier with ideal constant input and output impedances. The balanced amplifier can also be built with 180° hybrids if two 90° lines are added, as shown in Figure 20-8.

This use of cables to allow substitution of 180° hybrids for 90° hybrids in this circuit can be taken farther; a 180° hybrid can be converted to a 90° hybrid by adding 90° sections of line to two of its ports. Likewise, a 90° hybrid can be converted into a 180° hybrid. These conversions, which you can verify, are shown in Figure 20-9.

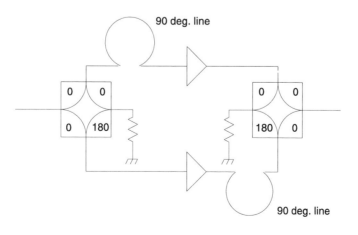

Figure 20-8. Balanced amplifier built with 180° hybrids.

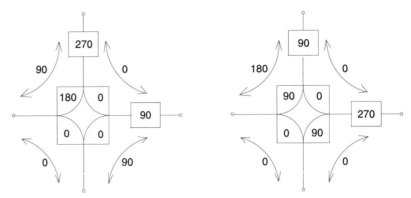

Figure 20-9. Conversions between 90 and 180° hybrids.

POWER COMBINING

To use a hybrid as a two-input power combiner, the forth port is simply terminated, as shown in Figure 20-10. The two signals to be combined must have equal amplitudes and the correct phase relationship to steer the total available power into the desired port. If the phase difference is changed by 180°, all the power will flow into the terminated port.

When a hybrid is used to combine two signals of different frequencies, as in Figure 20-10b, half the power of each signal will always be lost in the terminated port. Circuits to combine signals of different frequencies without loss are known as duplexers. (How would you make one?)

Figure 20-10. Hybrids used as power combiners.

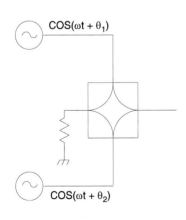

Combining equal frequencies

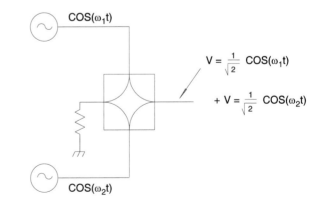

Combining different frequencies

OTHER HYBRIDS

There are many ways to make hybrids: you can collect them like stamps. When an application demands a hybrid with a certain bandwidth, phase relation, mechanical configuration, and so on, you will probably find just the right device in your file. A few more well-known hybrids are described below.

WILKINSON POWER DIVIDER (OR COMBINER)

This hybrid, shown in Figure 20-11 (in a 50 ohm version) uses two quarter-wave pieces of 70.7 ohm ($50\sqrt{2}$) cable. It has only three external ports; the fourth port is internally terminated, that is, connected to a load equal to its characteristic impedance. It is easy to see that power applied to port 1 will divide between ports 2 and 3. By symmetry, the voltages at ports 2 and 3 must be identical so no power is dissipated in the internal termination. Fifty ohm loads at ports 2 and 3 are transformed by the 90° cables to 100 ohms. The parallel connection of 100 and 100 ohms at port 1 produces the desired 50 ohm input impedance. The Wilkinson divider is usually classified as a 180° hybrid because its outputs have the same phase, even though this phase is 90° rather than zero.

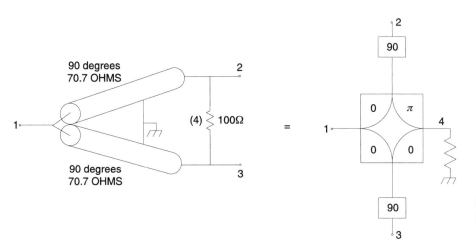

Figure 20-11. Wilkinson power divider.

RING HYBRID

The ring hybrid, shown in Figure 20-12, uses four pieces of transmission line. A 50 ohm hybrid requires 70.7 ohm cable.

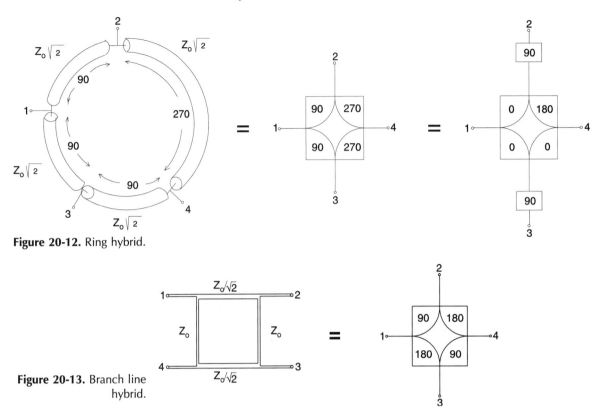

Figure 20-12. Ring hybrid.

Figure 20-13. Branch line hybrid.

BRANCH LINE HYBRIDS

These are ladder networks made of quarter-wave lengths of transmission line. The simplest is shown in Figure 20-13. More complicated versions have more branches and provide more bandwidth.

Two examples of 50 ohm lumped element hybrids are shown in Figure 20-14. These two circuits and many others are obtained by replacing the Z_0 and $Z_0/\sqrt{2}$ arms of the simple branch line hybrid with the pi (or T) lumped circuit impedance inverters discussed earlier.

All the hybrids discussed above use transformers, lumped elements, or simple transmission line elements. Many hybrids, especially those for use at microwave frequencies, use more complicated elements such as coupled transmission lines and waveguide structures. The backward coupler, shown in Figure 20-15, is a 90° hybrid [1].

The "magic T" of Figure 20-16 is a four-port waveguide junction that combines an E-plane tee with an H-plane tee. (In Volume 8 of the "MIT Radiation Laboratory Series" [1], any 180° hybrid is called a magic T.)

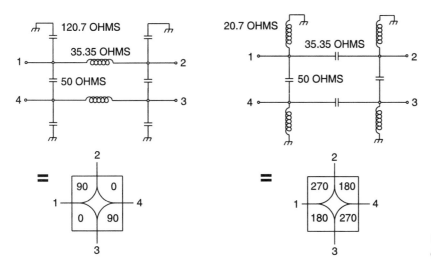

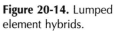

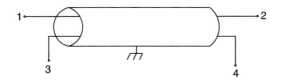

Figure 20-14. Lumped element hybrids.

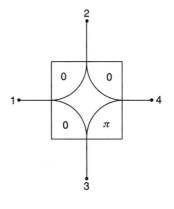

Figure 20-15. Backward coupler (often made with stripline).

Figure 20-16. Waveguide magic-T hybrid.

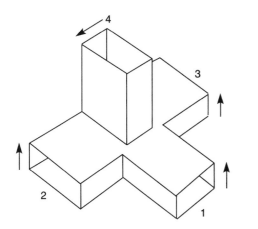

GENERAL DIRECTIONAL COUPLERS

Earlier we pointed out that the hybrid is a 3 dB directional coupler (half the power incident on a given port emerges at each of the two adjacent ports). It turns out that all directional couplers (including hybrids) can be represented by the circuit shown in Figure 20-17. The center element in

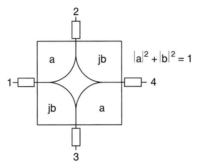

Figure 20-17. General directional coupler.

Figure 20-17 is a generalized hybrid with arbitrary power division (not just 3 db). The voltage transfer coefficients are a and jb. Conservation of energy requires that $|a|^2 + |b|^2 = 1$. When reciprocity is also taken into account, it turns out that a and b can always be real numbers [1]. You can see that if the line lengths are zero and if $a = b = 1/\sqrt{2}$, this general coupler becomes the 90° hybrid of Figure 20-6.

If you read Volume 8 of the "MIT Radiation Laboratory Series" you will learn that nearly any arbitrary asymmetric four-port structure (e.g. a metal box with four ports and arbitrary interior wiring or structure) will show some directivity and that there will exist impedances, Z_1 through Z_4 (generally complex), for which it will be a perfect directional coupler. Conversely, any four-port device that can be matched perfectly will then be a perfect directional coupler. You will also learn why hybrids are preferred to three-port junctions as combiners and splitters (even though only three ports are needed); no three-port junction can be completely matched unless it is a nonreciprocal device – a circulator.

BIBLIOGRAPHY

C. G. Montgomery, R. H. Dicke and E. M. Purcell (1948), "*Principles of Microwave Circuits.*" *MIT Radiation Laboratory Series*, Volume 8. New York: McGraw Hill. (Reprinted by Peter Peregrinus, London 1987).

1. Verify for the transformer hybrid of Figure 20-2 that port 2 is isolated from port 3, that is, show that with a generator connected to port 3, the voltage at port 2 will be zero provided the impedance terminating port 1 is twice the impedance terminating port 4.

2. Explain why the duplex telephone circuit and the two-way telephone repeater could be built with either 90 or 180° or any other hybrids.

3. **(a)** Calculate the overall gain of the balanced amplifier shown below if the (identical) individual amplifiers each have gain G_0. (*Answer:* G_0.)

(b) Calculate the overall gain if one of the interior amplifiers is dead, that is, has zero gain. (*Answer:* $G_0/4$.)

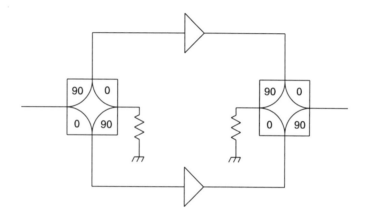

4. The 50 ohm hybrid shown below is properly terminated by 50 ohm resistors on two of its ports. The third port is terminated by a 25 ohm resistor. If a 50 ohm generator is connected to the remaining port, what fraction of the incident power will be reflected back into the generator? (*Answer:* 1/36.)

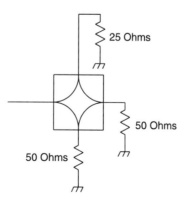

5. In the figure below, a 50 ohm hybrid is fed power from two amplifiers. The signals from these amplifiers have the same frequency and the same phase but the upper amplifier supplies 1 W while the lower amplifier supplies 2 W. In this unbalanced configuration, how much power reaches the load resistor? (*Answer*: 2.91 W.)

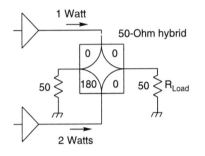

6. Four identical 1 W amplifiers and six hybrids are used to make a 4 W amplifier. Assume the input drive is such as to produce 1 W output from each amplifier. If one of the interior amplifiers fails, how much power will be delivered to the load? (*Answer*: 2.25 W.)

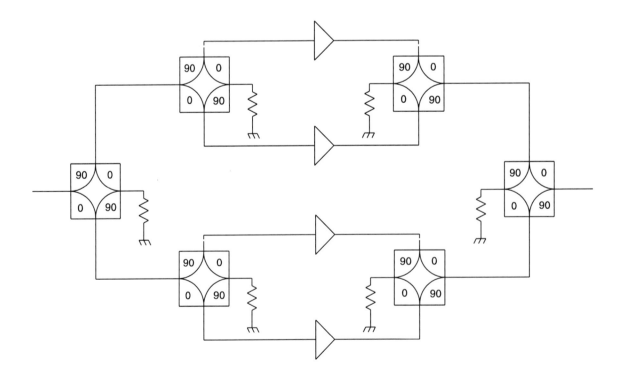

AMPLIFIER NOISE I

21

The output signal from any amplifier will always include some random noise generated within the amplifier itself. The noise of a phonograph amplifier can sometimes be heard as a steady hiss even when the needle is stationary. Most of the hiss from a radio receiver is due to noise generated by atmospheric electricity. But if the antenna is disconnected, the noise does not entirely disappear. The remainder is being generated within the receiver. Physical mechanisms that cause this noise include thermal noise (discussed below) and shot noise (due to the randomness in the flow of discrete charges – electrons and holes). The first stage in most receivers is an RF amplifier, and its noise usually dominates any other receiver-generated noise. This is easy to see; since this stage usually has considerable gain, its output power will be much greater than the noise power contributed by the second stage, so the noise of the second stage will hardly change the signal-to-noise ratio. In the same way, noise contributed by the third stage is even less important, and so forth.

THERMAL NOISE

Thermal noise is such a universal phenomenon that it provides the very vocabulary for the definition of terms such as *receiver noise temperature* and *antenna noise temperature*. Let us examine the fundamentals of thermal noise. Any object, being hotter than absolute zero, radiates electromagnetic radiation (mostly in the infrared) to a degree determined by its temperature and its emissivity. A black body has the highest possible emissivity, unity. Thermodynamics requires that high emissivity implies high absorptivity, and vice versa. Since any body that will absorb heat will radiate heat, it might be expected that a resistor, which can absorb electrical power, should also produce electrical power. A resistor can deliver its noise power to another resistor just as a black body can radiate into space. How much power can a resistor generate by virtue of being hot? Answer: any resistor can deliver kT watts per hertz, that is, kT is the spectral density of the power that a resistor will deliver to a matched load (a load of impedance $R + j0$). Here k is Boltzmann's constant $(1.38 \times 10^{-23} \, \text{J/K})$ and T is the absolute temperature. It is useful to remember that for $T_0 = 290\,\text{K}$, which is universally taken as standard reference temperature, $kT_0 = -114\,\text{dBm/MHz}$, since $10\log(1.38 \times 10^{-23} \times 10^3 \times 290 \times 10^6) = -114$.

To demonstrate that a resistor should produce this power, kT watts per hertz, an argument appropriate to radio engineering considers an antenna surrounded by a black body, that is, an antenna within a cavity whose walls are at a temperature T_1. We will be concerned with the spectral density at a particular spot frequency so we can specify that the antenna be resonant at that frequency, that is, it has a purely resistive impedance R. Let a transmission line connect the antenna to a resistor R that is outside the black body but also at temperature T_1. We know the antenna will intercept black body radiation and that power will be transmitted through the line to the external resistor. We also know from thermodynamics that, in this isolated system, it is impossible for the resistor to get hotter than T_1; heat cannot flow from a colder to a hotter object. The only way to resolve this is for the resistor to produce an equal amount of power that travels back to the antenna and is radiated back into the cavity. We can use some antenna theory to calculate the power. All antennas are directive; when used to receive, they have more effective area to intercept power from some directions than from other directions. But for any antenna, the average area turns out to be $\lambda^2/(4\pi)$ where λ is the wavelength. Black body radiation flux, at long wavelengths, is given by the Rayleigh–Jeans law, brightness $= 2kT/\lambda^2$ watts per square meter per hertz per steradian. This includes power in two polarizations. Since any antenna can respond to only one polarization, we use half the Rayleigh–Jeans brightness to calculate the power the antenna puts on the transmission line:

$$P = \int \frac{B(\theta,\phi)}{2} A(\theta,\phi)\,\mathrm{d}\Omega = \frac{kT}{\lambda^2}\frac{\lambda^2}{4\pi}4\pi = kT. \tag{21-1}$$

This value, kT is then also the power a resistor of R ohms will deliver to another resistor of R ohms. (The open circuit noise voltage from a resistor is therefore $\langle V_n^2 \rangle = 4kTR$ volts per hertz). Figure 21-1, below, shows the equivalent circuits of a resistor as a noise power generator.

Figure 21-1. Equivalent circuits for a resistor as a noise source.

R $(V_{rms})^2 = 4kTR$ volt^2/ Hz R $(I_{rms})^2 = 4kT/R$ amp^2 / Hz

NOISE FIGURE

At any given frequency, a figure of merit for a receiver, an amplifier, a mixer, and so forth, is its noise figure whose definition is as follows.

Noise Figure is the ratio of the total output noise power density to the portion of that power density engendered by the resistive part of the source impedance, with the condition that the temperature of the input termination be 290 degrees K.

Noise figure is a function of frequency and of source impedance but (as we will see later) is independent of output termination. Consider Figure 21-2. The voltage source V_n represents the thermal noise voltage from R_s, the resistive part of the source impedance, and the source, V_s, is the actual signal voltage, if any. The internal noise of the amplifier can be considered to result from another equivalent input noise source, V_a, at the input of the amplifier. With this model the noise figure, as defined above, can be written in terms of V_n and V_a: $\mathrm{NF} = (V_n^2 + V_a^2)/V_n^2$. Note that, because the amplifier noise is represented by an equivalent generator at the input side, this expression does not contain G, the gain. Since V_n^2 is proportional to the source temperature T_0, it is natural to assign the amplifier an equivalent noise temperature by writing the noise figure as $\mathrm{NF} = (T_0 + T_a)/T_0$. This amplifier noise temperature is just $T_a = (\mathrm{NF} - 1)T_0 = (\mathrm{NF} - 1) \times 290°$. Conversely, the noise figure is given by $\mathrm{NF} = (290 + T_a)/290$. An amplifier can have a noise temperature less than its physical temperature. The dish-mounted amplifiers used for home satellite reception have typical noise temperatures of 30 K. (Refrigeration, however, can help; FET (field effect transistor) amplifiers for radio astronomy are often physically cooled to about 10 K, and produce noise temperatures of only a few degrees kelvin).

So far we have not mentioned the signal voltage, V_s. The equation below shows that the noise figure also specifies the ratio of the input-to-noise ratio to the output signal-to-noise ratio:

$$\frac{\text{input SNR}}{\text{output SNR}} = \frac{V_s^2/V_n^2}{V_s^2/(V_n^2 + V_a^2)} = (V_n^2 + V_a^2)/V_n^2 = \mathrm{NF}. \qquad (21\text{-}2)$$

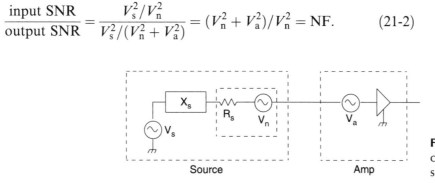

Source   Amp

Figure 21-2. Equivalent circuit of an amplifier and signal source.

CASCADED AMPLIFIERS

The noise from an amplifier of only modest gain will not totally dominate the noise added downstream, so it is useful to know how noise figures add. Suppose amplifier 1, with noise figure NF_1 and gain G_1, is followed by amplifier 2 with NF_2 and gain G_2. Suppose further that they are matched at their interface so that $G = G_1 G_2$ and that the output impedance of amplifier 1 is equal to the source impedance corresponding to the specified NF_2. Figure 21-3 shows how to compute the overall noise value. The noise figure of the cascade is $NF_{12} = NF_1 + (NF_2 - 1)/G_1$. It is interesting to calculate the noise figure of an infinite cascade of identical amplifiers as it is a lower limit to the noise we would get from any shorter cascade. You can verify that T_∞, the equivalent noise temperature of the infinite chain, is given by $T_a/(1 - 1/G)$, where T_a and G refer to the individual identical amplifiers.

Finally, let us look at the overall noise figure of an amplifier preceded by an attenuator (Figure 21-4). Suppose the gain of the attenuator is G_{attn}. (The gain of an attenuator is less than 1; the gain of a 6 dB attenuator, for example, is 1/4.) Referred to the amplifier input, the noise power engendered by the source resistance is $T_0 G_{attn}$. Since the amplifier still sees its standard source impedance, its total noise, referred to the input, is still $(NF_{amp}) T_0$. The overall noise figure is therefore

$$NF_{tot} = NF_{amp} \, T_0/(G_{attn} T_0) = NF_{amp}/G_{attn} \tag{21-3}$$

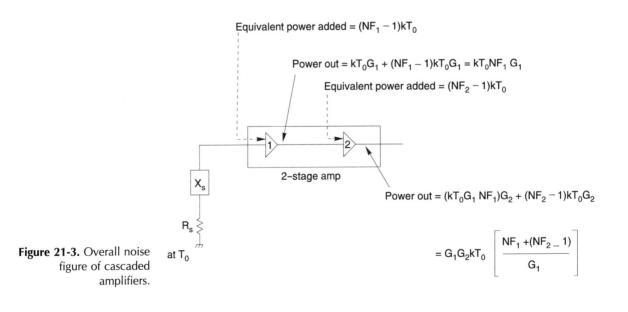

Figure 21-3. Overall noise figure of cascaded amplifiers.

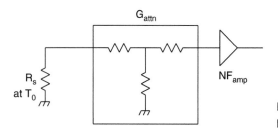

Figure 21-4. Amplifier preceded by an attenuator.

so if an amplifier preceded by an attenuator of "M decibels" (i.e. $G_{attn} = 10^{-M/10}$), the noise figure of the combination is M decibels higher than the noise figure of the amplifier alone. We could just as well have derived this result by using the relation for cascaded devices. The noise figure of the attenuator, from the definition of noise figure, is $NF_{attn} = kT_0/(G_{attn}kT_0) = 1/G_{attn}$. The noise figure of the cascade becomes $NF_{tot} = 1/G_{attn} + (NF - 1)/G_{attn} = NF/G_{attn}$ as before.

OTHER NOISE PARAMETERS

In what we have considered so far, the noise produced by a device, a transistor, amplifier, and so on, is specified by a single parameter, the noise figure. But the noise figure depends on the source impedance from which the device is fed, which makes this parameter seem to be something less than a complete noise description of the device. We will see later that a total of *four* noise parameters are sufficient to describe a device. The noise figure for any given source impedance can then be calculated from these four parameters, which are R_{opt}, X_{opt}, F_{min}, and R_n. The (complex) impedance $Z_{opt} = R_{opt} + X_{opt}$ is the source impedance that yields the minimum noise figure, F_{min}. The "noise resistance", R_n is a parameter that determines how fast the noise figure increases as the source impedance departs from Z_{opt}. We will see later that the noise figure for an arbitrary source impedance is given by

$$NF = NF_{min} + (R_n/G_s)|Y_{source} - Y_{opt}|^2. \tag{21-4}$$

(Here Y_{opt} is just $1/Z_{opt}$ and G_s is the real part of the source admittance.) We will also see that noise figure is somewhat deficient as a figure of merit. A piece of wire has $NF = 1$ but is not a valuable amplifier because it has no gain. With a given transistor, circuit A might produce a lower noise figure than circuit B, but circuit A may have less gain. We will see that T_∞, defined above, is the proper figure of merit.

NOISE FIGURE MEASUREMENT

A straightforward determination of the noise figure of an amplifier is possible if one knows its gain and has a spectrum analyzer suitable for measuring noise power density. Consider the common situation where we have an amplifier to be used in a 50 ohm environment. We connect a 50 ohm load to its input and use the spectrum analyzer to measure the output power density, S_{out} (watts per hertz), at the frequency of interest. We know that the portion of this power density engendered by the input load is kTG, where G is the amplifier gain. The noise figure is therefore given by

$$\text{NF} = S_{\text{out}}/(kTG). \tag{21-5}$$

This assumes we have done the measurement at $T = 290\,\text{K}$. If T was not 290, you can verify that

$$\text{NF} = [S_{\text{out}} - Gk(T - 290)]/(Gk \cdot 290). \tag{21-6}$$

For low-noise amplifiers, a comparison method is used. This method requires a cold load and a hot load, that is, two input loads at different temperatures (T_{hot} and T_{cold}). The amplifier is connected to a bandpass filter (whose shape is not critical) and then to a power meter (which needs to have only relative, not absolute, accuracy). The ratio of power meter readings, hot to cold, is called the Y-factor. The noise temperature of the amplifier is then given by

$$T_{\text{a}} = (T_{\text{h}} - YT_{\text{cold}})/(Y - 1). \tag{21-7}$$

PROBLEMS

1. Show that the noise figure of an infinite cascade of identical amplifiers is given by $F_\infty = (F - 1/G)/(1 - 1/G)$. Assume that the amplifiers have a standard input and output impedance such as 50 ohms, that G is the gain corresponding to this impedance, that F is the noise figure corresponding to a source of this impedance, and that F_∞ is also to be with respect to this standard impedance. *Hint:* Use the formula for a cascade of two amplifiers and the standard "infinite-chain-of-anything" technique (adding another link does not change the answer).

(This problem is not just academic. With only a short cascade the gain will be high enough to make the noise figure very close to F_∞, which is the best possible combination of the given amplifiers.)

2. Consider the balanced amplifier circuit shown below. The 3 dB, 90° hybrids are ideal. The amplifiers are identical and all impedances are matched. The individual amplifiers have power gain G and noise figure F_0. The hybrids are perfect, that is, they have no internal loss and are perfectly matched.

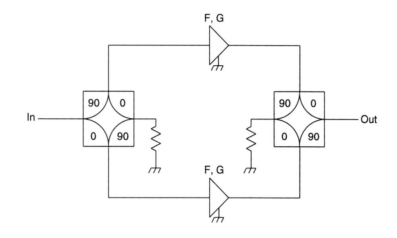

(a) Show that the overall noise figure of this circuit is equal to the noise figure of the individual amplifiers.

(b) If one amplifier dies, that is, provides zero output, what is the overall noise figure?

3. We derived the overall noise figure for an amplifier preceded by an attenuator when the physical temperature of the attenuator was T_0, the standard 290 K reference temperature. Assume now that the attenuator is at some different physical temperature, T_1, and find the overall noise figure. (Verify that your result gives the result $\mathrm{NF_{tot}} = \mathrm{NF}/G_{\mathrm{attn}}$ when $T_1 = T_0$). *Hint:* Model the attenuator as a lossless directional coupler with terminating resistors at the temperature T_1.

22 TRANSFORMERS AND BALUNS

Here we will first look at the conventional transformers used in power supplies, switching power supplies, and low-frequency matching networks. We will then examine transmission line transformers, which have extended bandwidth, and baluns, used to connect *bal*anced circuits to *un*balanced circuits.

Figure 22-1 shows a conventional transformer consisting of two windings on a toroidal core. Current flowing in either winding produces magnetic field lines – closed loops. When the toroid is made of a material such as iron, having a high magnetic permeability, the field lines are contained almost entirely within the core and provide nearly perfect coupling between the two windings. In a perfectly coupled transformer it makes no difference whether the windings are side by side, as shown, or wound one on top of the other. Any capacitance between the windings, however, produces electric coupling. This capacitance, as well as the distributed capacitance between turns of the windings, will affect the high frequency performance. At low frequencies the capacitances can be ignored.

The total magnetic flux, Φ, is the integral of the magnetic field, B, over the cross-section of the toroid. Faraday's law of induction states that a changing flux, $d\Phi/dt$, will induce a voltage $N\, d\Phi/dt$ into an encircling windings of N turns. If the windings are made of good conductors and/or the currents are small, the IR drops will be small, and induced voltages will account for nearly all the voltage across each winding. The voltages on the primary and secondary windings will be in phase, and their ratio will be equal to the turns ratio, N_1/N_2.

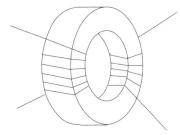

Figure 22-1. Prototype transformer.

TRANSFORMER CURRENTS AND THE IDEAL TRANSFORMER

The ratio of primary to secondary voltages is truly equal to the turns ratio if the transformer is perfectly coupled and lossless. The situation with the currents is not quite as simple. In a simplified model of a real transformer known as the *ideal transformer*, the ratio of the currents is the reciprocal of the turns ratio. The ideal transformer therefore transforms any impedance, Z_{load}, into $(N_1/N_2)^2 Z_{load}$, a simple multiplication. It is easy to see, however, that even perfect coupling and zero loss do not produce an ideal transformer. If the secondary of a real transformer is left open circuited, that is, Z_{load} equals infinity, then the primary impedance is just the inductance of the primary winding. An ideal transformer, on the other hand, would transform the open circuit into another open circuit. Another "unreal" feature of an ideal transformer is that it has no net magnetic field! The field (or flux) from the primary current is exactly cancelled by the field from the secondary current. And with no magnetic field there would be no $d\Phi/dt$ and therefore no voltage across the windings. Clearly the ideal transformer is not a good starting point for understanding transformer operation. We will see next that an equivalent circuit for a perfectly coupled lossless transformer (which might be called a "perfect real transformer") consists of an ideal transformer and a shunt inductance.

LOW-FREQUENCY EQUIVALENT CIRCUIT OF A PERFECTLY COUPLED LOSSLESS TRANSFORMER

A transformer is just a pair of coupled inductors. Depending on the circuit in which the transformer is used, these windings may be designated "primary" and "secondary". In Figure 22-2 the windings and their respective voltages and currents are labeled simply "1" and "2". The equivalent circuit of a real transformer is contained in the fundamental relations relating the voltages and currents in a pair of coupled windings. Using the conventional $e^{j\omega t}$ time dependence for ac circuit analysis, these equations, which follow from Ampère's law and Faraday's law, are

$$V_1 = j\omega L_1 I_1 + j\omega k \sqrt{L_1 L_2} I_2 + R_1 I_1 \qquad (22\text{-}1a)$$

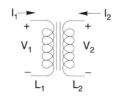

Figure 22-2. Transformer symbol with voltage and current assignments.

$$V_2 = j\omega k \sqrt{L_1 L_2} I_1 + j\omega L_2 I_2 + R_2 I_2 \qquad (22\text{-}1b)$$

We will assume for now that R_1 and R_2, the resistances of the windings, are negligible. Note that L_1 is the inductance of winding 1 when winding 2 is open ($I_2 = 0$), and L_2 is the inductance of winding 2 when winding 1 is open ($I_1 = 0$). The ratio of these self inductances, L_1/L_2, is equal to the square of the turns ratio, $(N_1/N_2)^2$. The factor k is called the *coupling coefficient*, and the quantity $k\sqrt{(L_1 L_2)}$ is the *mutual inductance*. For perfect (maximum) coupling, the coupling coefficient, k, is unity. With k equal to unity, and with R_1 and R_2 equal to zero, Equations (22-1a) and (22-1b) are equivalent to the following:

$$V_2 = V_1 \sqrt{\frac{L_2}{L_1}} \qquad (22\text{-}2a)$$

$$I_1 = \frac{V_1}{j\omega L_1} - I_2 \sqrt{\frac{L_2}{L_1}}. \qquad (22\text{-}2b)$$

These two equations exactly describe the circuit shown in Figure 22-3, an ideal transformer with a shunt inductor, the "magnetizing inductance." The effective turns ratio of this ideal transformer is the same as the turns ratio of the perfectly coupled transformer. The value of the shunt inductance is L_1, the inductance of the winding on the same side of the transformer. We could just as well have put the magnetizing inductance on the other side of the transformer, making its value L_2 instead of L_1. You can see this by interchanging the subscripts 1 and 2 in Equations (22-2a) and (22-2b) or by noting that the shunt inductance can be transferred to the opposite of the transformer if its value is multiplied by $(N_2/N_1)^2$. This gives $L_1(N_2/N_1)^2 = L_1(L_2/L_1) = L_2$, the same result. For that matter, L_1 could be represented by two parallel inductors and then one could be moved to the other side of the ideal transformer. However, that kind of symmetric representation could mistakenly imply that magnetizing inductances must be ascribed to both sides of the transformer.

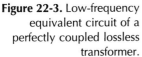

Figure 22-3. Low-frequency equivalent circuit of a perfectly coupled lossless transformer.

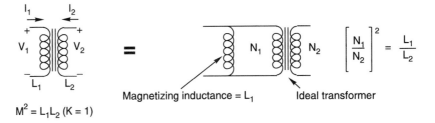

Magnetizing inductance = L_1 Ideal transformer

OPERATION OF THE PERFECTLY COUPLED LOSSLESS TRANSFORMER

Suppose we have a load, Z, on the right-hand winding (the secondary) and apply a voltage to the left-hand winding (the primary). If the reactance of the magnetizing inductance is much higher than the transformed value of Z, the magnetizing current will be negligible with respect to the total current, and the transformer can be considered ideal. Note, however, that it is the magnetizing current which produces all the magnetic flux; the currents caused by the load produce equal and opposite B-fields that cancel. The magnetizing current is present even when the transformer has no load and does not change when a load is applied. No average power is supplied by the magnetizing current since it is 90° out of phase with respect to the voltage. The magnetizing current in a typical power transformer might be as large as the in-phase current when the transformer is delivering full-rated power to a resistive load, R. This corresponds to a power factor of 0.707.* But usually the power factor is better than 90%, that is, $\cos[\tan^{-1}(R/X_{\mathrm{L}})] > 0.9$ or $X_{\mathrm{L}} > 2.06R$. Ordinary power transformers, therefore, have $X_{\mathrm{L}} > R$ rather than $X_{\mathrm{L}} \gg R$.

For circuit analysis the ideal transformer is sometimes an adequate model of a real transformer. However, we have already seen at least one circuit whose large-signal operation cannot be explained without a better model. That circuit, a transformer-coupled class-A amplifier, is shown in Figure 22-4. Since the transformer windings have almost no dc resistance, the average voltage at the collector must be V_{cc}. Under maximum signal

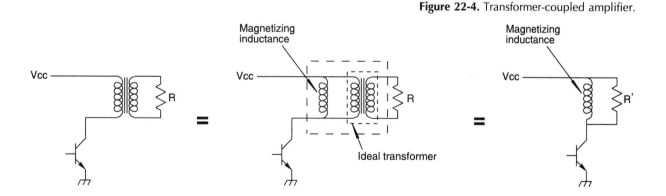

Figure 22-4. Transformer-coupled amplifier.

*The power factor of load (here a transformer with a resistive load on its secondary) is the cosine of the angle between the applied voltage and the resulting current. The average power delivered to the load is therefore the root mean square (r.m.s.) voltage times the r.m.s. current times the power factor.

conditions the collector voltage swings between zero and $2V_{cc}$, applying a peak-to-peak voltage of $2V_{cc}$ to the transformer primary. If we do not include the magnetizing inductance, the "ideally transformed" load is a pure resistance. We would mistakenly conclude that the quiescent collector voltage must be $V_{cc}/2$ rather than V_{cc} and that the largest peak-to-peak collector signal would be V_{cc} rather than $2V_{cc}$. We would also conclude incorrectly that the frequency response would extend all the way down to dc, rather than being shorted out at low frequencies by the magnetizing inductance.

MECHANICAL ANALOG OF A PERFECTLY COUPLED TRANSFORMER

A transformer multiplies (steps up or steps down) ac voltage. Figure 22-5 shows how a lever might be used to step down the velocity of a sinusoidally reciprocating arm. The resistive load on the right-hand side is a dashpot (damper). It produces a reaction force proportional to velocity. A transformer that steps down voltage provides increased current. This lever steps down velocity (and amplitude) and provides increased force. For an ideal transformer (infinite magnetizing inductance) or an ideal lever (zero mass) the input power (primary voltage times primary current or primary velocity times primary force) is equal to the output power at every instant. But for a real transformer, with finite magnetizing inductance, there is also an additional "excitation" current, lagging the voltage by 90°, that pumps magnetic energy in and out of the core. Likewise, for a real lever, with nonzero mass, there is an additional component of input force, leading the velocity by 90°, that pumps mechanical kinetic energy in and out of the lever. For both the transformer and the lever the average reactive power is zero, but the additional current or force can be considerable.

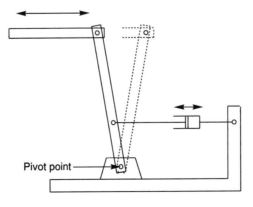

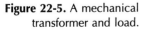

Figure 22-5. A mechanical transformer and load.

Pivot point

THE IMPERFECTLY COUPLED TRANSFORMER

We will see below that a transformer with $k < 1$ is equivalent to a perfectly coupled transformer with an inductor in series with one of its windings. This series inductor is known as the *leakage inductance*. The perfectly coupled transformer can then be represented, as shown above, by an ideal transformer with an external magnetizing inductance. These equivalent circuits are shown in Figure 22-6. Here the leakage inductance has arbitrarily been placed on the right-hand side of the transformer. We could just as well put some (or all) of the leakage inductance on the other side of the transformer, but, for circuit analysis, a single inductor is easier (and does not mistakenly imply that some leakage inductance must be ascribed to each side of the transformer). The value of the leakage inductance, $L_2(1 - k^2)$, and the effective turns ratio, n_1/n_2, of the perfectly coupled real transformer of Figure 22-6b can be deduced from Equations (22-1a) and (22-1b). As before, we neglect the winding resistances. Equation (22-1a) can represent a perfectly coupled real transformer with primary and secondary inductances given by $L_1' = L_1$ and $L_2' = k^2 L_2$. The effective turns ratio is therefore given by $n_1/n_2 = k^{-1}\sqrt{L_1/L_2}$. The secondary voltage, Equation (22-1b), becomes

$$
\begin{aligned}
V_2 &= j\omega\sqrt{k^2 L_1 L_2}I_1 + j\omega k^2 L_2 I_2 + j\omega(1 - k^2)L_2 I_2 \\
&= j\omega\sqrt{L_1' L_2'}I_1 + j\omega L_2' I_2 + j\omega(1 - k^2)L_2 I_2.
\end{aligned}
\tag{22-3}
$$

The first two terms on the right correspond to the same perfectly coupled real transformer. The last term can be attributed to an inductor – the leakage inductance – in series with the secondary.

Figure 22-6. Low-frequency equivalent circuit of a lossless transformer: arbitrary coupling.

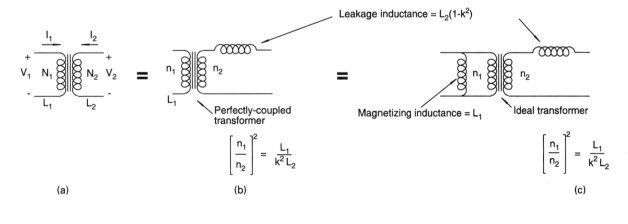

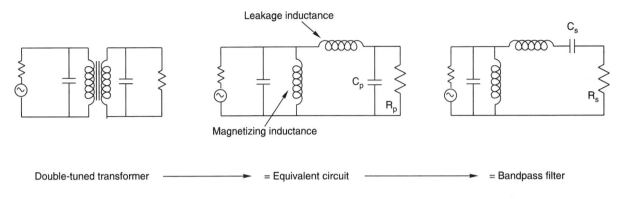

Double-tuned transformer ⟶ = Equivalent circuit ⟶ = Bandpass filter

Figure 22-7. Loosely coupled transformer used in a two-section bandpass filter.

DOUBLE-TUNED TRANSFORMER

Leakage inductance and magnetizing inductance limit the performance of transformers used in audio and other baseband applications. But in RF work these parasitic inductances can be tuned out with capacitors. In fact, the leakage and magnetizing inductance can be intentionally used, as in the double-tuned circuit shown in Figure 22-7. This circuit is a bandpass filter, commonly used in IF amplifiers. (Whenever you see capacitors across both windings of a transformer, you can guess that the coupling is less than unity – otherwise the two capacitors would be effectively in parallel, and a single capacitor could be used.) The double-tuned circuit works as follows: In Figure 22-6b the transformer has been replaced by its equivalent circuit. Only the leakage and magnetizing inductances are shown; the ideal transformer is either one to one, or the resistor and capacitor on the right-hand side have been multiplied by the transformer ratio. Figure 22-6c shows the parallel load resistor and capacitor replaced by their (narrow-band) series equivalents. The resulting circuit is just a simple two-section bandpass filter! The transformer, to have enough intentional leakage inductance, may have no permeable core (air-wound) or may be wound on a permeable rod or a low-permeability toroid.

CONVENTIONAL TRANSFORMERS WITH MAGNETIC CORES

Iron core transformers are used as low frequency (line current) power transformers and also for wide-band signals such as audio. The high-permeability ferromagnetic core greatly increases the inductance. (Remember that the magnetizing inductance should have a large reactance compared to impedance of the transformed load because otherwise the source bogs down supplying magnetizing current.) The permeable core also keeps the flux lines within the windings to minimize the leakage inductance. The *BH* curves of ferromagnetic materials are notoriously nonlinear, exhibiting saturation and hysteresis. These effects cause the magnetizing current to be a nonlinear function of the applied voltage. Nevertheless, the secondary voltage remains faithfully proportional to the voltage applied to the primary. The magnetization current will be nonlinear, but nonlinear in the way that makes $N_1\, d\Phi/dt$ equal to V_1. Since the secondary voltage, V_2, is given by $N_2\, d\Phi/dt$, it will be equal to V_1 times N_2/N_1. Severe saturation, however, will upset the proportionality between V_1 and V_2; the greatly reduced magnetizing inductance results in excessive magnetization current, and a significant voltage drop develops across the resistance of the primary winding. The primary voltage is now the sum of the induced voltage and this *IR* drop. Hysteresis causes an energy loss on every cycle equal to the area of the *BH* loop. This *core loss* is a consequence of the magnetizing current and is therefore independent of the load – unlike ohmic losses in the windings (*copper losses*), which increase with load.

EDDY CURRENTS AND LAMINATED CORES

In addition to suffering hysteresis loss, a core made of a conductive material such as iron will dissipate energy as ordinary $I^2 R$ loss because closed paths in the iron core will act as shorted turns around magnetic flux lines. To minimize this dissipation, any such closed paths are kept short by making the core a stack of thin sheet iron laminations separated by insulating varnish or oxide. The core of toroidal transformer can be a bundle of insulated iron wire rings.

DESIGN OF IRON CORE TRANSFORMERS

A transformer designer strives to find the smallest, lightest, and least expensive (usually all synonymous) transformer that conforms to a set of electrical specifications. To see the issues involved, let us consider the design of a 60 Hz power transformer. Suppose the primary voltage is 220 V r.m.s. and the power delivered to the load is 500 W. The efficiency is to be 96% and the magnetizing current must be no greater than the "working" (in-phase) current.

We will pick a silicon steel core material for which the maximum flux density before saturation is $1.5 \, \text{Wb/m}^2$. The core shape is shown in Figure 22-8. For minimum copper loss (neglecting the excitation current) the primary and secondary windings will have equal loss and will each occupy

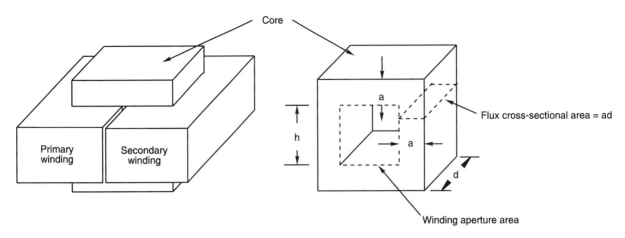

Figure 22-8. Iron core transformer geometry.

half of the winding aperture (as shown in Figure 22-8). The transformer will be specified by four parameters: the number of turns on the primary, N, and the three linear dimensions of the core, a, h, and d. To determine these four parameters we must write equations for the maximum B-field, the loss, and the inductance of the primary winding. Faraday's law of induction gives us the maximum B-field:

$$V_{\max} = N \frac{d\Phi}{dt} = N\omega a d B_{\max}. \tag{22-4a}$$

Since the primary voltage is 220 r.m.s., $V_{\max} = 220\sqrt{2}$, and we have

$$B_{\max} = \frac{220\sqrt{2}}{\omega N a d} < 1.5 \, Wb/m^2. \tag{22-4b}$$

The copper loss in the primary winding will be $I^2 R_{\mathrm{p}}$, where I is the r.m.s. current in the primary and R_{p} is the resistance of the winding. The number of turns on the primary, N, is given by

$$N = \frac{(h^2/2)}{\sigma} \tag{22-5}$$

where $h^2/2$ is the winding area for the primary and σ is the cross-sectional area of the wire. The mean length per turn is given by $2(a + d + h)$, so the primary resistance is found to be

$$R_{\mathrm{p}} = \rho \frac{\text{length}}{\sigma} = \frac{4\rho N^2 (a + d + h)}{h^2} \tag{22-6}$$

where ρ is the wire resistivity. As for core losses, the 60 Hz loss for the selected core material (when $B_{\max} = 1.5 \, Wb/m^2$) is 0.6 W/lb (11,000 W/m^3). The overall loss is the sum of the winding losses and the core loss. Since this loss is to be $500(1 - 0.96) = 20 \, W$, we can write

$$\text{Loss}_{\text{watts}} = 11,000[4ad(a+h)] + \left(\frac{500}{220}\right)^2 \frac{4\rho N^2(a+d+h)}{h^2} = 20.$$

(22-7)

Finally, the specification on the magnetizing current is equivalent to specifying that the reactance of the primary, ωL, be greater than the equivalent input load resistance or

$$\omega L \geq \frac{220^2}{500}.$$

(22-8)

The inductance of the primary, L, can be written as

$$L = \frac{\mu N^2 \times \text{flux area}}{\text{mean flux path length}}.$$

(22-9)

The mean flux path length, from Figure 22-8, is $4(h+a)$, so

$$L = \frac{\mu N^2 ad}{4(h+a)}.$$

(22-10)

We must use Equations (22-4b), (22-7) and (22-8) to find transformer parameters, a, d, h, and N, that will satisfy the given specifications and minimize the size of the transformer. This is not quite as simple as solving four equations in four unknowns. The equations are really inequalities and, in general, there will not be a solution that simultaneously produces the maximum allowable maximum flux density, the maximum allowable loss, and the minimum allowable inductance. Instead, this problem in linear programming is most often solved by cut-and-try iterative methods, conveniently done using a spread sheet program. In this particular example, such a procedure led to the following set of parameters: $d = 5\,\text{cm}$, $a = 2\,\text{cm}$, $h = 5\,\text{cm}$, and $N = 580$ turns. These dimensions give a core weight of 5.1 lbs (2.3 kg), a loss of 19.3 W and B_{max} of 1.42 Wb/m^2. The reactance of the primary, assuming a relative permeability of 1,000, is 5.9 times the input load resistance – five times more than the minimum required reactance. *Note*: in the equations presented above, no consideration was made for the space occupied by wire insulation and lamination stacking but these can be accounted for by simply increasing the value of the winding wire resistivity, ρ, and decreasing the permeability.

MAXIMUM TEMPERATURE AND TRANSFORMER SIZE

The heat generated by a transformer makes its way to the outside surface to be radiated or conducted away. The interior temperature build-up must not damage the insulation or reach the Curie temperature where the fer-

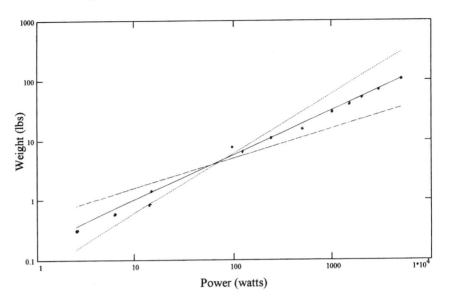

Figure 22-9. Weight versus power rating for 60 Hz power transformers.

romagnetism quits (a consideration with high-frequency ferrite cores). "Class-A" insulation materials (cotton, silk, paper, phenolics, and varnishes) are limited to a maximum temperature of 105°C. If reliable theoretical or empirically determined equations are available to predict internal temperatures, they can be included in the iterative design procedure described above. Rules of thumb can be used, at least as a starting point to determine core sizes for conventional transformers. One such rule for transformers up to, say 1 kW is that the flux cross-sectional area of the core (*ad* in Figure 22-8) in square inches should be about $0.25\sqrt{}$(power in watts). Transformer weight versus power rating (from catalog specifications) is plotted in Figure 22-9 for fourteen 60 Hz power transformers, ranging from 2.5 to 5000 W. The solid line, which fits the data, is given by

$$Wt_{lbs} = 0.18(power_{watts})^{3/4}. \qquad (22\text{-}11)$$

The dotted and dashed lines in Figure 22-9 correspond to exponents of 1.0 and 0.5. Transformer manufacturers seem to use the rule of thumb that makes core area proportional to power$^{1/2}$, because this results in the volume and weight being proportional to power$^{3/4}$.

TRANSMISSION LINE TRANSFORMERS

Leakage inductance and distributed capacitance eventually determine the high-frequency limit of conventional transformers. For wide-band applications we cannot simply resonate away these parasitics. Wide-band *transmission line transformers* [2] are built like ordinary core-type transformers

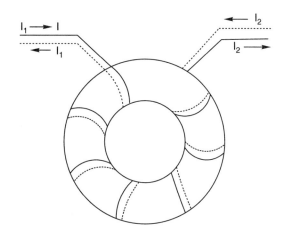

Figure 22-10. Transmission line transformer.

except that the windings are made wtih transmission line – either a bifilar winding, as shown in Figure 22-10, or coaxial cable. The core must have high permeability, but modest loss is acceptable (cores in chokes and transformers store little energy so high Q is not necessary). The effect of the core is to choke off any common-mode current in the transmission line, leaving only differential currents. We will therefore label the currents on the left as I_1 and $-I_1$ and the currents on the right as I_2 and $-I_2$. With ordinary transformers I_1 and I_2 are assumed equal, but here we treat the winding as a proper transmission line with a finite propagation velocity so the differential current at one end is not identical to the differential current at the other end. Figure 22-11 shows the transmission line transformer used in a reversing circuit, that is V_{out} is nominally equal to $-V_g$. The two circuits of Figure 22-11 are identical, but the form of Figure 22-11b makes the transmission line transformer appear more as a transmission line, with signal flow from left to right. Analysis of this circuit (or any other circuit containing this transmission line transformer) consists of replacing the transformer by a transmission line whose length is equal to the length of the transformer winding and assuming that no common-

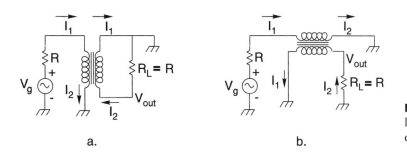

a.

b.

Figure 22-11. Transmission line transformer reversing circuit.

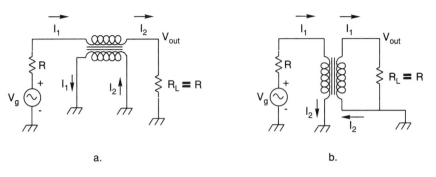

Figure 22-12. Transmission line transformer nonreversing circuits.

a.

b.

mode current flows. When $R_L = Z_o = R$, the result for Figure 22-11 is that $V_{out} = -1/2 V_g e^{j\theta}$, where θ is the electrical length of the line. Since $|e^{-j\theta}| = 1$, the power response is perfectly flat, but the reversal is perfect only as the frequency goes to zero ($\theta = 0$), leaving the 180° phase shift from the minus sign.

When the right-hand terminals of the transformer in Figure 22-11b are reversed, we get the circuit of Figure 22-12a. The response of this circuit, not surprisingly, is $V_{out} = 1/2 V_g e^{-j\theta}$. The negative sign is now positive. The amplitude and phase shift are identical to that of the reversing circuit. Note that this circuit is exactly equivalent to an unbalanced transmission line of the same impedance and same length (with or without a core). Note also that the circuit of Figure 22-12b is *not* the same as Figure 22-12a. Analysis of Figure 22-12b shows that its frequency response is not flat – it has a dip when $\theta = \pi/2$ – and that its phase response is not just a simple time delay.

Transmission line transformers extend the range of ordinary transformers by two octaves or more. In addition to this 1:1 transformer, many other transformers can be made with this transmission line technique [2,3]. Commercial hybrids good from 0.1 to 1000 MHz use transmission line transformers. Miniature transmission line transformers are commercially available as standard components.

BALUNS

A balun is any device that converts a *bal*anced (double-ended symmetric) signal into an *un*balanced (single-ended) signal. Figure 22-13 shows a common household appliance: a balun used with television sets and video recorders to splice 300 ohm balanced ribbon line to 75 ohm coaxial line. The circuit inside this balun is discussed below. Baluns are also commonly used to feed symmetric antennas (e.g. dipoles) from unbalanced coaxial feed lines. Push–pull amplifiers driven from or driving unbalanced

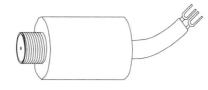

Figure 22-13. Common 75-to-300 ohm balun.

circuits require some kind of balun device. Figure 22-14 illustrates the requirement for a balun; with equal Z_1 and Z_2, V_1 and V_2 must be equal and opposite with respect to ground. The dotted ground symbol indicates that this point of symmetry will have zero voltage (when $Z_1 = Z_2$) and can be grounded if necessary or desirable. Figure 22-15 shows the equivalent situation with the load at the unbalanced side. when V_1 and V_2 are in phase (common mode) there must be no excitation of Z. But when V_1 and V_2 are 180° apart (differential mode) the load, Z, is fully excited. Baluns are normally reciprocal devices, so the name "unbal" is not needed.

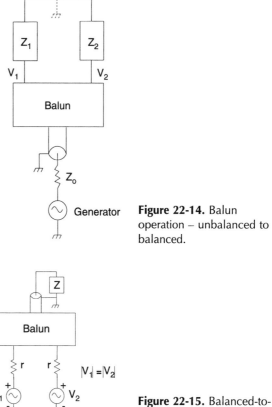

Figure 22-14. Balun operation – unbalanced to balanced.

Figure 22-15. Balanced-to-unbalanced conversion.

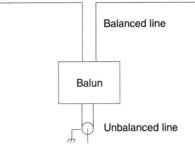

Figure 22-16. Balun connecting a dipole antenna to coaxial transmission line.

The dipole antenna shown in Figure 22-16 needs a balun so that (undesired) radiation picked up as common-mode current on the feedline will not get through to the receiver. From a transmitting standpoint, the balun eliminates common-mode current on the feedline that otherwise would radiate and affect the pattern of the antenna. Figure 22-17 shows a commonly used balun for this application. The phase shift through the half-wave piece of transmission line transforms V into $-V$. Note that this balun is also a 4:1 impedance transformer because the voltage across the dipole is 2 times V. (Note also that the half-wave line can have any Z_o). This transmission line balun is, of course, a narrow-band device because it depends on wavelength. A transformer can be used to make the wide-band 4:1 balun shown in Figure 22-18. An even wider-band balun uses transmission line transformers to overcome the high-frequency limitations of a conventional transformer. The 1:1 transmission line transformer discussed above can be used to provide the voltage reversal. A second 1:1 transformer is used in the noninverting configuration (Figure 22-12a) to provide an equal phase shift but without the reversal. The combination of these transformers, connected in parallel at one end and in series at the other end, as shown in Figure 22-19, makes a very wide-band 4:1 balun. This is the circuit most often used in the television balun of Figure 22-13. The transformers are often wound on a common "binocular core" (see Figure 22-17b). This core operates as two separate cores, that is, there is nominally

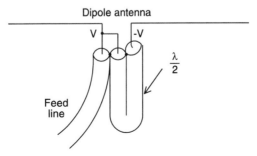

Figure 22-17. Transmission line balun.

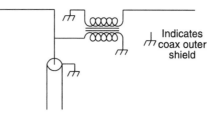

Figure 22-18. Transformer balun.

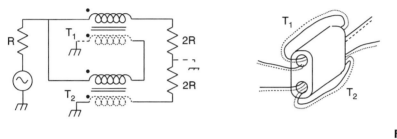

a.

b.

Figure 22-19. Wide-band 4:1 balun made from two 1:1 transmission line transformers.

no magnetic coupling between the two transformers, T_1 and T_2. For clarity, Figure 22-19 shows the transformer wound with only two turns; in practice, several turns are used.

BIBLIOGRAPHY
1. W. M. Flanagan (1993), *Handbook of Transformer Design & Applications*, 2nd Edition. New York: McGraw-Hill.
2. C. L. Ruthroff (1968), "Some broadband transformers," *Proc. IRE*, August 1357–42.
3. J. Sevick (1987), *Transmission Line Transformers* Newington, CT: American Radio Relay League.

PROBLEMS

1. Use Equations (22-1a) and (22-1b) to show that when the primary and secondary windings of a transformer are connected in series the total inductance is given by $L = L_1 + L_2 \pm 2M$, where M, the mutal inductance, is given by $M = k\sqrt{L_1 L_2}$ and the $\pm$ changes when one of the windings is reversed.

2. Find the low-frequency cut-off – 3 dB frequency – for a resistive load coupled by a 1:1 transformer to a generator. The transformer has perfect coupling. The source and load impedance are both 100 ohms and the reactance of the transformer primary is 50 ohms at 20 Hz.

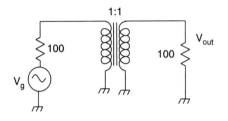

3. Upgrade your ladder network analysis program (see Chapter 1, Problem 3) to handle conventional transformers. Let the transformer be specified by its primary inductance, secondary inductance, and coupling coefficient.

4. When a power transformer is first turned on, that is, connected to the line, there is sometimes an initial inrush of current strong enough to dim lights on the same circuit and produce an audible "grunt" from the transformer itself. Decide whether this effect is strongest when the circuit is closed at a zero crossing of the line voltage or at a maximum of the line voltage. (This involves the magnetizing inductance of the transformer, so simply analyze the transient when an inductor is connected to an ac line.)

5. Extend the lever as a mechanical analog of the transformer to include leakage inductance. (The analogy does not have to be perfect; just as the mass of the lever gives a high-frequency roll-off instead of the low-frequency roll-off of magnetizing inductance, your mechanical "leakage inductance" can give a low-frequency rather than high-frequency roll-off.)

WAVEGUIDE CIRCUITS

23

In this chapter we will take an introductory look at waveguides and waveguide versions of low-frequency components. We will see how the concepts developed for two-conductor transmission lines apply to waveguides.

WAVEGUIDES

The ability of a hollow metal pipe to transmit electromagnetic waves can be demonstrated by holding it in front of your eye (you can see through it). While coaxial transmission lines are often used up to at least 20 GHz, waveguides have significantly less loss and more power-handling capacity. They need no insulating material to support a center conductor, so they are almost always filled only with air or some other low-loss gas.

SIMPLE EXPLANATION OF WAVEGUIDE PROPAGATION

A common RF engineering argument for the plausibility of transmitting electromagnetic waves through a hollow single conductor is shown in Figure 23-1 in which a twin-lead transmission line evolves into a waveguide. Quarter-wave shorted stubs are added to the line. Since a shorted quarter-wave line presents an open circuit, these stubs do not impair transmission. More stubs are added to both sides until a rectangular pipe is formed.

Figure 23-1. Transmission line-to-waveguide evolution.

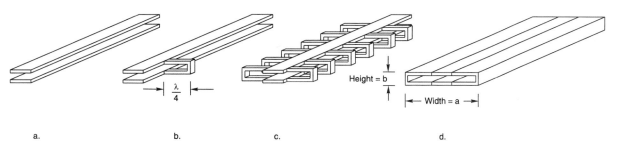

This plausbility argument, while oversimplified, does point out some important properties of waveguide propagation in the fundamental mode. The electric field, which is essentially vertical between the conductors in Figure 23-1a, becomes perfectly vertical in the waveguide, though its magnitude must fall to zero at the sides of the waveguide (the metallic walls short out any tangential electric field). Just as the conductors of the transmission line of Figure 23-1a can have any separation and still support wave propagation, the waveguide of Figure 23-1d can have any height. The width, however, is critical. The total width of the guide must be at least $\lambda/2$ to accommodate a quarter-wave stub on each side and still have nonvanishing conductor strips, as shown in Figure 23-1c. This means that there is a low-frequency cutoff; wave propagation is not possible if the wavelength is greater than $2a$ where a is the waveguide width. Standard waveguide designations indicate the shape and size of the guide. WR430, for example, denotes "waveguide, rectangular", 4.3 inches (10.9 cm) wide. The standard width-to-height ratio is 2:1. (While the height of the guide can be made arbitrarily small, the waveguide will become increasingly lossy.) One of the largest waveguide sizes, WR2300, with a width of 23 inches (58.4 cm), has a low frequency cutoff of 257 MHz. One of the smallest, WR3, with a width of about 1/32 inch, has a low frequency cutoff of 197 GHz.

For a standard (2:1 aspect ratio) waveguide, the fundamental mode (called the $TE_{1,0}$ mode) is the only possible mode for frequencies above the low-frequency cutoff and below twice the low-frequency cutoff. Above this range other modes become possible. One of the first higher modes is the $TE_{0,1}$ mode, in which the electric field lines span the width, rather than the height, of the waveguide. When higher modes are possible they are often unintentionally excited at bends and discontinuities, and divert power from the desired mode. This power then does not couple properly to circuit elements designed for the fundamental mode and is effectively lost. Whenever possible, a microwave system designer therefore tries to use only the fundamental mode. The essential details of this most important mode are stated below. (Abbreviated derivations are included for the reader familiar with electromagnetic field theory.)

PROPAGATION OF THE FUNDAMENTAL MODE IN A RECTANGULAR WAVEGUIDE

The designation TE means *transverse electric*. In the fundamental mode ($TE_{1,0}$), the electric field is everywhere perpendicular to the direction of propagation (the z-direction in Figure 23-2). The electric field is also perpendicular to the y-axis, that is, E_x is the only component. Its functional

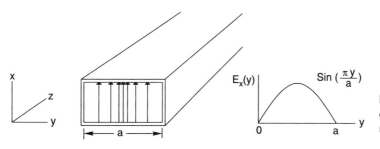

Figure 23-2. Electric field configuration in the $TE_{1,0}$ mode.

form is $\sin(\pi y/a)$, which falls to zero at the walls, $y=0$ and y-a, satisfying the boundary condition that the electric field tangential to a conductor must be zero.

Using this assumed $\sin(\pi y/a)$ dependence, we can find the wavelength in the guide and the form of the magnetic field. The magnetic field will satisfy the boundary condition that its normal component must be zero at the walls. This shows that the overall electromagnetic field is indeed a solution of Maxwell's equations. Finally, the current distribution (which flows on the inside waveguide walls) is obtained directly from the magnetic field.

GUIDE WAVELENGTH

Just as with waves on two-conductor transmission lines, such as coaxial cable, the wave-like dependence of the E- and B-fields will be contained in a factor $e^{j(\omega t - \beta z)}$ (for a wave traveling in the $+z$ direction) or $e^{j(\omega t + \beta z)}$ (for a wave traveling in the $-z$ direction). The propagation constant, β, is given by

$$\beta^2 = (\omega/c)^2 - \frac{\pi^2}{a^2} = (2\pi)^2 \left(\frac{1}{\lambda^2} - \frac{1}{(2a)^2} \right) \tag{23-1}$$

where c is the speed of light and $\lambda = 2\pi c/\omega$ (the free-space wavelength). This equation shows that if λ is greather than $2a$ there can be no propagation (β would be imaginary and the wave would decay exponentially). The guide wavelength (the spatial period in the z-direction) is given by $\lambda_g = 2\pi/\beta$, so, as the frequency approaches the low-frequency cutoff, the guide wavelength increases without limit. (The attenuation also increases drastically near the cutoff, and waveguides are normally operated at least 25% above their cutoff.) At frequencies far above the cutoff, the guide wavelength approaches the free-space wavelength, λ.

Equation (23-1) results from the substituting the assumed form of E, $\sin(\pi y/a)e^{j(\omega t - \beta z)}$, into the wave equation for free space. (The inside of the waveguide is free space.) The wave equation, which results from combining the curl E and curl H Maxwell equations is just

$$\frac{d^2E}{dx^2} + \frac{d^2E}{dy^2} + \frac{d^2E}{dz^2} + \frac{\omega^2}{c^2}E = 0. \tag{23-2}$$

FORM OF THE MAGNETIC FIELD

For this mode, the magnetic field has components in both the y- and z-directions, whereas the electric field only has an x-component. (In waveguide propagation, E and B cannot both be transverse to the direction of propagation.) For the assumed E-field (the real part of $\sin(\pi y/a)e^{j(\omega t - \beta z)}$, the corresponding B-fields (real parts) are given by

$$B_y = \frac{\beta}{\omega}\sin\left(\frac{\pi y}{a}\right)\cos(\omega t - \beta z) \tag{23-3}$$

and

$$B_z = \frac{\pi}{a\omega}\cos\left(\frac{\pi y}{a}\right)\sin(\omega t - \beta z). \tag{23-4}$$

The form of this B-field is shown in Figure 23-3b. Note that the magnetic field lines are concentric loops in the y–z plane with no component normal to the walls. You can use equations (23-3) and (23-4) to find the exact shape of these loops (see Problem 4). Equations (23-3) and (23-4) follow from the curl B Maxwell equation. In free space we have

$$By = -\frac{1}{j\omega}\frac{\partial E_x}{\partial z} \tag{23-5a}$$

$$B_z = \frac{1}{j\omega}\frac{\partial E_x}{\partial y}. \tag{23-5b}$$

Application of Equations (23-5a) and (23-5b) to the E-field produces Equations (23-3) and (23-4).

Figure 23-3. (a) Electric field lines and (b) magnetic field lines. (c) The electric lines are bundles of vertical vectors while the magnetic lines are stacks of nested loops.

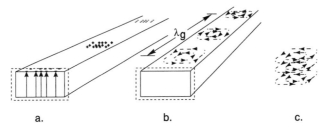

a.   b.   c.

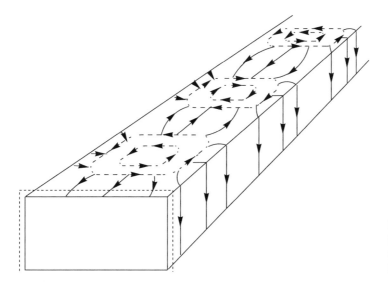

Figure 23-4. Wall currents (solid lines) in relation to the magnetic field (dashed lines).

WALL CURRENTS

Wall currents, which flow on the inside surfaces, are determined by the tangential magnetic field. The currents are perpendicular to the B-field, and their magnitude (in amperes per meter) is given by H, where $H = B/\mu_0$ (The permeability of free space, μ_0, is equal to $4\pi \times 10^{-7}$.) The wall currents are shown in Figure 23-4. These currents converge or diverge from areas on the broad wall where charge is being placed or removed. The E-field lines start and end at these charge patches. Note that the currents on the narrow walls are perfectly vertical because the tangential magnetic field has no x-component.

The fields and currents shown in Figures 22-2, 22-3, and 22-4 are, of course, snapshots at an instant in time. As the wave propagates, these patterns move uniformly along the z-axis with a (phase) velocity given by $v_{\text{phase}} = \omega/\beta$.

WAVEGUIDE VERSUS COAXIAL CABLE FOR LOW-LOSS POWER TRANSMISSION

Consider a situation requiring a low-loss transmission line. Let us compare a standard 2:1 aspect ratio waveguide to a cylindrical coaxial line. To minimize the loss we will make both as large as possible, but here we will impose the restriction that they are also small enough so that modes higher than the fundamental mode cannot propagate. Appendix 1 shows that the diameter of this lowest-loss coaxial line and the height of the lowest-loss waveguide are very close to $\lambda/2$ and that the coaxial line will have 2.4 times the loss of the waveguide and will carry only 23% as much power before breakdown.

WAVEGUIDE IMPEDANCE

There are several ways to define an impedance for a waveguide. One way is to define the voltage to be the potential difference between the top and bottom walls at the middle of the guide and the current to be the integrated current across the top wall. The ratio of voltage to current gives an impedance. Another definition uses voltage and power flow. Still another method uses the ratio of electric field to magnetic field at the center of the guide. The various definitions give $Z_0 = 377$ ohms (impedance of free space) within a factor of two for standard 2:1 waveguides. But regardless of how impedance is defined, there is no ambiguity in the concept of reflection coefficient. Recall that a shunt capacitance on an ordinary (transverse electromagnetic, TEM) transmission line produces a reflection coefficient on the negative j-axis of the Smith chart. The same kind of reflection is produced in a waveguide by a short vertical post or a horizontal iris. These obstructions are therefore called "capacitive posts" or "capacitive irises." An iris across the narrow dimension of the guide causes a reflection on the positive j-axis, so is called an "inductive iris." Figure 23-5 shows examples of inductive and capacitive irises. (The equivalent circuit for a thin iris is just a single shunt susceptance.)

The combination of an inductive and a capacitive iris (a thin wall with a hole) is equivalent to a parallel resonant circuit. You can see how these resonant irises could be spaced at quarter-wave intervals in a waveguide to make a coupled-resonator filter.

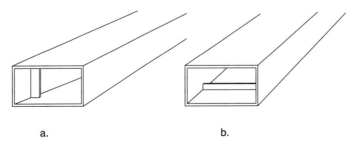

Figure 23-5. Waveguide irises: (a) inductive iris and (b) capacitive iris

a.

b.

MATCHING IN WAVEGUIDE CIRCUITS

Impedance matching in waveguide circuits can be done with the same techniques used for ordinary transmission lines. Suppose we are using a waveguide to supply power to some device, maybe a horn antenna, and that we have an instrument – a reflectometer or network analyzer – that can measure the reflection coefficient looking into the waveguide. We can locate the reflection coefficient on the complex reflection plane (Smith chart) as shown in Figure 23-6. As we move down the guide, away from the load, the reflection coefficient circles the center of the chart and eventually arrives at the unity conductance circle. We locate this position on the guide and install the appropriate inductive or capacitive iris. In practice, the tuning process is sometimes very simple: we find the point at which we need to add shunt capacitance. If the reflected wave is small (not a severe mismatch) we will not have to add much capacitance so we get out the ball-peen hammer and dent the broad side of the guide. An expert knows just how hard to swing the hammer.

A NOTE ON MATCHING   Suppose we join two dissimilar waveguides (perhaps of different sizes) at a junction, which could be some kind of elbow, coupling, or butt joint. Assume that the system is nominally lossless, that is, all metal. We want to match the junction so that a wave coming from either direction will suffer no reflection. We carry out the above procedure on one side of the junction. Do we have to then match the other side? No, the job has been done. Time reversal produces an equally good solution to Maxwell's equations in which all the power flows in the opposite direction. Of course this applies just as well to ordinary (TEM) transmission lines as it does to waveguides. The argument fails for lossy junctions because the time-reversed solution requires the absorptive material to *produce* power.

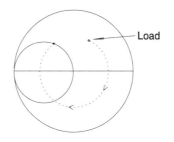

Figure 23-6. The reflection coefficient of a load located on the Smith chart.

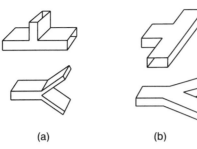

Figure 23-7. Waveguide tee junctions.

(a) (b)

THREE-PORT WAVEGUIDE JUNCTIONS

Two kinds of waveguide tee junctions (three-port junctions) are shown in Figure 23-7. The series tee gets its name from the fact that the voltage of the input guide divides between the two output guides. This works out well because the half-height output guides have half the impedance of the input guide, and the junction is inherently matched. (The half-height guides could be increased to full width in a gradual taper that would not cause much reflection.) The shunt tee applies to the full input voltage across each of the output arms – not so natural as the series tee.

FOUR-PORT WAVEGUIDE JUNCTIONS

The *magic-T* hybrid can be built using a procedure that itself seems like magic. We start with the bare waveguide junction (nothing hidden inside) as shown in Figure 23-8. First port 1 is matched. To do this we start by putting matched loads on ports 2 and 3. (We do not have to put a load on port 4 because, by symmetry, it is isolated from port 1.) With the loads in place we measure the reflection at port 1 and install the necessary iris (or dent) somewhere down line 1. Then we do the same process on port 4.

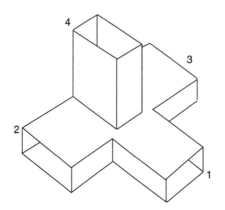

Figure 23-8. Waveguide magic T

That is it. The two matches and the isolation by virtue of symmetry are sufficient. We now have a perfectly matched magic-T hybrid.

Transitions from coaxial cable to waveguide have mostly been built with empirical methods. (Three-dimensional finite-element design programs are only just now coming onto the market.) For any kind of microwave transition, the designer first looks at the fields on both sides and finds a mechanical structure that causes the main features of the fields to line up. The remaining reflection should be small, and can be tuned out with a small iris.

Rectangular waveguides, like TEM lines, can carry only one signal in each direction. But square or round guides can have two independent waves; they are both fundamental mode waves but they have different polarizations. To launch or recover these two waves requires an *orthomode coupler* (which has no TEM counterpart). The simplest orthomode couplers use coaxial or waveguide connections mounted at right angles on the sides of the square or round guide. Some couplers produce circular rather than rectangular polarizations. Wide-band orthomode couplers are always needed for radio astronomy, and their development is an active field.

APPENDIX 1: LOWEST-LOSS WAVEGUIDE VERSUS LOWEST-LOSS COAXIAL LINE

For lowest loss we will make the waveguide and the coaxial line as large as possible, but, as explained above, with the restriction that each be capable of supporting only its fundamental mode. Our lowest-loss $TE_{1,0}$ waveguide will be made with its width equal to the wavelength. (If it is any wider, the second mode, $TE_{2,0}$, becomes possible.) We will make the height equal to half the width, that is, the usual aspect ratio. For air-filled coaxial line at the frequency where non-TEM modes become possible, the inner and outer radii, r_i and r_o, satisfy the inequality $(r_0 + r_i)\pi \leq 1.03\lambda$ (see Montgomery et al. [2], p. 42). The equal sign applies when $r_i/r_0 = 1/3.6$ This ratio also provides the lowest-loss air-filled coaxial line for a given outer diameter (see Appendix 2). Note that the characteristic impedance, $Z_0 = 60 \ln (r_0/r_i)$, will be 77 ohms for this lowest-loss coaxial cable. Using this ratio of diameters, the maximum outer diameter is given by $r_0 = 1.03\lambda\pi^{-1}/(1 + 1/3.6) = 0.26\lambda$. These relative waveguide and coaxial cable cross-sections are shown in Figure 23-9.

Let us compare the losses. The amplitude of a wave propagating in the $+x$ direction on any lossy line is proportional to $\exp(-\alpha x)$ where α, the loss factor, has units of inverse meters. The power is therefore proportional to $\exp(-2\alpha x)$, and the fractional power loss per meter is 2α. Note

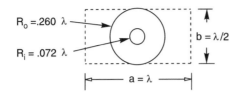

Figure 23-9. Relative cross-sections of lowest-loss waveguide and coaxial cable.

that $20 \log(e)\alpha = 8.686\alpha$ dB/m. Because of the skin effect, the loss of a line is proportional to its surface resistance, which is given by

$$R_s = \sqrt{\frac{\omega\mu}{2\sigma}} \text{ ohms per square} \tag{23-6}$$

where σ is the bulk dc conductivity and ω is the (angular) frequency. (For copper, $R_s = 2.61 \times 10^{-7}\sqrt{f}$ ohms/$\sqrt{Hz}$. For aluminium $R_s = 3.26 \times 10^{-7}\sqrt{f}$ ohms/$\sqrt{Hz}$.) The loss factor for air-filled coaxial line is given by

$$\alpha_{\text{coax}} = \frac{R_s}{2Z_0}\left(\frac{1}{2\pi r_0} + \frac{1}{2\pi r_i}\right). \tag{23-7}$$

Our lowest-loss coaxial line has $Z_0 = 77$, $r_0 = 0.26\lambda$ and $r_i = 0.072\lambda$, so its loss becomes

$$(\alpha_{\text{coax}})_{\min} = 0.0183\frac{R_s}{\lambda}. \tag{23-8}$$

For an air-filled rectangular waveguide in the fundamental mode the loss factor is given by

$$\alpha_{\text{WG}} = \frac{R_s}{377}\frac{\left[1 + \frac{2b}{a}\left(\frac{\lambda}{2a}\right)^2\right]}{b\sqrt{1 - \left(\frac{\lambda}{2a}\right)^2}} \tag{23-9}$$

where 377 ohms is $\sqrt{(\mu_0/\epsilon_0)}$,.the "impedance of free space". In our case $a = 2b = \lambda$, so

$$(\alpha_{\text{WG}})_{\min} = \frac{R_s}{\lambda}\frac{5}{377\sqrt{3}} = 0.0076\frac{R_s}{\lambda}. \tag{23-10}$$

The loss of the coaxial line is therefore higher than that of the waveguide by a factor of 0.0183/0.0076 or about 2.4. What about power handling capacity? The breakdown of either the waveguide or the coaxial cable depends on the maximum E field, $E_{\max}$. (For air at sea level pressure, $E_{\max}$ is about 30,000 V/cm.) For a rectangular waveguide in the fundamental mode the power is related to the maximum E-field by

$$\frac{\text{Pwr}}{E_{\max}^2} = \frac{1}{4 \times 377}ab\frac{\lambda_0}{\lambda_g} = 6.63 \times 10^{-4}ab\frac{\lambda_0}{\lambda_g} \tag{23-11}$$

where Pwr is in watts, E_{max} is in volts per centimeter, a and b are in centimeters, λ_0 is the free-space wavelength, and λ_g, the guide wavelength, is given by

$$\lambda_g = \frac{\lambda_0}{\sqrt{1 - \left(\frac{\lambda_0}{\lambda_{cutoff}}\right)^2}}. \tag{23-12}$$

For our waveguide $\lambda_0/\lambda_{cutoff} = 1/2$, so $\lambda_g = 2\lambda_0/\sqrt{3}$, and

$$\frac{Pwr}{E_{max}^2} = 5.74 \times 10^{-4} ab = 5.74 \times 10^{-4} \lambda \times \lambda/2 = 2.37 \times 10^{-4}\lambda^2. \tag{23-13}$$

Turning to the coaxial cable, the ln(r) dependence of voltage and the characteristic impedance, $Z_0 = (377/2\pi) \ln(r_0/r_i) = 60\ln(r_0/r_i)$, allows us to find the power in terms of the maximum E-field:

$$\frac{Pwr}{E_{max}^2} = \frac{Z_0 r_i^2}{2 \times 60^2}. \tag{23-14}$$

In our case $Z_0 = 77$ and $r_i = 0.072\lambda$, so

$$\frac{Pwr}{E_{max}^2} = \frac{77}{2 \times 60^2}(0.072\lambda)^2 = 0.55 \times 10^{-4}\lambda^2. \tag{23-15}$$

We see that the waveguide can handle $2.37/0.55 = 4.3$ times the power of the minimum-loss coaxial line. The waveguide is clearly better both for loss and power-handling capacity. In high-power applications the waveguide has the additional advantage that there are no interior surfaces needing cooling and mechanical spacers for centering. (Insulating spacers in high-power coaxial lines must fit tightly; high voltage develops across any gap. This problem generally reduces the power-handling capacity of the coaxial line by something like an order of magnitude.)

APPENDIX 2: COAXIAL LINE DIMENTIONS FOR LOWEST LOSS, HIGHEST POWER, AND MAXIMUM VOLTAGE

LOWEST LOSS

For a given outer diameter, the characteristic impedance of a coaxial line is increased by making the inner diameter smaller. For a given power, the current is decreased. But the smaller inner conductor has more resistance. The I^2R product, that is, the dissipation, has a maximum when the ratio of diameters is 3.6. This follows from Equation (23-7), which can be rewritten as

$$\alpha_{\text{coax}} = \frac{R_s}{2 \times 60 \times 2\pi r_0} \frac{1}{\ln(x)} (1 + x) \qquad (23\text{-}16)$$

where $x = r_0/r_i$. The minimum of $(1+x)/\ln(x)$ occurs at $x = 3.6$, so the characteristic impedance of lowest-loss air-filled coaxial line is $Z_0 = 60\ln(3.6) = 77$ ohms.

HIGHEST POWER

From the equation (23-14) we see that to maximize the power-handling capability of the coaxial line we must maximize the expression $Z_0 r_i^2$, that is, we must maximize $\ln(x)/x^2$. The maximum occurs when $\ln(x) = 1/2$, so the characteristic impedance of the maximum power line is $Z_0 = 60\ln(1.65)/2 = 30$ ohms.

MAXIMUM VOLTAGE

If the line is to withstand maximum voltage the optimum value of $\ln(x)$ is 1 and the characteristic impedance is 60 ohms. This also follows from Equation (23-14): If we express power as $V_{\max}^2/(2Z_0)$ then $V_{\max}$ is proportional to $Z_0 r_i$ or $\ln(x)/x$, which reaches a maximum at $x = e$.

RELATIVE PERFORMANCE OF 50 OHM COAXIAL LINE

The 50 ohm line commonly used in RF work ($x = r_0/r_i = 2.3$) strikes a compromise beween lowest loss, highest power, and highest voltage. For loss, we compare $(1+x)/\ln(x)$ for $x = 2.3$ and $x = 3.6$, to see that the 50 ohm line will have only 10 percent more loss than a 77 ohm line with the same outer diameter. For power handling, we compare $\ln(x/x^2)$, and find that the 50 ohm line can carry 62 percent as much power as a 30 ohm line with the same outer diameter. Finally, for voltage we compare $\ln(x)/x$, and find that the 50 ohm line can handle 98 percent as much voltage as a 60 ohm cable with the same outer diameter.

BIBLIOGRAPHY
1. R. E. Collin (1992), *Foundations for Microwave Engineering*, New York: McGraw-Hill.
2. C. G. Montgomery, R. H. Dicke and E. M. Purcell (1987) *Principles of Microwave Circuits*, London: Peter Peregrinus. (Originally Volume 8 of the *MIT Radiation Laboratory Series*, McGraw Hill, New York).
3. T. Moreno (1948), Microwave Transmission Design Data, Sperry Gyroscope Corp. (Reprinted by Dover, New York, 1958).
4. S. Ramo, J. R. Whinnery and T. Van Duzer (1994), *Fields and Waves in Communication Electronics*, 3rd Edition. New York: John Wiley. (Original edition was *Fields and Waves in Modern Radio*. Wiley, 1944).

1. Suppose a car enters a long tunnel that is essentially a rectangular metal tube 10 m wide by 5 m high. The car radio becomes silent inside the tunnel. Was the radio more likely tuned to an AM station or an FM station?

2. Examine the waveguide current distribution shown in Figure 23-4 (for the fundamental mode) and draw a sketch showing the position(s) in which a narrow slot could be cut through the waveguide wall without affecting its operation.

3. Describe an experimental set-up that could be used to demonstrate the waveguide *E*-field and *B*-field distributions shown in Figure 23-3.

4. Use Equations (23-3) and (23-4) to find the mathematical shape of the magnetic field loops. *Hint*: The slope of a field line, dz/dy, is given by B_z/B_y.

24 TELEVISION SYSTEMS

IMAGE DISSECTION

Standard television systems dissect the image and transmit the pixels serially. The image is divided into horizontal lines that are scanned sequentially for brightness information. The image is reconstructed at the picture tube or other display device by the inverse process; the pixels (picture elements) are illuminated, one after the other, left to right, line after line. With a cathode ray tube (CRT) display, the video signal controls the beam current and thereby determines the brightness of the phosphorescence at each pixel position on the screen. If the entire screen is scanned at a rate more than about 20 times per second, the viewer perceives a steady image due to the "persistence of vision" (the eye has a low-pass frequency response out to about 10 Hz). All broadcast television used analog modulation until 1994, when a commercial direct-broadcast satellite system went into service using digital modulation. A set-top box provides conversion back to the existing analog standard. A digital television standard, described further below, was proposed to the U.S. Federal Communictions Commission in 1996.

THE NIPKOW SYSTEM

Electron image dissection and reconstruction was proposed in the Nipkow disk system (patented in 1884), which used a pair of synchronized rotating disks; the camera disk dissected the image while the receiver disk reconstructed it. The receiver screen, a rectangular aperture mask, was covered, in effect, by an opaque curtain containing a pinhole illuminated from behind by an intensity-modulated gas discharge lamp. The position of the pinhole was analogous to the position of the illuminated spot on a CRT. This scanning pinhole was actually a set of N pinholes, arranged in a spiral on an opaque disk that rotated behind the aperture mask (see Figure 24-1). Only one hole at a time was uncovered by the aperture. As this active hole rotated off the right-hand side of the aperture, the next hole

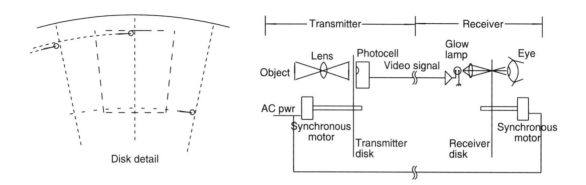

Figure 24-1. Nipkow rotating disk television system.

arrived at the left-hand side, but displaced downward by one scanning line. Corresponding holes in the transmitter disk allowed light from the original image to hit a photocell. Of course the rotating disks had to be synchronized.

This crude system was finally demonstrated in 1923 after the invention of the photoelectric cell, vacuum tube amplifier, and neon glow lamp. Some experimental broadcasting was done in the USA on 100 kHz wide channels in the 2–3 MHz range. All-electronic television broadcasts were first made in Germany in 1935 and in England in 1936. The all-electronic system used in the USA was proposed by the National Television Standards Committee (NTSC) of the Radio Manufacturers Assocation (RMA). Once this group had agreed on what is essentially still today's standard, the Federal Communications Commission (FCC) allowed commercial broadcasting to begin on July 1, 1941. NBC and CBS both started television service that day in New York City. These stations and a handful of others (Philadelphia, Schenectady, Los Angeles, and Chicago) continued to broadcast several hours per week throughout World War II. The existing receiver base numbered some ten or twenty thousand. Other than this, the war postponed the commercial development of television.

NTSC TELEVISION STANDARD

The NTSC standard specifies a horizontal-to-vertical aspect ratio of four-to-three for the raster (German for screen) with 525 horizontal lines, each scanned in 62.5 μs, including the retrace time. About 40 of these lines occur during the vertical retrace interval, so there are some 525–40 or 485 lines in the picture. If the horizontal resolution were equal to the vertical resolution the number of horizontal picture elements would be $4/3 \times 485 = 646$. The NTSC standard specifies somewhat less horizontal resolution, 440 picture elements. The horizontal retrace requires about $10 \, \mu s$, so the active portion of each line is $62.5\text{-}10 = 52.5 \, \mu s$. The maximum video frequency is therefore given by $\frac{1}{2} \times 440/52.5 = 4.2 \, \text{MHz}$. (A video sine wave at 4.2 MHz would produce 220 white stripes and 220 black stripes.)

The NTSC frame rate is $1/30 \, \text{s}$, that is, the entire image is scanned 30 times each second. The line rate is therefore $525 \times 30 = 15750 \, \text{Hz}$. To provide higher vertical resolution, a twofold interlaced scanning is specified. This is shown in Figure 24-2. Lines 1 through 262 and the first half of lines 263 make up the first field. The second half of line 263 plus lines 264

Figure 24-2. NTSC scanning sequence.

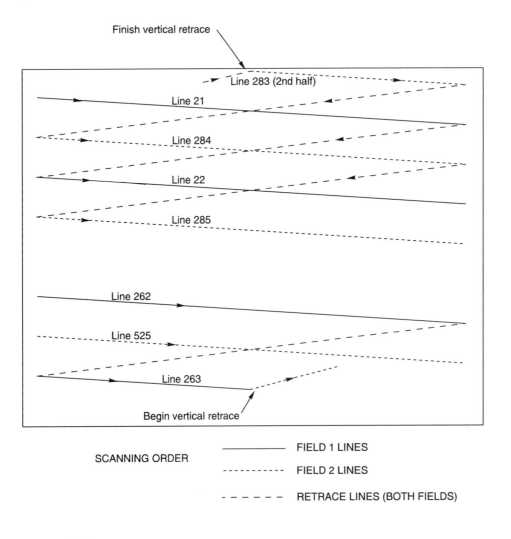

through 525 make up the second field. The lines in the second field fit between the lines of the first field. Interlacing improves the resolution for fixed scenes, but can create artifacts for moving objects (see Problem 6). Lines 1 through 20 occur during the vertical retrace period, and this entire period, like the horizontal retrace periods, is *blanked* (the electron beam is turned off).

THE VIDEO SIGNAL

The video signal amplitude modulates the video transmitter. Synchronization pulses are inserted between every line of picture information. (Just as the Nipkow disks had to be synchronized, the all-electronic image dissection and reconstruction must be synchronized; the illuminated dot on the receiver raster must be at the same relative position as the pixel being sensed at the camera tube.) The NTSC system uses negative video modulation, so that less amplitude denotes more brightness. The principle reason for using this polarity is that impulse inference creates black dots rather than more visible white dots on the screen of the receiver. (The French chose positive modulation so that noise pulses would not cause false synchronization, but modern synchronization systems are practically immune to noise.)

HORIZONTAL SYNCHRONIZATION

A horizontal synchronization pulse is inserted in the retrace period between each scanning line, and is distinguished by having a higher amplitude than the highest-amplitude picture information, that is, the synchronization pulse is "blacker" than the black level already blanking the beam during the retrace. The composite video (picture information plus synchronizing pulses) is shown in Figure 24-3. This waveform, which would be observed at the output of the video detector (after low-pass filtering to remove the 4.5 MHz sound) shows three successive scan lines.

Television receivers have a threshold detector in the synchronization circuitry in order to look only at the tips of the synchronization pulses, that is, the portion that is above the black level (and therefore totally independent of the video information). An *RC* differentiator produces a pulse coincident with the front (reference edge) of the synchronization tips. The burst of eight sinusoidal cycles at 3.579545 MHz on the "back porch" of each synch pulse provides a reference for the color demodulator (to be described later).

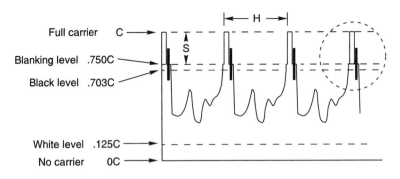

A. THREE LINES OF VIDEO

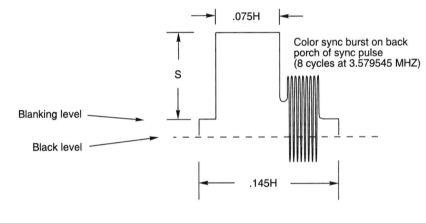

B. HORIZONTAL SYNC PULSE DETAIL

Figure 24-3. NTSC video waveforms.

VERTICAL SYNCHRONIZATION

A vertical synchronization reference is provided by a series of wider synchronization pulses that occur near the beginning of the vertical blanking period (i.e. every 1/60 s at the end of every field). An *RC* integrator produces a (rounded) pulse each time these wide pulses come by. A threshold detector, triggered by this pulse, signals the start of the vertical retrace period. The vertical synchronization scheme is complicated slightly because of the interlaced scanning. Since each field has 262.5 horizontal lines, every other vertical sync interval is shifted by half the horizontal scan period. Figure 24-4 shows how the pulse sequence preceding and following the vertical synchronization pulse interval differs between the even-numbered fields and the odd-numbered fields. But to maintain correct inter-

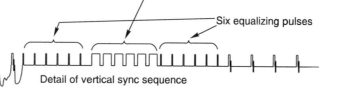

Figure 24-4. NTSC synchronizing pulses.

lace, the integrator output should be identical for each field in order that the vertical retrace be triggered exactly every 1/60 s. To buffer the integrator from seeing different sequences on odd and even fields, a series of six *equalizing pulses* precedes and follows the wide pulses. The time constant of the *RC* integrator is made short enough that when the vertical synchronization pulse inverval begins, the integrator output voltage has always been determined by an identical sequence of equalizing pulses.

Note that the equalizing pulses and the wide pulses provide uninterrupted horizontal synchronization in and around the vertical sync interval. In field 1 the horizontal timing references are the leading edges of the *even*-numbered equalizing and wide pulses, while in field 2 it is the odd pulses. In Figure 24-4 the small tick marks indicate these horizontal synchronization edges.

MODULATION

Radio transmission of video information (television) requires that we modulate a carrier wave with the composite video signal. The NTSC system uses AM modulation for the video. (Satellite links use FM.) Because

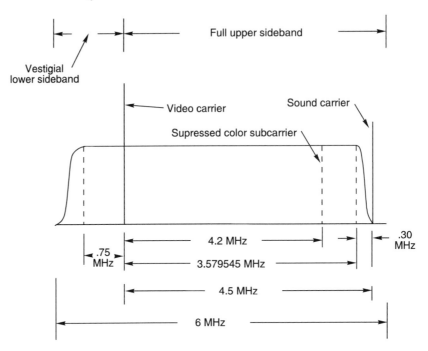

Figure 24-5. FCC standard 6 MHz television channel.

the NTSC video signal extends to 4.2 MHz, ordinary double-sideband AM would require a bandwidth of 8.4 MHz. To save bandwidth, allowing a 6 MHz channel spacing rather than a 8.4 MHz channel spacing, the lower part of the lower sideband is removed at the transmitter by filtering. The resulting signal consists of the entire upper sideband, the carrier, and the "vestigial" lower sideband, as shown in Figure 24-5.

At the receiver, a low-frequency video component, because it is present in both sidebands, would produce twice the voltage that would be produced by a high-frequency video signal, present only in the upper sideband. This problem is corrected by using an intermediate frequency (IF) bandpass shape that slopes off at the lower end, such as shown in Figure 24-6. At the video carrier frequency the amplitude response is $\frac{1}{2}$. (This response is as good as and simpler to obtain than the dotted curve where all of the double sideband region is reduced to $\frac{1}{2}$). For this IF equalization to work correctly, this IF filter should also have linear phase response. Fortunately, the surface acoustic wave (SAW) filters now commonly used to determine the IF bandpass shape can provide the linear phase (constant delay) response.

Figure 24-6. IF response to equalize the vestigial sideband.

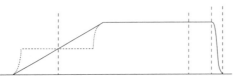

SOUND

The audio or "aural" signal is transmitted on a separate carrier, 4.5 MHz higher in frequency than the video carrier. The NTSC system uses FM modulation for the audio component, though some other standards use AM. The maximum deviation is 25 kHz, that is, the maximum audio amplitude shifts the audio carrier 25 kHz. Normally the audio transmitter is separate from the visual transmitter, and their signals are combined with a duplexer at the antenna. Early receivers had what amounted to an independent internal FM receiver for the sound. This was superseded by the "intercarrier" system, where the IF amplifier has enough bandwidth to pass the sound signal. The video detector then produces the 4.5 MHz sound signal as if it were a subcarrier of the video. It is passed through a 4.5 MHz bandpass filter, further amplified, and then FM detected. A corresponding 4.2 MHz low-pass filter removes the sound signal from the video channel.

OTHER TELEVISION STANDARDS

The NTSC is only one of some 13 world standards defined by the CCIR (International Radio Consultative Committee). These standards vary according to number of lines (525 or 625), sound carrier offset, FM or AM sound, and positive or negative video. These systems also differ slightly in the way they have been adapted to provide compatible color.

COLOR TELEVISION

Like color photography, color television is based on a tricolor system. When two or more colors (monochromatic spectral lines, sets of monochromatic lines, or continuous spectral shapes) are presented simultaneously to the eye, their relative intensities determine the perceived hue. When three complete images in three suitable chosen *primary* colors are superimposed, the eye perceives an accurate full-color image. The number of primaries and their spectral composition is not unique, though three primaries are enough to produce the required range of hues for color television. The particular red, green, and blue standards specified as FCC standards are based on practical phosphors used for the dots on the faceplate of the CRT. Combinations of these three FCC primaries can produce most hues except for very dark greens and bluish greens. Projection television sets use three high-intensity picture tubes, each with a color filter. Large direct-view receivers use a single tricolor tube with three electron guns. The inside of the glass face plate is tiled with red, green, and blue phosphor dots (i.e. dots that produce light of these colors when hit by electrons). Just behind the face plate is a *shadow mask*, which is a metal plate with thousands of precisely positioned holes. Electrons from the "red" gun, for example, are prevented from hitting any blue or green dots.

THREE COLORS THROUGH A SINGLE CHANNEL

Since color television requires the simultaneous transmission of three images, it is interesting to see how the color system was designed to get all three into the 6 MHz channel orginally allocated for the single monochrome image. Not only was this done, but it was done in a way that made color broadcasting compatible with existing monochrome receivers. The NTSC standard for compatible color television was adopted in 1953.

The solution to the bandwidth problem takes advantage of the fact that the monochrome video signal does not use the entire video bandwidth. There is a lot of redundancy in a typical picture. In particular, any given line is usually very much like the line preceding it (lots of correlation at $62.5\,\mu s$) so the video signal is similar to a repetitive waveform with a frequency of 15,570 Hz, the horizontal scan frequency. If the lines were truly identical, the spectrum would be a comb of delta functions at 15,750 and its harmonics. Since one line differs somewhat from the next, these delta functions are broadened but the spectral energy is still clumped around the harmonics of the horizontal scanning frequency, leaving relatively empty windows between each clump that can hold color information. (Note that if the entire picture is stationary, then the spectrum is a comb of delta functions spaced every 30 Hz, and essentially all the bandwidth is unused.)

COMPATIBILITY

Before getting the color signal into the empty windows we should mention how compatibility is maintained with monochrome receivers. Instead of transmitting the red, blue, and green (RGB) signals on an equal basis, three linear combinations are used. One of these, the *luminance* signal, Y, is chosen to be the brightness signal that would have been produced by a monochrome camera: $Y = 0.299R + 0.587G + 0.114B$. The other two linear combinations are $I = 0.74(R\text{-}Y)\text{-}0.27(B\text{-}Y)$ and $Q = 0.48(R\text{-}Y(+0.41(B\text{-}Y).$

The luminance signal directly modulates the carrier, just as in monochrome television, and monochrome receivers respond to it in the standard fashion. Each of the other two signals, I and Q, modulates a subcarrier (just as the right-minus-left audio signal modulates a 38 kHz subcarrier in FM stereo.) The color subcarriers have the same frequency, 3.579545 MHz, but they differ in phase by 90°. Figure 24-7 shows how two independent signals are transmitted and recovered from a single-frequency channel using so-called quadrature modulation. (Note that this type of multiplexing would make it possible to have two AM radio stations on each assigned frequency, but it would require receivers with phase-locked product detectors rather than simple envelope detectors.) In color television the subcarrier is suppressed, so the I and Q (for in-phase and quadrature) signals are double-sideband suppressed carrier signals. The receiver must supply local carriers (BFO signals) with the precise frequency and correct phases to demodulate (detect) these signals. The reference signal necessary to regenerate these local carriers is sent as a burst of about eight sinusoidal cycles at 3.579545 MHz on the back porch of each horizontal synchronization pulse, as shown in Figure 24-4. This color burst is used in the receiver as the reference for a phase-locked loop. The color subcarrier frequency is held to a tight 10 Hz tolerance, so the VCO is usually a VCXO (voltage-controlled crystal oscillator).

The color information, like the luminance information, is similar from line to line, so its power spectrum is also a comb whose components have a spacing equal to the horizontal scanning frequency. The subcarrier frequency is chosen at the middle of one of the spectral slots left by the luminance signal, so the comb of color sidebands interleaves with the comb of luminance sidebands.

Originally color television receivers had no way to separate the luminance and chrominance signals; the necessary comb filters were beyond the state of the art. Compatibility depended on the fact that the spectral

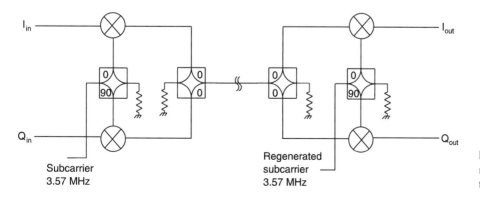

Figure 24-7. Quadrature modulator/demodulator lets two signals share one band.

interlacing greatly reduces visible "cross-talk" between chrominance and luminance information. To see this, consider a very simple signal, a uniform color field such as an all-yellow screen. Since this field has a color, that is, it is not black, white, or gray, there will be nonzero I- and Q-signals. In this example, since the color information is constant, the I- and Q-signals together are just a sine wave at the color subcarrier frequency. Their relative amplitudes determine the hue while their abolute amplitudes determine the saturation. One would expect this 3.58 MHz video component to produce vertical stripes. And the beam, as it sweeps across the screen, does indeed get brighter and dimmer at a 3.58 MHz rate, trying to make some 186 stripes in the 52.5 μs scan. But, on the next line, these stripes are displaced by exactly one-half cycle. The result is that the entire screen, rather than having 186 vertical stripes, has a fine-gridded checkerboard or "low-visibility" pattern. Colored objects viewed on a monochrome receiver can be seen to have this low-visibility checkerboard pattern. The pattern can also be discerned in colored objects viewed on an older color receiver; new receivers have a comb filter (described below). Low-priced monochrome receivers often have less than the full 4.2 MHz of video bandwidth; this suppresses the checkerboard pattern as well as the fine picture detail.

Note that there are some unusual situations where spectral interleaving does not work. If the image itself is like a checkerboard with just the right grid spacing, the luminance signal will fall into the spectral slots allocated for the chrominance signal and vice versa. A herringbone suit, for example, will often have a gaudy sparkling appearance when seen on a color receiver.

The low-visibility principle is applied not only to avoid luminance-chrominance cross-talk but also to reduce the effect of the beat between the 4.5 MHz sound carrier and the color subcarrier. To take advantage of the low-visibility principle, television standards were modified slightly when color television was introduced. The sound carrier remained the same at 4.5 MHz above the video carrier. The relation between the horizontal scanning frequency, f_h, and the color subcarrier frequency, f_{sc}, was picked to be $f_{sc} = 227.5 f_h$. Then the sound subcarrier-minus-color subcarrier beat was likewise made an odd number of half multiples of the horizontal scanning frequency: 4.5 MHz-$f_{sc} = 58.5 f_h$. Putting these two relations together determines the horizontal frequency, 15,734.264 Hz, and the color subcarrier frequency, 3,579545 MHz. The number of scanning lines remained at 525, so the vertical frequency changed from 60 Hz to $262.5 f_h = 59.940$ Hz. With these choices, the sound carried is at 286 times the horizontal frequency. This would produce a high-visibility pattern, but the sound carrier is above the nominal video band and can be filtered out easily.

COMB FILTERS

Comb filters to separate the luminance spectrum from the chrominance spectrum can use a delay line (ultrasonic, charged-coupled device (CCD), or digital). The CCD delay line, a sort of bucket brigade analog shift register, has been the most widely used. The circuit shown in Figure 24-8 uses a 1 H delay line (a delay line of one horizontal period) to provide comb filters for both the luminance and chrominance signals.

Figure 24-9 shows how the frequency response is calculated for these filters. (Frequency is normalized to the horizontal scan frequency.) The amplitude response has the desired interleaving of the combs such that the luminance filter passes signals around the harmonics of the horizontal scan frequency while the chrominance filter covers the in-between regions. The graphs cutoff at only three times the horizontal frequency, but the periodic response of the filter extends all the way through the 4.2 MHz wide video band.

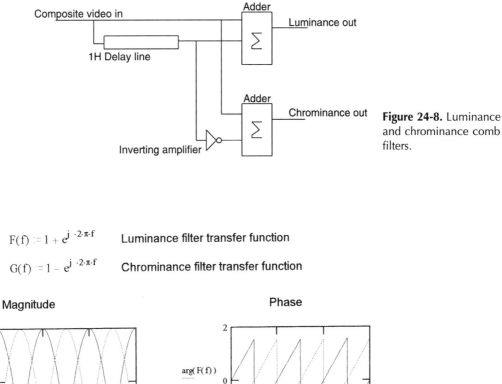

Figure 24-8. Luminance and chrominance comb filters.

$$F(f) := 1 + e^{j \cdot 2 \cdot \pi \cdot f} \qquad \text{Luminance filter transfer function}$$

$$G(f) := 1 - e^{j \cdot 2 \cdot \pi \cdot f} \qquad \text{Chrominance filter transfer function}$$

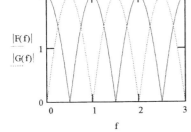

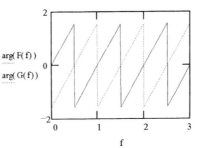

Figure 24-9. Comb filter amplitude and phase responses.

Note that the luminance filter is a true linear phase filter, so it will delay but not distort the luminance information. The chrominance filter would also have linear phase except for a constant 90° offset. But since the chrominance information is a relatively narrow-band signal (the color information is clustered around the color reference frequency), a simple all-pass equalizer can correct this offset. The operation of this comb filter is quite simple when viewed in the time domain; when two successive horizontal lines are superimposed, the chrominance signals have opposite polarities while the luminance signals have the same polarity. In as much as one line is not very different from the next, the sum of the two lines leaves only the luminance signal while the difference leaves only the chrominance signal.

TELEVISION TRANSMITTERS

Klystron amplifiers are used for UHF transmitters while VHF amplifiers usually use ordinary gridded tubes. Television transmitters usually use low-level signal generation followed by linear power amplifiers. Sometimes, though, a final amplifier tube is grid modulated (the high voltages needed for plate modulation are too difficult to use to generate video frequencies, that is, thousands of volts with several megahertz of bandwidth). The amplifier produces the most power at the synchronization tips where linearity is not too important. The need to produce the high-power synchronization pulses contributes to low efficiency. Special modulator circuits can be used to step up the supply voltage just for the synchronization pulses, to improve the efficiency. The separate sound (aural) transmitter usually has about a quarter of the carrier power of the visual transmitter.

TELEVISION RECEIVERS

Television receivers are single-conversion superheterodynes with an IF band that puts the video carrier at 45.75 MHz. The front-end amplifier will often use a gallium arsenide FET (field effect transistor) and have a noise figure lower than 3 dB. The local oscillator is usually a frequency synthesizer locked to a crystal reference; a fine tuning control is no longer needed. The shape of the IF pass band must be quite accurately set to equalize the vestigial sideband, to provide full video bandwidth, and to reject adjacent channels. This once required careful factory adjustment of many tuned circuits, but is now determined by the geometry in a single SAW filter. Present-day receivers use digital circuitry to provide stable phase-locked synchronization, and to perform various corrections. Broadcasters transmit secondary and calibration signals on the blanked lines that occur during the vertical retrace period. The calibration signals allow premium receivers to correct color balance, set up multipath cancelers, recover closed caption text, and so forth. Switching power supplies are integrated with high-efficiency horizontal deflection circuits (see Chapter 16).

COLOR TELEVISION RECEIVER

The block diagram of Figure 24-10 shows the overall organization of a typical color television receiver. The first block is simply an AM radio receiver for VHF, UHF, and cable frequencies (about 52–400 MHz). The carrier of the selected channel is translated to an IF frequency of 45.75 MHz, and the IF bandwidth is about 6 MHz. The IF signal contains the composite video (luminance and synchronization) plus the sound and color information. The sound and color signals, which are essentially narrow-band signals around 4.5 MHz and 3.57 MHz, respectively, ride on top of the luminance signal. A 4.5 MHz bandpass filter isolates the sound signal in the block labeled "4.5 MHz FM receiver." The output of this FM receiver either is fed directly to an audio power amplifier or, for stereo sound, is fed into the stereo sound block, where the left and right channels are separately demodulated.

STEREO SOUND The stereo system used for television sound is essentially the same as is used for FM broadcasting. At the transmitter the left (L) and right (R) audio signals are added, to form an L + R signal, and subtracted, to form an L − R signal. The L − R signal, which is the stereo difference signal, is multiplied by a subcarrier at 31.4686 kHz. We have seen that a multiplier is a balanced mixer, so the product is just the set of upper and lower modulation sidebands, centered on the subcarrier frequency. This double-sideband suppressed carrier AM signal plus the L + R sum signal and a low-level pilot signal at half the subcarrier frequency forms the modulating signal for the FM transmitter. At the receiver, the output of the FM demodulator contains the L + R audio already in place, which can be used directly for monaural sound, but the L − R signal, if used, must be shifted back down from the inaudible range around the subcarrier frequency. This down-conversion is just another multiplication. The local oscillator, which must have the right frequency and the right phase, is produced by a phase-locked loop that locks to a pilot signal at 15.75 kHz. A frequency doubler then produces the 31 kHz multiplicand. (Note that the frequency of the pilot is the same as the horizontal scan rate; the horizontal synchronization pulses could provide the reference.

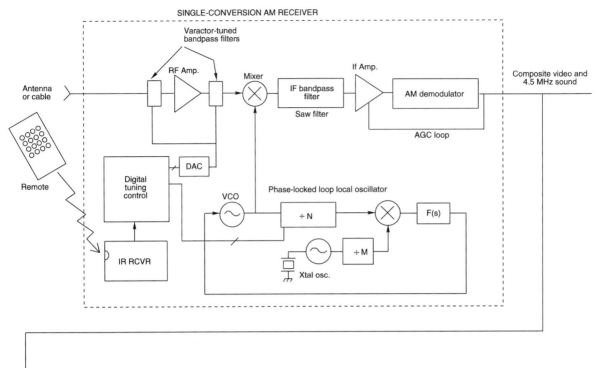

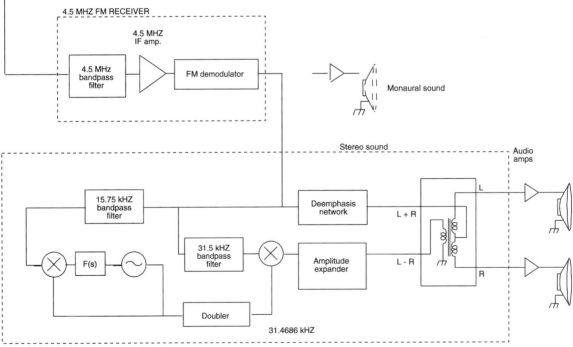

Figure 24-10. Television receiver block diagram.

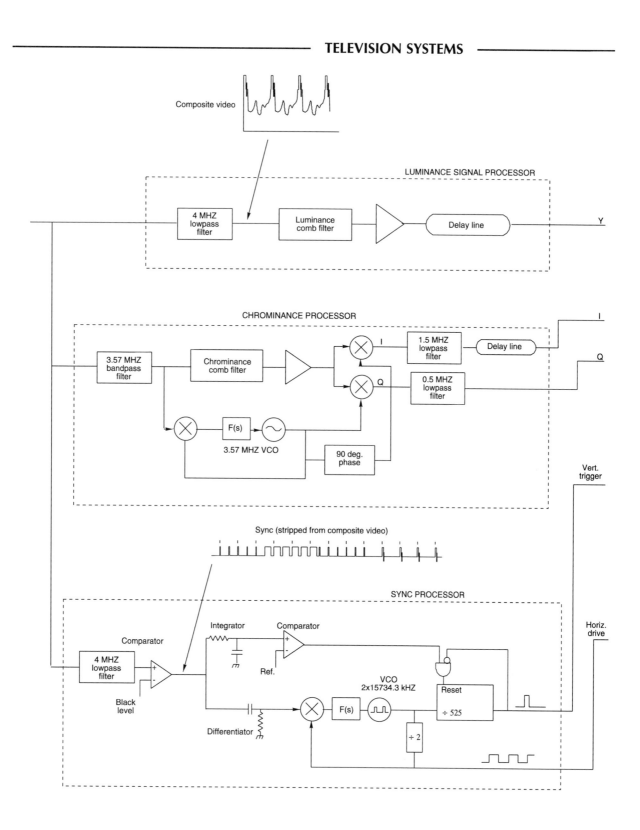

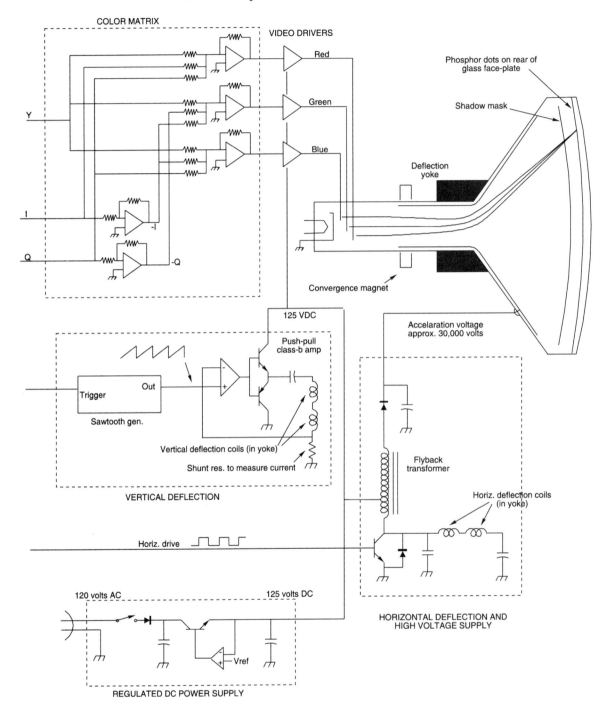

Figure 24-10. *Continued*

But including a pilot in the 4.5 MHz sound channel makes it simpler to demodulate stereo television sound in small audio-only receivers for AM, FM, and television sound.) A matrix then adds the L – R signals to form the left channel, and subtracts them to form the right channel. (Any network that forms linear combinations is known as a matrix.) In the block diagram this matrix operation is done with a transformer. More often it would be done with a resistor network or a network of resistors and operational amplifiers (op amps) (as shown in the color matrix block).

The amplitude expander is a nonlinear signal processor that inverts an amplitude compression done at the transmitter on the L – R signal. This is a Dolby-like system that improves the signal-to-noise ratio of the L – R signal. The deemphasis network, a linear "equalizer", is explained later in Chapter 28.

FM stereo broadcasting is the same except that the subcarrier is at 38 kHz, the pilot is at 19 kHz, and no compression–expansion is used for the L – R channel.

LUMINANCE PROCESSOR In this block a 4 MHz low-pass filter eliminates the sound signal and a comb filter (described above) eliminates the chrominance signal. The resulting video signal, the "Y" signal contains only the brightness information (and synchronization pulses) and would produce the correct picture if sent to a monochrome (black and white) picture tube. The delay line is needed because bandpass filters in the chrominance processor inevitably delay the I- and Q-signals. If a corresponding delay were not included in the luminance channel, the coloring would not be in exact registration with the black and white luminance picture but, arriving late, would be shifted to the right.

CHROMINANCE PROCESSOR This block is very much like the stereo sound processor. Here the double-sideband suppressed carrier (DSBSC) chrominance signal must be multiplied by a locally generated carrier at the subcarrier frequency. The pilot signal is the color burst on the back porch of the horizontal synchronization pulses. This burst provides the reference signal for a phase-locked loop, which then provides the local carrier. Not shown on the diagram is an electronic switch, the burst gate, which is controlled by the synchronization circuitry to apply the reference signal only during the burst period. This improves the signal-to-noise ratio of the loop. The local carrier is fed to a product detector (multiplier) to demodulate the I signal. A 90° phase shift network provides a second local carrier, shifted in phase for a second product detector to demodulate the Q signal. The Q-signal has less bandwidth than the I-signal, so the bandpass filter after the Q-demodulator is narrower than the bandpass filter after the *I*-demodulator. To equalize the delays, a delay line is needed in the *I*-channel.

SYNC PROCESSOR A comparator with its threshold set at the black level strips away the video signal, producing a clean train of synchronization pulses. As explained above, a simple *RC* differentiator then provides horizontal reference pulses and an *RC* integrator provides vertical reference pulses. A VCO, phase locked to the horizontal reference pulses, provides a flywheel-stabilized horizontal time base. The VCO actually operates at twice the horizontal frequency and is divided by two to provide 15,734 kHz for the phase-locked loop phase detector and to drive the horizontal deflection circuitry. The VCO is also divided by 525 to provide an equally stable vertical time base. This divider must operate with the right phase for the picture to have the correct vertical alignment, so the divide-by-525 counter has a reset input that will be triggered when the counter output has failed to coincide with several consecutive vertical synchronization pulses.

COLOR MATRIX AND PICTURE TUBE This block forms the appropriate linear combinations of *Y*, *I* and *Q* to produce the *R*, *G*, and *B* video signals. Some of the coefficients are negative, so inverters are included to provide -*I* and -*Q*. The *R*, *G*, and *B* signals are amplified to levels of many tens of volts for application to the control grids of the respective electron guns in the tricolor picture tube. Not shown is the blanking circuitry that turns off the electron beams during the horizontal and vertical retrace periods. In normal operation this locally generated blanking is redundant because the composite video signal provides blanking.

DEFLECTION CIRCUITRY The sawtooth current waveform for the vertical deflection coils is produced by a linear class-B push–pull amplifier. A sawtooth voltage generator, triggered from the synchronization circuitry, provides an accurate reference waveform. A low-resistance shunt provides a voltage proportional to the output current, and this voltage provides negative feedback to make the current follow the reference waveform. The horizontal deflection circuit was discussed earlier with switching power supplies.

POWER SUPPLY A simple half-wave capacitor-input circuit provides about 140 V dc. A series regulator lowers this to 125 V, which is appropriate for the transistors used in the high-power sections of the receiver (deflection, video drivers, and audio power amplification). The receiver circuits provide essentially a constant load, so the regulator needs only to correct power line voltage variations. A low-voltage supply, say 15 V, for the low-power signal-processing circuitry is often generated from a secondary winding on the flyback transformer. Another secondary winding provides several volts for the filaments in the CRT. This kind of power supply ties one side of the ac line to ground, usually a metal chassis. With this "hot chassis" arrangement it is important that the cabinet provide a barrier against outside contact with the chassis. A 60 Hz power transformer at the input leaves the chassis safe, but is heavy and expensive. A small transformer may be used to provide a cold subchassis for parts of the circuitry that must connect to the outside – audio and video jacks for VCRs, and so forth.

DIGITAL TELEVISION

At the time of writing, a successor to the more than fifty-year-old NTSC television standard appears close at hand. Variously known as high-definition television (HDTV), advanced television (ATV), or digital television, the emerging standard is based on digital processing and memory (frame storage). In the United States of America, a consortium known as the Grand Alliance (GA) has proposed a system in which processing and storage achieve a data compression of about sixty times. This standard includes a video compression technique (MPEG-2) based on the temporal and spatial redundancy of the video program material. The same redundancies are exploited to a much smaller degree in the NTSC system with its temporally interlaced fields and its frequency-interleaved luminance and chrominance signals. The proposed GA system would carry two standard definition programs or one high-definition program on a single 6 MHz television channel with a net data rate of about 18 Mbps. For terrestrial (radio) transmission, all the video and audio information is encoded in 3-bit words. An alternate transmission standard (for cable) uses 4-bit words. The 3-bit words are converted by an analog-to-digital converter into an 8-level analog signal which amplitude-modulates the carrier wave. The carrier is suppressed except for a low-amplitude pilot. At the receiver, the signal is digitized immediately after synchronous detection to recover the 3-bit words. As in the NTSC system, a bandpass filter in the transmitter eliminates most of the lower sideband to save spectrum – a vestigial sideband system. As in any other narrow-band RF system, the transmitted signal is essentially always a sine wave. Here the amplitude distribution is peaked around each of the eight modulation levels (an 8-ary digital system instead of a binary digital system). GA and NTSC signals require entirely different receivers, but the signals are at least compatible in that they can coexist on adjacent channels (unused now to avoid interference between adjacent NTSC broadcasts) without causing mutual interference. It is anticipated that broadcasters will simulcast with GA and NTSC channels for perhaps 15 years while phasing out the NTSC standard. A simplified description of the GA proposed system is given below.

VIDEO COMPRESSION

Temporal and spacial redundancies in the succession of images that make up the moving picture are exploited to reduce the data rate by a factor of fifty or higher. The picture is updated in blocks of pixels (8×8 pixels for the GA terrestrial format and 16×16 pixels for the GA cable format). From one frame to the next, the video encoder determines the change that has occurred in each block, and this difference information is used at the receiver to update the blocks, that is, modify the data in that block's area of the digital frame storage memory. Except for scene changes, the differences from one frame to the next are due to motion of elements within the picture (subject motion) or motions of the picture as a whole (camera motion). From frame to frame, a given block will therefore mostly just shift its position somewhat. In the GA system, for each block in a new frame, the encoder determines the displaced block in the previous frame that provides the best match (by minimizing the sum of the absolute values of the differences of the pixel brightness values). The position of the displaced block is specified as a *motion vector*, for example 1,1 would indicate that the best match block is shifted one pixel up and one pixel to the right. These motion vectors are part of the video update information and allow the receiver to construct the new frame mostly from the pixels of the last frame, which is stored in memory. The rest of the information blocks and the pixels in the new blocks. As a result of motion compensation, these differences will be small numbers; temporal correlation provides the first stage in data compression. The next step is to exploit the spatial coherence of the picture, that is the generally high correlation between adjacent pixels. (A screen of uncorrelated pixels would appear as fine-grained snow.) At the encoder, the 8×8 block of pixel differences is subjected to a two-dimensional Fourier Transform (in particular, a discrete cosine transform of DCT). Because of spatial correlation, most of the resulting 8×8 transform coefficients have very small amplitudes. For example, if a certain block of pixels represents an area with a linear increase of brightness in the horizontal direction, only two coefficients would have significant values. When most of the coefficients in a block are negligible, the resulting stream of digital numbers consists mostly of zeros, and run-length encoding (specification of the number of consecutive zeros) gives a high degree of compression. The number of consecutive zeros is increased by using a certain zig-zag read-out ordering of the transform differences. In addition, variable-length coding is used for the non-zero transform differences; the viewer can tolerate coarser quantization for the high spacial frequencies. Not all the frames can be predicted from previous frames; changes of scene require all-new data. Occasional *refresh frames* (tagged as such) are included in the stream of prediction frames. The proposed standard provides for two resolution standards, $1,280 \times 720$ pixels, with progressive scanning at frame rates of 60, 30, and 24 frames per second, and $1,920 \times 1,080$ pixels, with progressive scanning at 24 frames per second and progressive or interlaced scanning at 30 frames per second. (Both resolutions have an aspect ratio of 16:9.)

COLOR, SOUND, AND PACKETS

The GA system, like the NTSC system, transmits a luminance signal and two reduced-definition chrominance signals. (The chrominance information is smoothed and sampled at half the spatial rate of the luminance signals.) The sound (CD quality), like the video, uses digital compression. Video and sound data blocks are time multiplexed into separate packets. Other packets can carry data for auxiliary purposes. Buffer memory at the receiver provides the means to concatenate successive video or audio packets and present the decoded and processed signals at the proper rates. The packetization scheme, that is, the mix of packets, lengths (maybe variable), headers, and so forth, is referred to as the *Transport System*. For transmission, synchronization symbols are added, as well as redundant information for error detection and correction

BIBLIOGRAPHY
1. K. B. Benson (Ed.) (1986), *Television Engineering Handbook*. New York: McGraw-Hill. (updated version of *Television Engineering Handbook*. D. G. Fink (Ed.), New York: McGraw-Hill, 1957).
2. K. Challapali, X. Lebegue, J. S. Lim, W. H. Paik, R. S. Girons, E. Petajan, V. Sathe, P. A. Snopko and J. D. Zdepski (1995), "The Grand Alliance System for USA HDTV," *Proc. IEEE* 83: 139–50.
3. K. G. Jackson and G. B. Townsend (Eds) (1991), *TV & Video Engineer's Reference Book*. Oxford: Butterworth Heineman. (Modern handbook from the U.K.).
4. *Engineering Handbook*, 8th Edition (1993), National Association of Broadcasters, Washington, DC.
5. A. Netravali and A. Lippmann (1995), *Digital Television: A Perspective*, Proc. IEEE, 83: 834–42.
6. E. Petajan (1995), "The HDTV Grand Alliance System," *Proc. IEEE*, 83: 1094–105.

PROBLEMS

1. Suppose the signal from a television receiving antenna consists of the direct signal (via the line-of-sight path to the transmitting station) and a weak secondary signal (via reflection from a metal tower off the line-of-sight). If the path taken by the secondary signal is 1 km longer than the path of the direct signal, what will be the position of the "ghost" image on screen of the receiver?

2. Motion pictures from film shot at 24 frames/s are transmitted as 60 field/s television images using a technique called *3:2 pull down*. A film frame is held in place while two television fields are transmitted. The subsequent film frame is then pulled down and held in place while three television fields are transmitted. Show that this scheme results in an average film rate of 24 frames/s.

3. Draw a block diagram for an NTSC television receiver with a "picture in picture" feature, that is, a window that will display a second channel. Explain the operation of each block.

4. Consider a high-definition television monochrome signal with 1,920×1,080 pixels and progressive scan (no interlace) at 60 frames/s. If the brightness of each pixel is specified by an 8-bit number, and if no compression is used, what is the data rate? (*Answer:* 995 Mbps.) What compression ratio is needed if this signal is to be transmitted over a standard-width television channel at a rate of 18 Mbps?

5. Suppose an NTSC test pattern consists of five vertical bars of equal width but different colors. Let the bars all have the same luminance (brightness) and be only lightly colored (unsaturated). Sketch the waveform for one line of video.

6. Interlaced scanning provides more resolution for fixed scenes but can produce artifacts with moving objects. Think of a situation where interlaced scanning could make a moving object vanish.

RADAR PULSE MODULATORS 25

Any continuous wave (CW) transmitter can generally be used as a pulse transmitter if a *pulse modulator* is added to provide the rapid turn on and turn off. Tube-type amplifiers can be operated with much higher instantaneous powers when they are pulsed. Tubes are primarily limited by their maximum anode dissipation (heat removal); the dissipation can be the result of either modest CW operation or high-power pulse operation. A CW amplifier can be converted for pulse operation by changing the output-matching circuit in order to present a lower load resistance to the tube. Some tubes are available in special pulse-rated versions; they are fitted with high-emission cathodes. Gridded tubes (triodes, tetrodes, and pentodes) can be pulsed by switching the grid bias from negative, for pulse-off, to positive, for pulse-on. The negative bias keeps the tube completely turned off between pulses. Since the grid voltage and current are much smaller than the plate voltage and current, grid control requires only low-power circuitry compared to anode control. At microwave frequencies, magnetrons and klystrons replace gridded tubes. Magnetrons have no control element and therefore require high-power anode pulsers. Klystrons may or may not have a modulating anode ("mod anode") by which the beam current can be cut off. If not, they need high-power pulsers.[*]

Transistor amplifiers, unlike tube amplifiers, cannot make much of a trade-off between duty cycle and peak power. Transistors suffer one type of breakdown or another when operated much past their maximum continuous ratings. A high-power transistor amplifier for pulse service might differ from a CW amplifier only in that it will dissipate less heat (from the reduced duty cycle) and can therefore get by with a smaller heat sink.

No matter how an amplifier is pulsed, the power supply must furnish high-power pulses with minimum voltage droop. Duty cycles of pulsed transmitters are usually much less than unity so, in addition to at least one switching element, pulse modulators (pulsers) contain some form of energy storage element(s). The simple pulser circuit shown in Figure 25-1 stores energy in a capacitor. In this circuit the tube (magnetron, klystron, or whatever) is shown as requiring negative voltage. Microwave tubes

---

[*]An air traffic control radar might have a peak power output of 2 MW and an efficiency of 50%. A klystron tube in this service could require 50 kV pulses at 80 A.

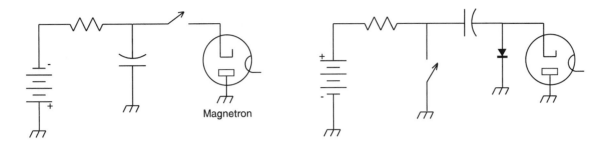

Figure 25-1. Capacitor discharge pulsers.

often use a negative supply voltage applied to their cathodes because it is convenient to ground the external heat-dissipating anode. The right-hand version of the circuit allows one side of the switch to be grounded, which is another convenience. The diode provides a charging path for the energy storage capacitor. The circuit of Figure 25-2 uses a thyratron (vacuum tube version of the silicon-controlled rectifier) as the switch.

The simple pulse modulators of Figures 25-1 and 25-2 have two main disadvantages:

1. The voltage droops during the pulse. The droop can be reduced by increasing the size (weight and cost) of the capacitor.
2. Not much of the stored energy is used. Even if a 10% voltage droop is permitted, only 20% of the stored energy is used for each pulse. This might be compared to a car, which would not run well if the fuel tank was less than 80% full.

Despite these drawbacks (they are not really limitations), capacitor banks are often used, as in the 430 MHz pulse transmitter used for ionospheric research at the Arecibo Observatory, because a more efficient circuit, the

Figure 25-2. Thyratron-switched pulser.

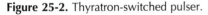

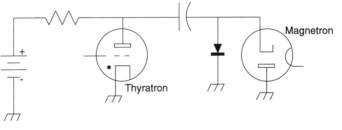

line modulator discussed below, does not easily provide the flexibility needed to change the pulse width. A capacitor bank cannot supply longer pulses without increased droop. (Normally inductors are not used as energy storage elements because, compared to capacitors, their maximum energy density is low.)

LINE MODULATORS

A length of transmission line (with the far end open) has capacitance and can therefore store electrostatic energy. When the line is discharged into a resistive load equal to its characteristic impedance, it will supply a perfect rectangular pulse rather than a drooping exponential pulse. The constant pulse amplitude during discharge is maintained by the distributed inductance of the line acting together with the distributed capacitance. In Figure 25-3 the line is a piece of coaxial cable, replacing the energy storage capacitor. As before, the tube is supplied with a negative pulse. A diode provides a path to recharge the line. Often the load has a higher impedance than the characteristic impedance of the line, and a pulse transformer is required.

The line supplies a pulse at half the charging voltage because, during the pulse, the charging voltage evenly divides between the load and the equivalent source resistance. The duration of the pulse is the time taken for the current to make a round trip through the line. At the end of the pulse the line is totally discharged; all the stored energy is delivered on every pulse. Waveforms of the line voltage and current are shown in Figure 25-4.

It is common to use an "artificial transmission line" or pulse-forming network (PFN), which is a ladder network of inductances and capacitances. A four-section network is shown in the modulator circuit of Figure 25-5. The network looks like a low-pass filter, and it is. Its cutoff frequency is given by $\omega^2 = 4/(LC)$. For frequencies well below cutoff, the network behaves like a transmission line with $Z_o = \sqrt{L/C}$. Here L and C are in

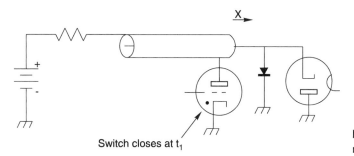

Switch closes at t_1

Figure 25-3. Line-type modulator.

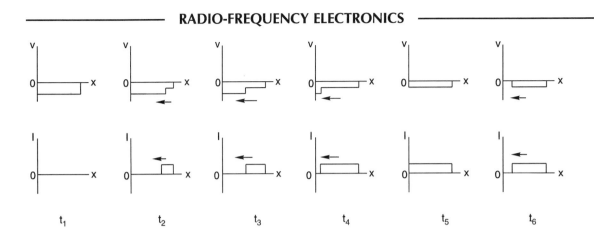

t_1 t_2 t_3 t_4 t_5 t_6

Figure 25-4. Line modulator waveforms.

henries and farads rather than henries per meter and farads per meter, as in the distributed element transmission line. The one-way time delay on this lumped line is $\sqrt{LC}$ seconds per section.

Let us consider a numerical example. A lumped line, as an approximation to a distributed line, does not require a great number of sections to produce a fairly rectangular pulse. Let us use four sections, as in Figure 25-5. Suppose we need a $1\mu s$ pulse at $10\,kV$ and $10\,A$. The voltage and current require that $Z_0 = \sqrt{L/C} = 1000$ ohms. The desired $1\mu s$ of delay in a round-trip through four sections requires that $8\sqrt{LC} = 10^{-6}$. These impedance and time delay equations are satisifed by $L = 125\mu H$ and $C = 125\,pF$. We can verify that the energy stored in the line is indeed equal to the energy delivered by the pulse. The latter is just $(IV)\tau = 10 \times 10,000 \times 10^{-6} = 0.1\,J$. The former, remembering that we must charge the line to $20,000\,V$, is $CV^2/2 = 4(125 \times 10^{-12}) \times 20,000^2/2$, which is also $0.1\,J$.

As often happens in filter design, these are not particularly practical values; real inductors of $125\,\mu H$ may well have distributed capacitances that are not negligible compared with $125\,pF$. We can build the line for a lower impedance and use a pulse transformer between the line and the magnetron. If we lower the line impedance to 100 ohms, the L- and C-

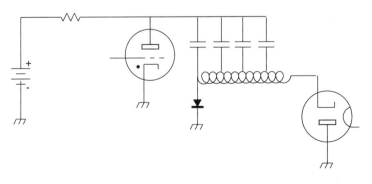

Figure 25-5. Pulser using an artificial transmission line (PFN).

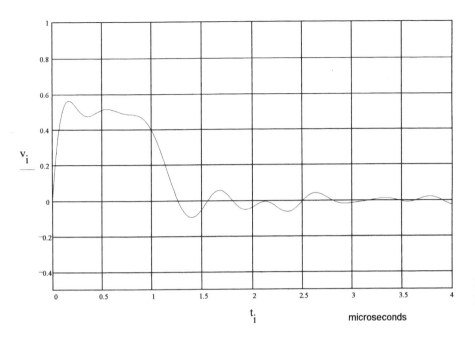

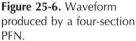

Figure 25-6. Waveform produced by a four-section PFN.

values become $12.5\,\mu H$ and $126\,pF$, respectively, values that are more practical. Using these values in a Spice simulation of the discharge produced the voltage waveform shown in Figure 25-6. The voltage scale is normalized, that is, the capacitors were charged to $1\,V$ so the nominal pulse voltage is $0.5\,V$. Lines with more sections provide better-shaped pulses.

The line modulator uses all the stored energy on each pulse but, precisely because of this virtue, deserves a more sophisticated charging circuit than the resistor shown in the circuits above. Remember that when a capacitor is charged through any resistive path from empty (no energy) to $CV^2/2$, the resistor will dissipate this same amount of energy, $CV^2/2$. Here the charging resistor, no matter what its value, would dissipate half the power consumed by the radar. The solution to this problem is to charge the line through an inductor instead of a resistor. Figure 25-7 shows the voltage waveform on a capacitor as it is *resonantly charged* through an inductor. The voltage is a sinusoid, building up to a maximum of twice the supply voltage. The modulator can be triggered just as the voltage reaches this maximum. The brief pulse discharges the line, and the charging curve begins anew. It would seem that the pulse repetition frequency is therefore determined rigidly by the charging time but, if a diode is put in a series with the inductor, the charging stops at the maximum voltage and the next pulse can occur anytime. The resonantly charged modulator, with the diode and a pulse transformer is shown in Figure 25-8. Note that the primary of the pulse transformer provides a charging path, eliminating the diode originally in parallel with the magnetron. Also

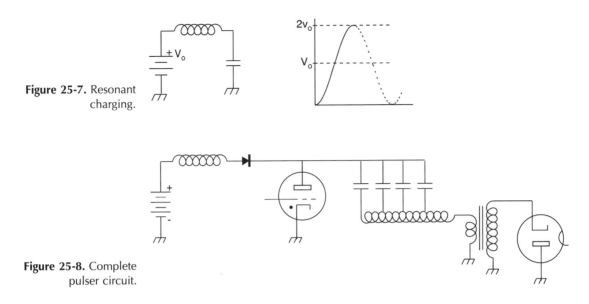

Figure 25-7. Resonant charging.

Figure 25-8. Complete pulser circuit.

remember that, because of the resonant charging, the supply voltage needs only to be half of the line charging voltage.

Line modulators present less risk to tubes than partial-discharge capacitor modulators because there is less stored energy available when an arc occurs in the tube.

BIBLIOGRAPHY

1. W. North (1994), *High-Power Microwave-Tube Transmitters*. Los Alamos National Laboratory, LA-12687. (Available from the US Dept. of Energy, Office of Scientific and Technical Information, PO Box 62, Oak Ridge, TN 37831, USA.)
2. M. I. Skolnik (1970), *Radar Handbook*. New York: McGraw-Hill.

PROBLEMS

1. (a) Show that when an uncharged capacitor is brought to potential V by connecting it through a resistor to a voltage source V, the energy supplied by the source is twice the energy deposited in the capacitor (CV^2 rather than $CV^2/2$).

(b) The charging efficiency in Problem 1(a) is only 50%. Find the efficiency when the capacitor initially has a partial charge, that is, when the capacitor is initially charged to a voltage αV, where $\alpha < 1$.

2. (a) Find the characteristic impedance of the artificial transmission line shown below. This impedance, Z_0 (which is complex), can be found by adding another LC section to the properly terminated line and noting that the new impedance must still be Z_0.

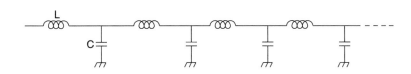

(b) Use the expression for Z_0 to show that the line has a cutoff frequency, $\omega_o=2/\sqrt{LC}$, above which signals are reflected rather than transmitted.

3. Show that when $\omega \ll \omega_c$, the propagation delay per section for the artificial transmission line of Problem 2 is given by $\tau=\sqrt{LC}$.

26 TR SWITCHING

Monostatic radars, which transmit and receive from the same location, can use either separate transmitting and receiving antennas[*] or a single antenna together with a TR (transmit–receive) switch. In most radar applications the desired echo arrives so soon after the pulse is transmitted that the TR switch (also known as a *Duplexer*) must be electronic rather than mechanical. Here we will first look at self-duplexing radar techniques based on the use of circular polarization and circulators, then at standard TR switch circuits, and finally at RF electronic switches in general.

SELF-DUPLEXING RADAR TECHNIQUES

If a radar transmits a circularly polarized signal, reflection by the target changes the sense of polarization from left hand to right hand or vice versa. Circular polarization can be produced by transmitting simultaneous crossed linear polarizations 90° out of phase. Figure 26-1 shows how a 90° hybrid not only produces circular polarization of one sense but also routes received circular polarization of the other sense (the return signal) into the receiver. Note that the x-dipole and the y-dipole together with the hybrid are really just equivalent to two separate antennas having opposite circular polarizations. In practice, the isolation between the transmitter and receiver in this scheme is usually no better than about 30 or 40 dB, so a limiter or SPST electronic switch (a *monoplexer*) is installed at the receiver input to protect it from burnout. A waveguide version of this circuit uses a *turnstile junction*, the microwave component shown in Figure 26-2. It is classified as a six-port junction because the round waveguide supports two independent modes: x and y or RCP and LCP or any other pair of orthogonal elliptical polarizations. When two opposite rectangular ports are fitted with shorts at $\lambda/8$ and $3\lambda/8$, the resulting four-port network is equivalent to the pair of dipoles and hybrid of Figure 26-1. The turnstile

[*]*Bistatic* radars (as opposed to monostatic radars) always use separate antennas because the transmitter and receiver are separated by a distance comparable to the distance from the target.

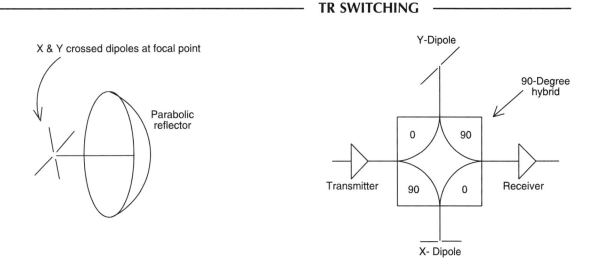

Figure 26-1. A self-duplexing radar using circular polarization.

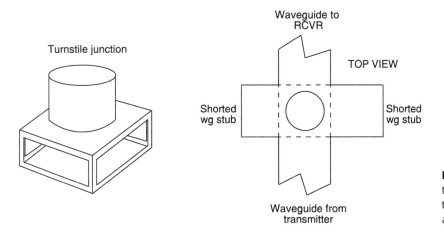

Figure 26-2. A waveguide turnstile junction combines the functions of the hybrid and crossed dipoles.

junction is described in Volumes 8 and 9 of the *Radiation Laboratory Series*. The transmitter and receiver are connected to the remaining rectangular ports (the pair without shorts) while the antenna, usually a feed horn, is connected to the round waveguide.

A true self-duplexing circuit, shown in Figure 26-3, uses a circulator. The circulator has the property that a signal injected at the transmitter port will exit via the antenna port while a signal injected at the antenna port will exit through the receiver port. (If a signal were injected at the receiver port, it would exit through the transmitter port.) This nonreciprocal action depends on transmission through a nonreciprocal medium which, for the circulator, is a ferrite material subjected to the field of a permanent magnet. This elegant TR system is limited by available circulators to powers of tens of kilowatts.

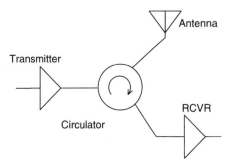

Figure 26-3. Circulator used as a TR switch.

TR SWITCHING DEVICES AND CIRCUITS

The classic TR circuits developed during World War II use gas discharge tubes or spark gaps and are self-activated by high-power radio frequency (RF) on the transmission line. Lower-power versions use p-i-n diode switches, turned on by an external bias circuit. (The radar has a timing generator providing pulses that (1) turn on the TR switch for transmitting, (2) pulse the transmitter, and (3) turn off the TR switch for receiving.) Gas discharge tube switches are usually built into a short piece of waveguide and come in two types: TR and ATR (anti-TR). The distinction is as follows: on transmit a TR device ionizes and presents a low-impedance shunt across the line; the ATR device also ionizes on transmit but it presents a low impedance in series with the line. (Volume 14 of the *Radiation Laboratory Series* [1] is devoted mostly to these tubes.) There are two general classes of TR circuits, branch line TR switches and balanced TR switches. The former uses segments of transmission line while the latter uses hybrids.

BRANCH LINE TR SWITCHES

Figure 26-4 shows some standard branch line TR switch circuits using TR, ATR, or both TR and ATR tubes (or p-i-n diode equivalents). In the circuit of Figure 26-4a, the switches are shown in the nonconducting (transmitter off) position. The open ATR switch is connected to the antenna by a half-wave line, so it presents the same open circuit to the antenna-to-receiver line. Likewise, the open TR switch does not disturb the antenna-to-receiver line. On transmit, the TR switch places a protective short circuit at the receiver input. This short circuit is transformed by the quarter-wave line into an open circuit, that does not affect the connection between the transmitter and the antenna. At high frequencies the switches contain some non-zero length that forms part of the half-wave or quarter-wave lines. For low-frequency designs the half-wave lines can

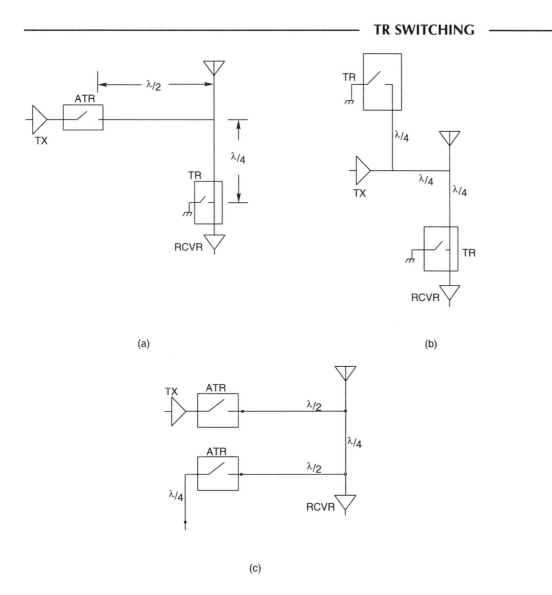

Figure 26-4. Branch line TR switches.

be reduced to zero length and the quarter-wave lines can be replaced by lumped-element impedance inverters. You can see from the ways that the TR and ATR elements are used that their names are somewhat arbitrary.

BALANCED DUPLEXERS

Balanced duplexers use hybrids and can have wider bandwidth than the branch line circuits (though more elaborate branch line circuits than those

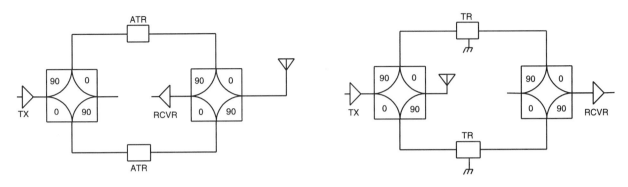

Figure 26-5. Balanced TR switches.

shown above can also have wider bandwidths). The balanced duplexer circuits shown in Figure 26-5 both use two 90° hybrids (see Problem 4).

DIODE SWITCHES

A single-diode shunt switch circuit is shown in Figure 26-6. Positive control voltage gives the diode a forward bias current and makes its dynamic resistance, dV/dI, low. Negative control voltage turns the diode off, making its dynamic resistance very high. Let these dynamic resistance values be denoted respectively by r and R. You can verify (see Problem 2) that the transmission values for the switch of Figure 26-6 are those given in Table 26-1, where Z_0 is the impedance of the load and of the generator and $r << Z_0 << R$. Note from these expressions that you could favor better isolation or lower insertion loss by transforming the line to have a larger or smaller Z_0 at the diode location. Better performance can be obtained with ladder networks analogous to multisection filters. Figure 26-7 shows how a shunt switch can be combined with a series switch to form a two-element

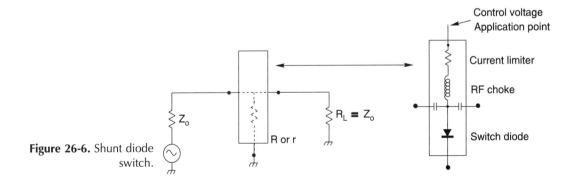

Figure 26-6. Shunt diode switch.

TABLE 26-1

SINGLE-DIODE SWITCH STATES

Diode state	Switch state	Power out/power available
Forward biased	Isolation	$4r^2/Z_0^2$
Reversed biased	Transmission	$1/(1 + Z_0/R)$

ladder network. Isolation is improved because any signal leakage across the open series switch is shorted to ground by the closed shunt switch.

It is often more convenient to use shunt diodes than series diodes, because shunt diodes are easier to bias and to heat sink. Impedance inverters can transform series elements into shunt elements, as we saw when designing coupled resonator filters. The switch circuit of Figure 26-8 uses two impedance inverters and three shunt switches.

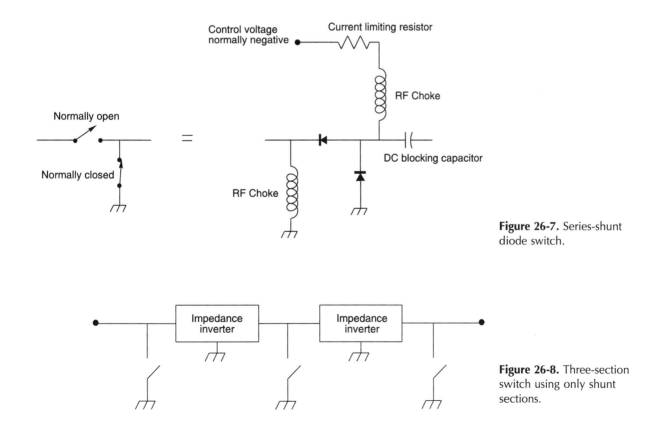

Figure 26-7. Series-shunt diode switch.

Figure 26-8. Three-section switch using only shunt sections.

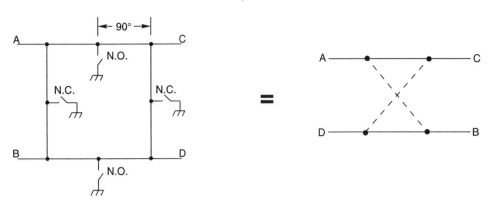

Figure 26-9. Transfer switching using quarter-wave transission lines.

The switches discussed above are all SPST switches. More complicated switches can be built up from the elementary SPST switch, but special designs can often be made such as the four-diode transfer switch shown in Figure 26-9. When the switches are in the indicated normal positions, the shorted quarter-wave line appear as infinite-impedance shunts and do not disturb the transmission paths from A to C and from B to D. When the switches are reversed, transmission is from A to B and from C to D.

DIODES FOR RF SWITCHING

When ordinary diodes are used in these switching applications, the biases must be enough to keep the diode in the desired state. In particular, when the diode is off, the reverse bias voltage must be greater than the peak RF voltage and, when the diode is on, the forward bias current must be greater than the peak RF current. However, the *p-i-n diode*, a sandwich of p-type, intrinsic, and n-type semiconductor material, has the remarkable property that, for RF switching, the bias current and bias voltage values can be less than the corresponding RF current and voltage by perhaps an order of magnitude. The operation of the p-i-n diode [2] depends on having a large (small) stored charge in the intrinsic region when the diode is on (off). At high frequencies, the time between electric field reversals is much less than the transit time through the intrinsic region, determined by diffusion and drift, so the diode remains in the commanded state. Finally, a word of caution: diodes, since they are nonlinear circuit elements, have the potential to distort a signal. In particular, they can create intermodulation products between the various signals in a complicated spectrum. For critical applications, such as receiver band changing, diode switches must be turned off and on hard enough to keep any generated intermodulation products at a negligible level.

BIBLIOGRAPHY

1. L. D. Smullin and C. G. Montgomery (1948), *Microwave Duplexers, Radio Laboratory Series*, Volume 14, New York: McGraw Hill.
2. J. F. White (1982), *Microwave Semiconductors Engineering*, New York: Van Nostrand Reinhold.

PROBLEMS

1. If we try to use the hybrid of the circular polarization duplexer as a circulator, we might consider the TR circuit shown below. Assume the antenna, transmitter, receiver, load, and hybrid all have the same characteristic impedance. This circuit at least protects the receiver from the transmitter. What are its disadvantages (a) when transmitting and (b) when receiving?

2. Derive the expressions for power out/power available given in Table 26.1.

3. Apply your circuit analysis program (see Chapter 1 Problem 3) to the transfer switch circuit of Figure 26-9. Assume a 50 ohm load is connected to port C, a 50 ohm generator is connected to port A, and the transmission line sections have a 50 ohm characteristic impedance. Assume also that the internal switches are ideal. Find the transmission coefficient (in decibels) over the frequency range from half the design frequency to twice the

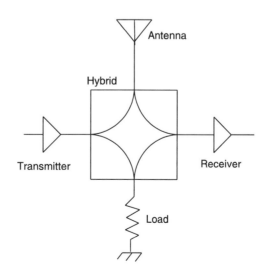

design frequency. *Hint*: The closed switches divide the circuit in two so you can ignore the bottom half.

4. Explain the operation of the balanced duplexers shown in Figure 26-5. What restrictions, if any, are there on the lengths of the interconnecting transmission lines?

DEMODULATORS AND DETECTORS

The terms AM detection and FM detection mean demodulation (recovering information from a carrier). The term *detector* is also used for circuits designed to measure power (e.g. square law detectors). Here we will discuss various AM detector, FM detector, and power detector circuits.

DIODE DETECTOR

The classic diode envelope detector circuit for AM is shown in Figure 27-1. The input signal voltage, V, is usually provided by a tuned transformer at the output of the IF amplifier. This tuned circuit forms part of the IF bandpass filter. Basically, the diode and the parallel RC form a fading memory peak detector. Except for the resistor, R, the output capacitor would remain charged to the maximum peak voltage of the input sine wave. In order to follow a changing peak voltage (AM), the resistor provides a discharge path. Since the input sine wave, a radio-frequency (RF) signal, has a much higher frequency than the amplitude modulation frequencies, the RC time constant can be made large enough so that the droop between charge pulses is much less than indicated in Figure 27-1. (Of course the time constant must also be small enough that output voltage can accurately follow a rapidly changing modulation envelope.) We will therefore denote the detector output at V_{dc}. Keep in mind, however, that this "dc" voltage follows the modulation envelope. This detector, or any other envelope detector, is known as a *linear detector* because its output voltage is linearly proportional to the amplitude of the input sine wave.

Figure 27-1. Diode envelope detector

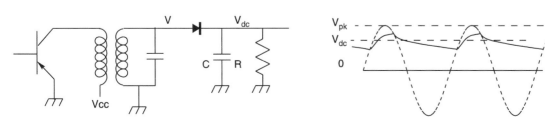

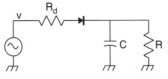

Figure 27-2. Diode envelope detector equivalent circuit.

ANALYSIS ASSUMING AN IDEAL RECTIFIER

Note that the detector of Figure 27-1 is identical to the simple half-wave rectifier capacitor-input power supply. As with the power supply, this circuit has poor regulation with respect to a changing load. But here we have a constant load resistance R (which we will asume includes the parallel input resistance of the subsequent audio amplifier). The equivalent circuit is shown in Figure 27-2. If the diode is modeled as a perfect rectifier (zero constant forward resistance and infinite reverse resistance) the analysis of this circuit is straightforward.

The value of the forward resistance of the diode, R_d can be increased to account for source resistance. But here the high-Q tuned circuit at the detector input acts as a zero-impedance voltage source (since the waveform is forced to stay sinusoidal). If we assume that C is large enough to make the output ripple negligible compared to the output voltage, the output voltage, V_{dc}, can be calculated by noting that V_{dc}/R must equal the average current through the diode, which is the average of $(V-V_{dc})/R_d$ during the part of the input cycle when this expression is positive. The result is that V_{dc} is proportional to the source voltage. (Curves showing output voltage versus ωCR for various values of R_d/R are found in the power supply chapters of many handbooks.) The ratio of V_{dc} to the peak source voltage is known as the detector efficiency. For a typical AM detector, $R \geqslant 10 R_d$ and $\omega CR \geqslant 100$. This gives a detector efficiency greater than 65% and an r.m.s. ripple less than about 1% of the dc output. With the assumed ideal rectifier, however, R could be any value. The analysis in the following section shows the limitations imposed on R by a real diode.

ANALYSIS WITH A REAL DIODE

Here we will discard the perfect rectifier in favor of the standard diode for which $I_{diode} = I_s \exp(V_{diode}/V_T - 1)$, where I_s is the reverse saturation current and V_T is the so-called thermal voltage, 0.026 V. In the equivalent circuit of Figure 27-2, we will now let R_d be zero, that is, we will assume that any voltage drop across the bulk resistance of the diode is negligible compared to the drop across the junction. As before, the analysis to find V_{dc} consists in equating V_{dc}/R to the average current through the diode:

$$\frac{V_{dc}}{R} = <I_{diode}> = \frac{1}{2\pi}\int_0^{2\pi} I(\theta)d\theta \tag{27-1}$$

where

$$I(\theta) = I_s\{exp[(V\cos\theta - V_{dc})/V_T] - 1\}. \tag{27-2}$$

This pair of equations is equivalent to

$$\frac{V_{dc}}{0.026} = \ln\left(\frac{V_s I(V)}{2\pi(V_{dc} + V_s)}\right) \tag{27-3}$$

where V_s, a "saturation voltage", is defined by $V_s = I_s R$, and

$$I(V) - \int_0^{2\pi} exp(V\cos\theta/0.26)d\theta. \tag{27-4}$$

Using a desktop computer math utility to solve Equation (27-3) results in a set of curves (Figure 27-3) showing V_{dc} versus V for various values of V_s.

Note that the detector output is very nonlinear for low-amplitude input signals when V_s is less than about 0.01 V. Suppose, then, that we pick $V_s = 0.01$ V. A germanium diode or zero-bias Schottky diode might have a saturation current of 10^{-6} A, which would then require that $R = 0.01/10^{-6} = 10k$ ohms, a convenient value. The low I_s of an ordinary silicon diode might require that R be more than 10^7 ohms, which would require the audio amplifier to have an inconveniently high input impedance. The power dissipated in the detector is V_{dc}^2/R. This detector would typically produce, say, 2 V (to operate up in the linear range), which corresponds to

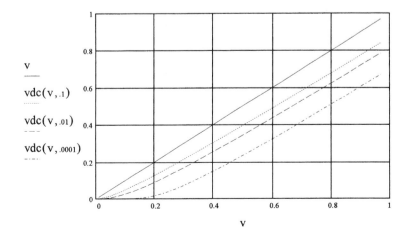

Figure 27-3. Detector output versus input voltage for several values of $V_s = I_s R$.

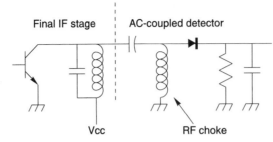

Figure 27-4. AC-coupled diode detector.

a power dissipation of $2^2/10^4 = 0.4\,\text{mW}$. For a given signal strength at the receiver input, the RF stages must have enough gain to produce $0.4\,\text{mW}$.

AC-COUPLED DIODE DETECTOR

Sometimes circuit considerations require that the detector input be ac coupled. In this case a dc return must be furnished for the detector diode. Such a circuit is shown in Figure 27-4 where an RF choke (large-value inductor) provides this dc return path, forcing the average voltage at the left side of the diode to be zero, as in the circuits of Figures 27-1 and 27-2.

SINGLE-SIDEBAND (SSB) AND MORSE CODE DETECTION

Demodulation of SSB and continuous wave (CW) signals (Morse code) is done by a *product detector*, which multiplies the IF signal by a locally generated carrier. The resulting difference frequency components become the demodulated signal while the sum frequency components are discarded. As shown in Figure 27-5, the local carrier is provided by a free-running oscillator known as a *beat frequency oscillator* (BFO).

For SSB reception of voice signals, the frequency of the BFO is manually adjusted until the audio sounds approximately natural. For Morse code reception, the BFO is deliberately offset to produce an audible tone or "beat note." An interesting early circuit to demodulate CW was the heterodyne detector of Fessenden. This predecessor to the superheterodyne used a special dynamic headphone with two electromagnets rather

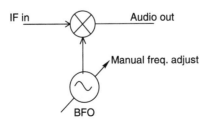

Figure 27-5. Product detector for SSB and CW.

than an electromagnet and a permanent magnet. The force on the diaphragm was then proportional to the product of the currents. A BFO signal from a small arc source (an early negative-resistance oscillator) was connected to one electromagnet while the signal fed the other, so, in operation, there was no audio output from this product detector headphone until the frequency difference between the incoming signal and the local oscillator was in the audio range. A beat note can also be produced by simply adding a strong BFO signal to the IF signal and then envelope detecting the sum (see Problem 2).

PRODUCT DETECTOR FOR AM

A product detector can also be used for AM demodulation. Again the signal is *multiplied* by a locally generated carrier. Here the local carrier must have the frequency and phase of the received carrier. An error in frequency creates a strong audio beat note with the carrier of the received signal and an error in phase reduces the amplitude. Nevertheless, the product detector overcomes the limitations of diodes; input signal levels do not have to be as high and there is no low-signal threshold below which the detector is useless. When the AM signal is consistently strong (usually the case for most listeners) the local carrier can be a hard-limited version of the input signal. This works because, in double sideband AM, the modulated signal has the same zero crossings as the unmodulated carrier. The product detector shown in Figure 27-6 uses this method. This detector is common in "radio on a chip" integrated circuits and as the video detector in television sets.

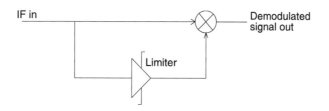

Figure 27-6. Product AM detector.

SYNCHRONOUS AM DETECTOR

The *synchronous detector*, shown in Figure 27-7, is an improved product detector circuit in which a phase-locked loop (PLL) is used to generate the

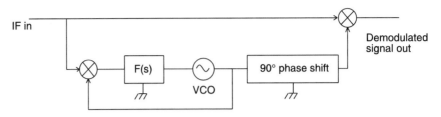

Figure 27-7. Synchronous
AM detector.

local carrier. The carrier of the input signal provides the reference signal for the loop. In a practical circuit, a limiting amplifier could be used at the reference input to make the loop dynamics independent of the signal level.

The PLL gives the synchronous detector a flywheel effect: the narrow-band loop maintains the regenerated carrier during abrupt selective fades common in short-wave listening. This prevents distortion, common in short-wave receivers, caused by momentary dropouts of the carrier. The PLL provides, in effect, a bandpass filter so narrow that its output cannot change quickly. Note the 90° phase shift network; if the phase detector is the standard multiplier (mixer), the voltage-controlled oscillator (VCO) output phase differs from the reference phase by 90°. Without the network to bring the phase back to 0°, the output of the detector would be zero.

FM DEMODULATORS

A variety of circuits have been used to demodulate FM. Most of these circuits are sensitive to amplitude variations as well as frequency variations so the signal is usually amplitude limited before it arrives at the FM detector. This reduces the noise output and also removes distortion that would result if the IF amplifier does not have a flat passband.

PLL FM DEMODULATOR

We have already pointed out that a PLL may be used as an FM discriminator (Figure 27-8). As the loop operates, the instantaneous voltage it applies to the VCO is determined by the reference frequency, which here is the signal frequency. The linearity of the VCO determines the linearity of this detector.

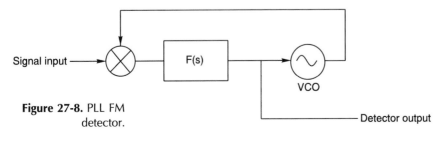

Figure 27-8. PLL FM
detector.

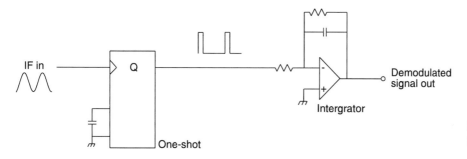

Figure 27-9. Tachometer FM detector.

TACHOMETER FM DETECTOR

A tachometer FM detector or "pulse-counting detector" is just a one-shot multivibrator that fires on the zero crossings of the signal (Figure 27-9). Each positive zero crossing produces a constant-width output pulse. The duty cycle of the one-shot output therefore varies linearly with input frequency, so, by integrating the output of the one-shot, we get an output voltage that varies linearly with frequency.

DELAY LINE FM DETECTOR

The *delay line discriminator* is often used in satellite television receivers to demodulate the FM-modulated video and sound. The IF frequency in these receivers is typically 70 MHz. Figure 27-10 shows a quarter-wave delay line (which could be a piece of ordinary transmission line) which delays the signal at one input of the multiplier.

If the input signal is $\cos[(\omega_o + \delta\omega)t]$, then the signal at the output of the delay line is $\cos[(\omega_o + \delta\omega)(t - \tau)]$ and the baseband component at the output of the multiplier is $-\sin(\tau\delta\omega)/2$. For small $\tau\delta\omega$ this is just $-\tau\delta\omega/2$. The output voltage is thus proportional to the frequency offset, $\delta\omega$. If the delay line is lengthened by an integral number of half-wave lengths, the sensitivity of the detector is increased, that is, a given shift from center frequency produces a greater output voltage.

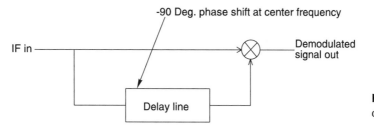

Figure 27-10. Delay line FM detector.

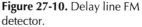

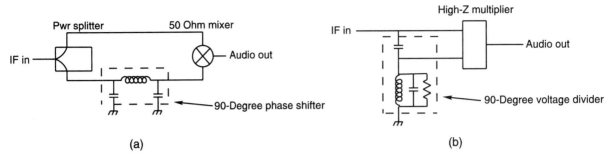

(a) (b)

Figure 27-11. Quadrature FM detectors.

QUADRATURE FM DEMODULATOR

This circuit (Figure 27-11) is the same as the delay line discriminator except that an *LC* network is used to provide the delay, that is, a phase shift that varies linearly with frequency. These circuits are commonly used in integrated circuits for FM radios and television sound; the *LC* network or an equivalent resonator is normally an off-chip component. The circuit of Figure 27-11a uses an *LC* inverter as a delay element. The multiplier provides the necessary resistive termination. The circuit of Figure 27-11b uses a multiplier with high input impedances such as a Gilbert cell multiplier. A voltage divider provides the necessary phase shift (see Problem 3).

SLOPE DETECTOR

Slope detection, in which FM modulation is converted to AM modulation, is the simplest method to demodulate FM. The amplitude response of the IF bandpass filter is made to have a constant slope at the nominal signal frequency. An input signal of constant amplitude will then produce an output signal whose amplitude depends linearly on frequency. A simple envelope detector can detect this amplitude variation. A receiver with an AM envelope detector can slope detect an FM signal if detuned slightly to put the FM signal on the upper or lower sloping skirt of the IF pass band filter. A refined slope detector (Figure 27-12) uses two filters with equal but opposite slopes. The filter outputs are individually envelope detected, and the detector voltages are subtracted. This makes the output voltage zero when the input signal is on center frequency, f_0, and also linearizes the detector by cancelling any curvature, that is, the $(f-f_0)^2$ term, in the filter shape.

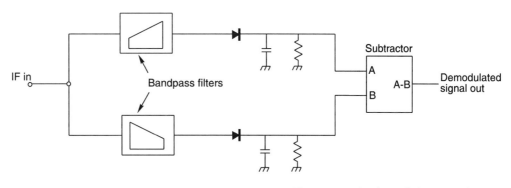

Figure 27-12. Balanced slope FM detector.

THE FOSTER–SEELEY DISCRIMINATOR

The Foster–Seeley discriminator circuit is an example of the balanced slope detector shown above. The sloping filters are made of a phase shift network and an adder. The network produces a $-90°$ phase shift at the center frequency, less than $-90°$ below the center frequency, and more than $-90°$ above the center frequency. Figure 27-13 shows how this combination produces a sloping filter; when the frequency is higher or lower than the center frequency, the magnitude of the vector sum changes.

Two of these sloping filters are used in the balanced circuit shown in Figure 27-14, a prototype for the standard Foster–Seeley circuit.

In the most common implementation of the Foster–Seeley circuit the phase shifts are produced by a loosely coupled transformer, as shown in Figure 27-15. The magnetizing inductance, L_2, and leakage inductance, L_1, of this transformer are shown explicitly. The voltage on the primary of the ideal transformer is derived from the voltage divider whose top branch is

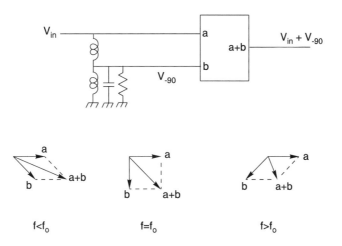

Figure 27-13. Sloping filter for the Foster–Seeley discriminator.

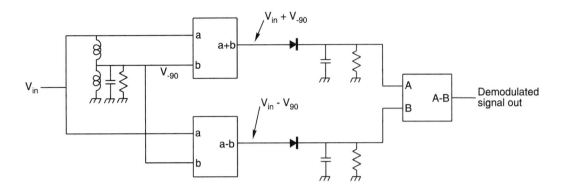

Figure 27-14. Foster–Seeley discriminator equivalent circuit I.

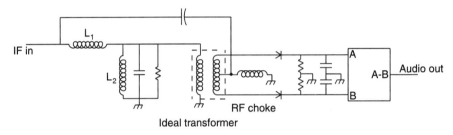

Figure 27-15. Foster–Seeley discriminator equivalent circuit II.

L_1 and whose bottom is L_2 in parallel with a resistor and a capacitor. The output of this divider is 90° out of phase with respect to the input. (The same divider network was used in the quadrature FM detector of Figure 27-11b. Here the resistor is provided by the loading of the diode detectors.) The IF input signal, V_{in}, is connected to the center tap of the transformer secondary so the voltage at the top of the secondary is $V_{in} + V_{90}$ while the voltage at the bottom of the secondary is $V_{in} - V_{90}$. Although the IF signal could be connected directly to the center tap, the next step will require dc isolation between the primary and secondary circuits, so at this point we have already connected it through a blocking capacitor. An RF choke now provides the necessary dc return for the diodes as in the circuit of Figure 27-4.

The Foster–Seeley circuit, as usually drawn, is shown in Figure 27-16. The magnetizing and leakage inductance are not shown explicitly. A capacitor is added across the primary so that the final IF stage sees a resistive load. If the magnetizing inductance of the transformer is low, this forms a high-Q resonant circuit that becomes a significant part of the bandpass filter. The reason for ac coupling of the primary signal to the center tap of the secondary is that we can now move the ground point in the secondary

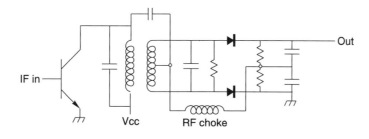

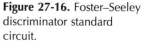

Figure 27-16. Foster–Seeley discriminator standard circuit.

circuit and eliminate the subtractor block; the desired $A - B$ signal then appears at the top of the output circuit.

POWER DETECTORS

Square law detectors are not used as demodulators but are used in laboratory instruments to measure power (in wattmeters or r.m.s. voltmeters). If we are measuring a sine wave we know that the r.m.s. voltage is equal to the peak voltage divided by $\sqrt{2}$. When we know the shape of the waveform, a true square law meter is not necessary. Even noise power can sometimes be estimated with other than a square law device. The power of a Gaussian random noise source can be calculated after averaging the output of a V^{2n} law device or a $|V|^n$ law device where $n = 1,2,\ldots$ The square law device, however, is always the optimum detector in that it provides the most accurate power estimate for a fixed averaging time. When we need the optimum power measuring strategy or if we need to measure the rms voltage of an unknown waveform we must average the output from a true square law device. "True r.m.s. voltmeters" built for this purpose use a variety of techniques to form the square of the input voltage. Some instruments use a network of diodes and resistors to form a piece-wise approximation of a square law transfer function. Other instruments use a thermal method where the unknown voltage heats a resistor. The temperature of the resistor is monitored by a thermistor while a servo circuit removes or adds dc (or sine wave ac) current to the resistor to keep it at a constant temperature. The diode network requires large signals, and the thermal method has a very slow response. Gilbert cell analog multipliers can be used to square the input voltage. They are limited to low-frequency signals. Generally, when the power is very low and/or the frequency is very high, and/or a very wide dynamic range is needed, the square law detector uses a semiconductor diode. A simple (but not particularly recommended) circuit is shown in Figure 27-17. In this circuit the average current through the diode is converted to a voltage by the operational amplifier (op amp). The virtual ground at the op amp input insures that the V_{in} is applied in full to the diode.

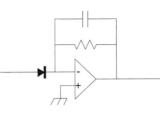

Figure 27-17. Simple diode power detector.

It is important to remember that a diode does not have an I versus V curve that follows a square law, even in a limited region. The diode junction law is the exponential $I = I_s[\exp(V/V_T)-1]$ where V_T, the thermal voltage, is 26 mV, which is KT_0/e. The diode is used at voltages much smaller than V_T, so it is permissible to write $I = I_s[V/V_T + 1/2(V/V_T)^2 + 1/6(V/V_T)^3 + 1/24(V/V_T)^4]$. Obviously we can restrict the input voltage enough to neglect the last term. But what about the first term, the linear term? This dominant term will provide a current component that has the same frequency spectrum as the input signal; if that spectrum extends down to zero, the output of the detector will be corrupted with extra noise. The third-order term will also have a baseband component. But if the input signal is a bandpass signal, the first-order and third-order terms are high-frequency signals that will be eliminated by whatever low-pass filter is applied to do the averaging (the capacitor in the above circuit). The simple square law diode detector, then, is appropriate for measuring signals whose frequency components do not extend down into the baseband output spectrum of the detector circuit. (You can think of various two-diode balanced circuits to cancel the linear component but remember that the two diodes must be matched very closely to start with and then maintained at the same temperature.) The diode and op-amp circuit shown in Figure 27-17 serves to explain why diode detectors are used for bandpass signals but not for baseband signals. But a better circuit – the preferred circuit – is shown in Figure 27-18. Here the sensitivity is not dependent on I_s, and the circuit has a much better temperature coefficient. In this circuit no dc current can flow in the diode. The capacitor, however, insures that the full ac signal voltage is applied to the diode. Expanding Equation (27-1) and taking only the dc components, we have

$$0 = -V_{dc}/V_T + 1/2 < V_{in}^2 > /V_T^2 \qquad (27\text{-}5)$$

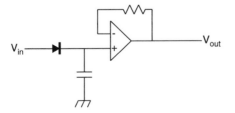

Figure 27-18. Preferred diode power detector.

$$V_{dc} = 1/2 < V_{in}^2 > / V_T \qquad\qquad (27\text{-}6)$$

which shows that the dc output voltage is indeed proportional to the average square of the input voltage.

BIBLIOGRAPHY
1. W. Gosling (1986), *Radio Receivers*, London: Peter Peregrinus.
2. R. W. Landee, D. C. Davis and A. P. Albrecht (1957), *Electronics Designers' Handbook*. New York: McGraw-Hill.

PROBLEMS

1. Assume that in the envelope detector circuit shown below, the diode is a perfect rectifier (zero forward resistance and infinite reverse resistance) and that the op amps are ideal

(a) Calculate the efficiency, V_{dc}/V.

(b) Calculate the effective load presented to the generator (the value of a resistor that would draw the same average power from the source).

2. Draw a vector diagram to show why an envelope detector will produce an audio tone when it is fed with the sum of an IF signal and a (much stronger) BFO signal.

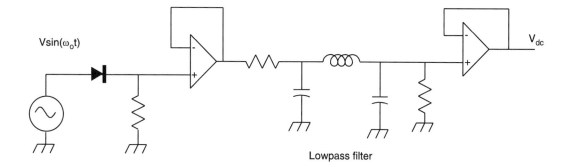

Lowpass filter

3. Show that the voltage divider network in Figure 27-11b can produce an output voltage shifted 90° from the input voltage when the input voltage is at a specified center frequency, ω_0. Determine the position of the output phase when the input signal is slightly higher or slightly lower than ω_0.

FREQUENCY AND PHASE MODULATION

28

Remembering again that radio signals are essentially narrow band and therefore sinusoidal, amplitude and phase angle are the only parameters available for modulation. We have already discussed conventional AM, where the amplitude of the audio signal controls the amplitude of the carrier wave and hence the power. In FM and phase modulation (PM) the audio voltage controls the phase angle of the sinusoid; the amplitude remains constant. Since frequency is the derivative of phase with respect to time, the two quantities cannot be varied independently; FM and PM are only slightly different methods of *angle modulation*. The advantage of FM (or PM) broadcasting over AM broadcasting is that under strong signal conditions, the audio signal-to-noise ratio (SNR) at the output of the receiver can be much higher for FM than for AM. The only price paid for this improvement is increased bandwidth.

BASICS OF ANGLE MODULATION

A simple unmodulated carrier wave is given by $V(t) = \cos(\omega_0 t + \phi_0)$ where ω_0 and ϕ_0 are constants. The total phase angle, $\phi(t) = (\omega_0 t + \phi_0)$, of this continuous wave (CW) signal increases linearly with time at a rate ω_0 radians per second. In FM the instantaneous frequency is made to shift away from ω_0 by an amount proportional to the modulating voltage, that is, $\omega(t) = \omega_0 + k_0 V_a(t)$. A linear voltage-controlled oscillator (VCO) driven by the audio signal, as shown in Figure 28-1, makes an FM transmitter. The excursion from the center frequency, $k_0 V_a$, is known as the (radian) *deviation*. In FM broadcasting the maximum excursion, defined as 100% modulation, is 75 kHz, that is, $k_0 V_{a\,max}/(2\pi) = 75 \times 10^3$. Consider the case of modulation by a single audio tone, $V_a(t) = A \cos(\omega_a t)$. The instantaneous frequency of the VCO is then $\omega(t) = \omega_0 + k_0 A \cos(\omega_a t)$. And, since the instantaneous frequency is the time derivative of phase, the phase is given by the integral

$$\phi(t) = \int \omega(t)\ dt = \int [\omega_0 + k_0 A \cos(\omega_a t]\ dt = \omega_0 t + \frac{k_0 A}{\omega_a}\ \sin(\omega_a t) + \phi_0.$$

$$(28\text{-}1)$$

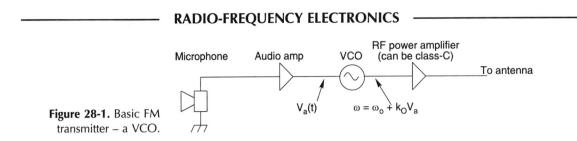

Figure 28-1. Basic FM transmitter – a VCO.

From Equation (28-1) we see that, in addition to the linearly increasing phase of the CW signal, there is a phase term that varies sinusoidally. The amplitude of this sinusoidal term is $k_0 A/\omega_a$. The inverse dependence on the modulation frequency, ω_a, results from phase being the integral of frequency. PM differs from FM only in that it does not have this ω_a denominator; it could be produced by a fixed-frequency oscillator followed by a voltage-controlled phase shifter. An indirect way to produce a PM signal is to use a standard FM modulator (i.e. a VCO), after first passing the audio signal through a differentiator. The differentiator produces a factor ω_a that cancels the ω_a denominator produced by the VCO. This scheme is shown in Figure 28-2. A PM transmitter can be similarly adapted to produce FM.

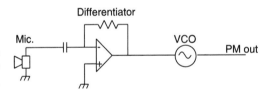

Figure 28-2. Phase modulation via frequency modulation.

FREQUENCY SPECTRUM OF FM

Using the expression for phase from Equation (28-1) (neglecting ϕ_0) and using the usual expansion for cos $(a + b)$, the complete FM signal (VCO output) becomes

$$V(t) = \cos[\phi(t)] = \cos(\omega_0 t)\cos(\phi_m \sin(\omega_a t)) - \sin(\omega_0 t)\sin[\phi_m \sin(\omega_a t)]$$
(28-2)

where $\phi_m = k_0 A/\omega_a$. This peak *phase* excursion is called the *modulation index*.

VERY NARROW-BAND FM OR PM

At very low modulation levels, both FM and PM produce a power spectrum similar to AM, that is, a constant carrier with an upper sideband ω_a above the carrier and a lower sideband ω_a below the carrier. Expanding Equation (28-2) for small ϕ_m, we have

$$V(t) \rightarrow \cos(\omega_0 t) - \phi_m \sin(\omega_a t)\sin(\omega_0 t). \tag{28-3}$$

Expanding the product of the sines gives the two sidebands. This differs from an AM signal in that the sidebands are equivalent to multiplication of the audio not by $\cos(\omega_0 t)$, the carrier, but by $\sin(\omega_0 t)$. Note that this narrow-band signal could be demodulated directly with a product detector that multiples the signal by a beat frequency oscillator (BFO) signal that is replica of the carrier but shifted $90°$ in phase. (Such a multiplicand could be generated by a narrow-bandwidth phase-locked loop referenced to the incoming signal.) Note also that the spacing between the two sidebands is $2\omega_a$, no matter how small the modulation index becomes.

WIDE-BAND FM SPECTRAL WIDTH

Looking at Equation (28-2) we see that the signal consists of quadrature components corresponding to $\cos(\omega_0 t)$ and $\sin(\omega_0 t)$ multiplied respectively by baseband signals, $\cos[\phi_m \sin(\omega_a t)]$ and $\sin[\phi_m \sin(\omega_a t)]$. When ϕ_m is not small, these product are similar, and each produces upper and lower sidebands around ω_0. The $\cos(\omega_0 t)$ and $\sin(\omega_0 t)$ carriers are suppressed, and the spectrum has no distinct central carrier spike. We can estimate the width of the spectrum from the width of these baseband modulating signals because they produce ordinary sidebands around their respective suppressed carriers. A rigorous analysis expands the last two expressions into a sum of Bessel functions. A simpler argument gives a good estimate of the spectral width of the FM signal. In a time equal to one-quarter of an audio cycle the expression $\phi_m \sin(\omega_a t)$ changes by ϕ_m (several radians) so the phase of the baseband-modulating signals changes by ϕ_m radians. Dividing this change by the corresponding time, $1/(4f_a) = \pi/(2\omega_a)$, the average sideband frequency is given by $2\phi_m\omega_a/\pi$. For FM we had $\phi_m = k_0 A/\omega_a$, so the average frequency is $2k_0 A/\pi$. The one-sided bandwidth of the FM signal when the modulation index is large is therefore roughly equal to the maximum deviation and the full bandwidth is about twice the maximum deviation. (If the deviation is reduced, the bandwidth goes down proportionally at first but then, in the narrow-band regime, stays constant at twice the audio bandwidth.)

FREQUENCY MULTIPLICATION OF AN FM SIGNAL

When a signal is passed through a times-N frequency multiplier its phase is multiplied by N. A square law device, for example, can serve as a frequency doubler because $\cos^2\phi = 1/2 + 1/2 \cos(2\phi)$. So if the phase of an FM signal, ϕ, the right-hand expression in Equation (28-1), is multiplied by two in a frequency doubler, both ω_0 and $k_0 A/\omega_a$ are multiplied by two, that is, the frequency and the modulation index are doubled. If a given VCO cannot be linearized well enough for the intended deviation, it can be operated with a low deviation and its signal can be frequency multiplied to multiply the deviation. If the resulting center frequency is too high, an ordinary mixer can shift it back down, preserving the increased deviation.

NOISE

The instantaneous received signal voltage, V_{SIG}, is accompanied by a noise, V_{N}, as shown in Figure 28-3. The vector sum of $V_{\text{SIG}} + V_{\text{N}}$ differs in phase from V_{SIG} by a "phase noise", ϕ_{N}, which will cause noise in the detected output of an FM or PM detector. Clearly the angle ϕ_{N} becomes smaller if the signal strength is increased. But there is another way to defeat the noise. The *SNR* at the detector output depends on the ratio of the modulation phase excursions of the signal to the phase noise. If the modulation level is increased, even without increasing the signal strength, the output SNR will be improved. If, for example, the r.m.s. phase noise is 1/10 radian and the modulation index is 1 radian, the phase SNR is 100. If the deviation is increased to produce a modulation index of 5 radians, the phase SNR increases to 2,500. The improvement has been obtained not by increasing the signal power but by increasing the signal bandwidth. In the case of amplitude modulation, the SNR depends on the noise modulating the length of the sector $V_{\text{SIG}} + V_{\text{N}}$. Since V_{N} is fixed in any given situation, the only way to improve the SNR in AM is to increase V_{SIG}, that is, increase the transmitted power.

Figure 28-3. Signal and noise voltages.

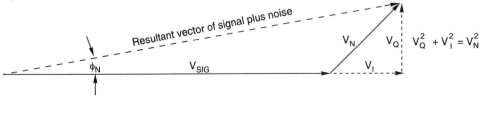

ANALYSIS OF THE SNR IMPROVEMENT IN FM

A quantitative analysis of the SNR improvement in FM is simpler if we take the noise to be the background noise produced by the detector when the signal is unmodulated, that is, the total power in the hiss coming from the loudspeaker (or whatever output device). For the signal we will take an audio sine wave with 100% modulation, that is, maximum deviation. We can represent the noise, V_N by in-phase (I) and quadrature (Q) noise components, V_I and V_Q, where $V_I^2 + V_Q^2 = V_N$, as shown in Figure 28-3. Both V_Q, and V_I are phasors rotating at ω_0, the frequency of the unmodulated carrier. Their amplitudes are random and independent. The I-component of the noise is most effective in causing amplitude fluctuations, and therefore contributes noise in AM demodulation. But it is mostly the Q-component, since it is perpendicular to V_{SIG}, that causes angle fluctuations and therefore contributes noise in the FM demodulation. For $V_N \ll V_{SIG}$, the instantaneous angle noise, $\phi_N(t)$, is just $V_Q(t)/V_{SIG}$ radians. Since V_{SIG}, the carrier amplitude, is constant, the power spectrum of ϕ_N (call it S_ϕ) is proportional to the power spectrum of V_Q. The spectral distribution of V_Q can be assumed uniform (white), so S_ϕ is also uniform. The integral of S_ϕ over the IF band gives the mean square phase fluctuation, $\langle[\phi_N(t)]^2\rangle$, so we can write $S_\phi = \langle\phi_N(t)^2\rangle/B_{IF} = \langle V_Q^2(t)\rangle/(V_{SIG}^2 B_{IF}) = \langle V_N^2(t)/2\rangle/(V_{SIG}^2 B_{IF})$, where B_{IF}, the IF bandband, is twice the maximum deviation. An FM demodulator produces an output spectral density proportional to the time derivative of the phase, $\langle(d\phi_N(t)/dt)^2\rangle$. The spectral density of the noise in the (one-sided) audio band at the detector output is therefore given by $2\omega_a^2\phi_N(t)$, and the total noise power is the integral of this spectral density over the output bandwidth of the detector, that is, the audio band (0 to B_a radians):

$$\text{output noise power} = \int_0^{B_a} 2\omega^2 S_\phi \, d\omega = S_\phi \int_0^{B_a} 2\omega^2 \, d\omega$$

$$= \frac{2}{3} S_\phi B_a^3 = \frac{B_a^3}{6kA_{MAX}} \frac{\langle V_N^2\rangle}{V_{SIG}^2}. \tag{28-4}$$

The maximum amplitude of the sine wave signal at the output of the detector is $k_0 A_{MAX}$, the maximum deviation, so the maximum signal power is just $P_{SIG} = k_0^2 A_{MAX}^2/2$. Taking the noise power from Equation (28-4), the SNR at the detector output is

$$\text{maximum output SNR} = 3\left(\frac{k_0 A_{MAX}}{B_a}\right)^3 \frac{V_{SIG}^2}{\langle V_N^2\rangle}. \tag{28-5}$$

Note that the output SNR improves as the cube of the ratio of the maximum deviation to the full audio bandwidth.

OUTPUT SNR FOR AN AM SIGNAL WITH THE SAME CARRIER POWER

Let us consider an AM system with the same carrier power and the same audio bandwidth, in order to compare its output SNR to that of the FM system. Again the modulation will be a single sine wave tone and the amplitude of the carrier will be V_{SIG}. At 100% modulation, the amplitude of the sinewave modulation envelope will also be V_{SIG}, so the audio signal power at the AM detector outut will be $V_{\mathrm{SIG}}^2/2$. For the AM system the IF bandwidth needs to be only $2B$, twice the audio bandwidth (wide enough to accommodate the highest audio frequency but, to minimize noise, no wider). The noise voltage at the detector output will be V'_{I}, the in-phase component of V'_{N}, where the primes distinguish the noise voltages in the AM IF band from the noise voltages in the wider FM IF band. The noise power will be given by $\langle V'^2_{\mathrm{I}}\rangle = \langle V'^2_{\mathrm{N}}\rangle/2$. The SNR at the AM detector output is therefore

$$\text{AM output SNR} = \frac{V_{\mathrm{SIG}}^2}{\langle V'^2_{\mathrm{N}}\rangle}. \tag{28-6}$$

COMPARISON OF NOISE, FM VERSUS AM UNDER STRONG SIGNAL CONDITIONS

All that remains in order to compare the FM and AM systems is to note that the ratio of the IF noise powers is just the ratio of the IF bandwidths, that is,

$$\frac{\langle V_{\mathrm{N}}^2\rangle}{\langle V_{\mathrm{N}}'^2\rangle} = \frac{2k_0A_{\mathrm{MAX}}}{2B_{\mathrm{a}}} = \frac{k_0A_{\mathrm{MAX}}}{B_{\mathrm{a}}}. \tag{28-7}$$

From Equations (28-5)–(28-7) we can write

$$\frac{\text{FM SNR}}{\text{AM SNR}} = \frac{3k_0{}^2A_{\mathrm{MAX}}^2}{B_{\mathrm{a}}^2} = 3\ (\text{deviation ratio})^2. \tag{28-8}$$

In FM broadcasting the deviation ratio is about 5, so the output SNR for a maximum amplitude audio tone will be higher with FM than with AM by a factor of $3 \times 5^2 = 75$ or about 19 dB. The FM sound in television uses a deviation ratio only one-third as large as that of FM broadcasting, so the SNR improvement is correspondingly less. Remember that the above analysis is only for the situations of high signal to noise. When the signal is comparable to or lower than the noise, the phase is almost totally determined by the noise, and the FM system is useless.

PREEMPHASIS AND DEEMPHASIS

We have seen that the phase excursions in FM modulation are inversely proportional to the audio frequency because phase is the integral of frequency and that the noise at the detector output is therefore greater at the high end of the audio band. This effect is compensated for in broadcasting by emphasizing the high audio frequencies at the transmitter, ahead of the modulator, and then compensating at the receiver with a deemphasis network after the FM detector. By convention, the deemphasis network is just an RC integrator with a time constant of 75μs. (Note that applying preemphasis actually amounts to changing FM to PM, as shown in Figure 28-2.)

FM, AM, AND CHANNEL CAPACITY

The improvement in the SNR that is possible with wide-band FM is an example of increasing the *channel capacity* of a communications channel by increasing the bandwidth. Nyquist's formula for channel capacity (bits per second) is given by

$$C = B \log_2\left(1 + \frac{S}{N}\right) \qquad (28\text{-}9)$$

where B is the bandwidth and S and N are the signal and noise powers (see, for example, Thomas [3]). It is not difficult to show that this formula makes sense at the high- and low-signal extremes. Nyquist showed that, with appropriate coding, totally error-free information transmission is possible at any rate not exceeding the channel capacity. Here we are concerned with narrow-band channels, so $N = N_0 B$, where N_0 is the spectral density of the noise, and the channel capacity can be written as

$$C = B \log_2\left(1 + \frac{S}{N_0 B}\right). \qquad (28\text{-}10)$$

Note that S/N_0 has units of bandwidth and can be considered a "natural" bandwidth for a given N_0 and S. N_0 is determined by the noise added along the channel, such as atmospheric noise and noise added by the receiver. The signal power, S, is determined by the transmitter power, transmitter and receiver antenna gains, and propagation loss. Channel capacity (from Equation (28-10)) versus bandwidth is plotted in Figure 28-4. Both are normalized to S/N_0.

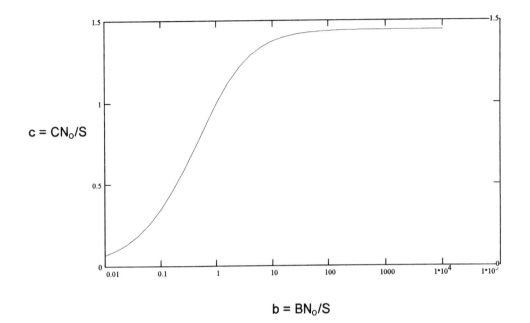

$c = CN_0/S$

$b = BN_0/S$

Figure 28-4. Channel capacity (Equation 28-10) versus bandwidth (both are normalized to S/N_o, the "natural bandwidth").

Note that, for $b > 1$, the channel capacity has essentially reached an asymptotic value of $1.44 S/N_0$. If we are below the knee of the curve we can increase the channel capacity significantly at no cost in transmitter power by (somehow) using more bandwidth. We have seen that FM broadcasting does just this. On the other hand, Equation (28-9) shows that it is expensive (and ultimately impractical) to increase channel capacity by increasing power since the log term increases slowly.

There is usually no reason to use a bandwidth higher than S/N_0 because, by that value, most of the achievable channel capacity has been obtained. An exception, however, is spread spectrum, where the bandwidth is deliberately made much greater than S/N_0. A common method of producing spread spectrum is to phase code the RF signal with a pseudorandom biphase sequence. The signal is multiplied by a digitally generated pseudorandom sequence of plus ones and minus ones. At the receiver, a locally generated replica of the code makes the phase of the local oscillator flip back and forth between zero and 180° with the same code sequence. The phase replica code must exactly match the encoded incoming signal. The output of the first mixer is then the same as if no encoding and decoding had been done, and a narrow IF bandpass filter will recover the signal from the noise. This seems like getting nothing for something (no SNR improvement for increased complexity), but spread spectrum can make a signal hard to detect and hard to jam (by spreading out the spectral energy density). It also lets many communication links use the same spectrum – as long as they have different codes (or the same code with different phases).

In AM the bandwidth is fixed by the highest modulation frequency; the total bandwidth in standard full-carrier double-sideband AM is twice the highest modulation frequency. When a weak AM station is received, the SNR is low enough to put us beyond the knee of the channel capacity versus bandwidth curve, and there would be no gain in going to a modulation scheme that increases the bandwidth. But if the station is strong, we are probably far below the knee of the curve. In this case, changing to FM modulation (without changing transmitter power) can bring us up the curve, where the higher channel capacity allows a higher SNR.

Channel capacity improvement through increased bandwidth is easy to understand in the case of straight digital modulation. If the noise is already small, widening the band allows the transmission of more bits per second, and the channel capacity is increased. But if the bandwidth is widened even more, the increased noise due to increased bandwidth eventually causes significant error rates. By this time we have climbed over the knee of the curve, and further widening of the bandwidth is to no avail.

BIBLIOGRAPHY
1. D. G. Fink (Ed.) (1975), *Electronic Engineer's Handbook*. New York: McGraw Hill.
2. *Engineering Handbook*, 8th Edition (1993), National Association of Broadcasters, Washington, DC.
3. J. B. Thomas (1969), *An Introduction to Statistical Communication Theory*. New York: John Wiley.

PROBLEMS

1. Show how a PM transmitter can be used to generate FM.

2. When an interfering AM station is close in frequency to a desired AM station, an audio tone "beat note" is produced, no matter whether the receiver uses an envelope detector or a product detector. (In the case of an envelope detector, the beat note is produced because the amplitude of the vector sum of the two carriers is effectively modulated by an audio envelope. In the case of the product detector, the carrier of the undesired station acts as a modulation sideband and beats wtih the BFO.) Will the same thing happen with FM? Suppose two carriers (i.e. CW signals), separated by say, 1 kHz, appear in the IF passband of an FM receiver. Let their amplitudes be in the ratio of, say, 1:10. Draw a phasor diagram of the sum of these two signals. Does the vector sum have phase modulation? Will the receiver produce an audio tone? What happens in the case when the amplitudes of the two signals are equal?

3. Try the following experiment with two FM receivers. Tune one receiver to a moderately strong station near the low-frequency end of the band. Use the other receiver (the local oscillator) as a signal generator. (This receiver must have continuous rather than digital tuning). Turn its volume down and hold it close to the first receiver so there will be a local oscillator pickup. Carefully tune the second receiver 10.7 MHz higher in frequency until an effect is produced in the sound from the first receiver. What is the effect? Can you use this experiment to confirm your answer to Problem 2?

4. Draw a block diagram for an FM transmitter in which a phase-locked loop keeps the average frequency of the VCO equal to a stable reference frequency.

5. Draw block diagrams for PM and FM generators based on the direct digital synthesizer (DDS) principle.

ANTENNAS AND RADIO WAVE PROPAGATION

While discussing transmitter and receiver circuitry we have not really had to know much about antennas and propagation. It was sufficient to know only that a voltage applied to the terminals of a transmitting antenna causes a voltage to appear some time later at the terminals of a receiving antenna. To be more exact, it was sufficient to know that everything between the terminals of the two antennas is equivalent to a linear two-port network. Here we will consider the transmission through this propagation link.

ANTENNAS

When an ac source (e.g. a transmitter) is connected to antenna (practically any metal structure) the resulting current has a component that is in-phase with the applied voltage. The impedance of the antenna therefore has a real part and draws power from the source. If the antenna is efficient, most of this power flows away from the antenna in the form of (energy-bearing) electromagnetic waves and only a small fraction of the power will be dissipated by ohmic heating of the antenna itself. The impedance will also generally have a nonzero imaginary part, that is, reactance. If the reactance is zero at the operating frequency the antenna is said to be resonant, just as an *RLC* circuit is purely resistive at its resonant frequency. An external tuning network can cancel the reactance and also transform the resistance to any value needed to match the output impedance of a transmitter or the input impedance of a receiver.

ELECTROMAGNETIC WAVES

As an electromagnetic wave propagates away from the transmitting antenna it takes on a spherical wave front, which, by the time it reaches a distant receiving antenna, is essentially a plane wave front. The *E* and *H* (electric and magnetic field) vectors lie in the plane of the wave front, that is, they are transverse to the direction of propagation, as shown in Figure 29-1. These fields are perpendicular to each other such that their vector

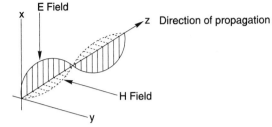

Figure 29-1.
Electromagnetic wave – *E*-
and *H*-fields are transverse
to the direction of
propagation.

cross-product, $E \times H$, points in the direction of propagation. A wave with a sinusoidal time dependence, $e^{j\omega t}$, also has sinusoidal spacial components, $E_0 e^{-jkz}$ and $H_0 e^{-jkz}$, as shown in Figure 29-1. The wave number, k, is equal to $2\pi/\lambda$. At any point in time and space the ratio of the electric field strength to the magnetic field strength is given by $E/H = \sqrt{\mu_0/\epsilon_0} = \sqrt{1.26\mu H/m \div 8.85\text{pF}/m} = 377$ ohms. This ratio is known as the "impedance of free space." The speed of propagation, ω/k is given by $c = 1/\sqrt{\mu_0 \epsilon_0} = 3 \times 10^8 m/s$ (the speed of light). Electromagnetic waves are produced by electric charges undergoing acceleration. A sinusoidal current distribution in an antenna is a system of accelerated charges and launches sinusoidal electromagnetic waves. The time-averaged power density of the waves is given by $S = 1/2(E \times H) = E^2/(2 \times 377)$ watts per square meter, where E and H are the (peak) field amplitudes. The fields from an incident wave accelerate charges in a receiving antenna, producing a voltage at the antenna terminals.

PROPAGATION IN A VACUUM

Static electric and magnetic fields are always associated with sources, that is, electric field lines terminate on electric charges and magnetic field lines encircle current filaments. But when an antenna launches an electromagnetic wave, the field lines break away from the sources, reconnecting into closed loops, and the waves becomes autonomous. For a vacuum, Maxwell's equations are

$$\nabla \times E = -\mu_0 \frac{\partial H}{\partial t} \text{ and } \nabla \times H = \epsilon_0 \frac{\partial E}{\partial t}. \tag{29-1}$$

The right-hand sides of these equations act like sources, that is, the E-field is produced by the changing H-field and the H-field is produced by the changing E-field. (To solve these equations and verify the statements made above about electromagnetic waves – that they are transverse and that they travel at the speed of light – you take the curl of one of the equations and then substitute from the other.)

ANTENNA DIRECTIVITY AND GAIN

The radiation from any antenna is always stronger in some directions than in others (Isotropic radiation is impossible for a transverse vector field.) Let $S(\theta, \phi, r)$ be the energy flux (in watts per square meter) produced in a given direction. The integral S over an enclosing sphere will be the total radiated power transmitted, P_{RAD}:

$$\int S(\theta, \phi, r)r^2 \, d\Omega = P_{RAD}. \qquad (29\text{-}2)$$

From any given direction, the ratio of the flux to the average flux is defined as the *directivity* of the antenna:

$$D(\theta, \phi, r) = \frac{S(\theta, \phi \, r)}{P_{RAD}/4\pi r^2}. \qquad (29\text{-}3)$$

Electromagnetic theory shows that the flux eventually falls off as r^{-2}. For large r, then, the far-field directivity, D, is a function only of θ and ϕ. Combining Equations (29-2) and (29-3), we see that the average directivity of an antenna is unity:

$$\frac{1}{4\pi} \int D(\theta, \phi)d\Omega = 1. \qquad (29\text{-}4)$$

Directive gain, $G(\theta, \phi, r)$, has the same definition as the directivity except that the radiated power is replaced by P_{INC} the transmission power incident on the antenna terminals:

$$G(\theta, \phi, r) = \frac{S(\theta, \phi, r)}{P_{INC}/4\pi r^2}. \qquad (29\text{-}5)$$

If an antenna has no ohmic losses and its feedpoint impedance matches the transmission line impedance (so that no power is reflected), all the incident power will be radiated, and the directive gain will be equal to the directivity. In most antennas used for transmitting, the losses are not more than a few percent and the distinction between directivity and gain is unimportant. However, many receiving antennas, especially those used for the AM broadcast band, have very low efficiency.[*] The maximum value of the directive gain of an antenna is simply called the gain. A transmitter connected to an antenna having a gain of 20 dB will produce a directed signal

---

[*]For a typical AM broadcast band or short-wave receiver the input signal-to-noise ratio is the ratio of the desired signal power to the noise power from static – atmospheric electricity. Even with an inefficient antenna, the total power delivered to the receiver is much greater than the thermal noise added by the antenna and by the receiver itself, so the output signal-to-noise ratio is not significantly increased.

100 times more powerful than if it were connected to a lossless isotropic radiator.

EFFECTIVE CAPTURE AREA OF AN ANTENNA

The distance between transmitting and receiving antennas is generally so large that a plane wave can be assumed incident on the receiving antenna. The energy extracted from the incident flux is, of course, proportional to the flux density. The proportionality constant is known as the *effective area* of the receiving antenna. (Power per unit area times area = power.) It turns out that the effective area is proportional to gain:

$$A_{\text{eff}} = G\lambda^2/4\pi. \tag{29-6}$$

This is an extremely important relation, which we stated earlier in order to show that a resistor can supply a noise power of kT watts per hertz. Since gain and effective area are proportional, there is really no distinction between transmitting antennas and receiving antennas; the best transmitting antenna (most gain) is also the best receiving antenna (most capture area). A standard derivation of Equation (29-6) applies the reciprocity theorem to a system of two arbitrary antennas. The two antennas, 1 and 2, need not be identical. We can suppose they are both matched to the same impedance value and that we have available both a generator and a receiver that match this impedance. First we connect the generator to antenna 1 and measure the power at antenna 2. If we now interchange the generator and receiver, the reciprocity theorem states that the power measured will be unchanged. Expressing this in terms of gain and effective area, we have $G_1 A_{\text{eff2}} = G_2 A_{\text{eff1}}$, from which we see that the ratio of gain to effective area has the same constant value for any and all antennas. We can pick any conveniently simple antenna and use electromagnetic theory to derive its gain and effective area. When this is done it is found that the ratio of gain to effective area is always $4\pi\lambda^2$. A half-wave dipole antenna has a maximum gain of about 1.6. Its effective area is therefore $1.6\lambda^2/4\pi$. If the dipole is made of thin wire, it has no real physical area, only a length, yet it has a nonzero effective area and can extract energy from an incident electromagnetic wave. The effective area of a microwave dish antenna is a large fraction, usually between 0.50 and 0.80, of its physical aperture area (like a rain gauge). This fraction, known as *aperture efficiency*, is usually not determined by ohmic losses but rather by the illumination pattern of the primary feed antenna. A perfectly uniform illumination pattern (and no ohmic loss) produces an aperture efficiency of unity. The Arecibo dish uses an aperture 700 feet (230 m) in diameter (area = 35,800 m²). Its aperture efficiency is about 70% for $\lambda = 12\,\text{cm}$ (a wavelength often used for radar astronomy), so its gain, using Equation (29-6), is $G = 4\pi$ $(0.70 \times 35,800)/(0.12)^2 = 22 \times 10^6$ or $73\,\text{dB}$.

A SPACECRAFT RADIO LINK

Consider the following example of a spacecraft telemetry link for which we wish to find the maximum range. Suppose we have a 1 W telemetry transmitter aboard a spacecraft and that the data rate requires a channel capacity corresponding to a signal-to-noise ratio of at least unity in a 1 Hz bandwidth. This link uses a frequency of 3 GHz (10 cm wavelength). The transmitting antenna on the spacecraft is a 2 m diameter dish. The ground station antenna is a 10 m diameter dish, as shown in Figure 29-2. Assume that both these dish antennas have an effective area equal to 60% of their physical apertures. Assume also that there is no pointing error, that is, the antennas always point directly at each other and that the system temperature of the ground station receiver is 25 K. (The system temperature is the sum of the equivalent receiver noise temperature, the antenna noise temperature, and the sky noise temperature.)

1. What is the equivalent input noise of the receiver? Boltzmann's constant, k is 1.38×10^{-23} $W/(\text{HzK})$, so the equivalent input noise power, kTB is $1.38 \times 10^{-23} \times 25 \times 1 = 3.45 \times 10^{-22}$.
2. What is the effective area of the receiving antenna? The physical aperture is πR^2, so the effective area is $0.60\pi R^2 = 0.60\pi 5^2 = 47.1\text{m}^2$.
3. What is the gain of the transmitting antenna? The physical aperture is πR^2 and the aperture efficiency is 0.60, so the effective aperture is $0.60\pi(1^2) = 1.88\text{m}^2$. The gain is $4\pi A_{\text{eff}}/\lambda^2 = 4\pi(1.88)/(0.1^2) = 2,369$.

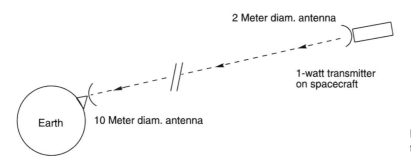

Figure 29-2. Spacecraft telemetry link.

4. What is the maximum range, R, in kilometers for the spacecraft to maintain the required signal-to-noise ratio? Here we simply set $P_{noise} = P_{received} = A_{rcvr} P_{trans} G_{trans}/(4\pi R^2)$, from which we have $R^2 = (4\pi)^{-1} P_{trans} G_{trans}/P_{noise}$. Using the parameters calculated above, we have $R = \sqrt{[(4\pi)^{-1} \times 1 \times 2,369/(3.45 \times 10^{-22})]} = 7.4 \times 10^{11} m = 740 \times 10^6 km$. This is roughly the mean distance to Jupiter.

TERRESTRIAL RADIO LINKS

The VHF and UHF two-way radios used by cellular phones, emergency vehicles, and so on, have transmitters with several watts of power, but their range is limited by the curvature of the Earth to only a few miles or tens of miles. (Radio waves do not propagate through the highly conductive Earth though they do diffract slightly so that the radio horizon is somewhat beyond the optical horizon.) These mobile radios also use nondirective antennas. Such antennas have gains only of the order of unity, so they do not have to be pointed accurately – or even at all. Finally, in ground-to-ground radio links, the signal usually arrives from an angle near the horizon; the receiving antenna will pick up noise from the ground (thermal radiation). In these systems, then, extremely low-noise receivers are of no benefit. Broadcasting stations for FM and television also use VHF and UHF frequencies, so their range is also essentially line of sight. Long-distance propagation in AM and short-wave broadcasting depends on reflection from the ionosphere.

THE IONOSPHERE

At altitudes above about 60 km the atmosphere is ionized by ultraviolet radiation from the Sun; electrons are stripped from the neutral particles (mostly oxygen atoms and O_2 and N_2 molecules) to produce a mixed electron and ion gas. During the day the density of this ionized gas is highest at around 250 km, the peak of the "F-region." Above the peak the ionization is less because the thinner atmosphere presents fewer particles to be ionized. Below the peak the ionization is less because the denser atmosphere exhausts the supply of ultraviolet photons; the electrons they produce quickly encounter nearby ions and recombine. At night, without sunlight, the ionization rate is zero. Recombination quickly neutralizes the ionization at the lowest altitudes, around 100 km, and erodes the F-region until sunrise.

WAVE PROPAGATION IN THE IONOSPHERE

An electromagnetic wave induces electric currents in the electron gas of the ionosphere. (The electrons, by virtue of their low mass, are accelerated by the incident wave to much higher velocities than the ions, so the ion contribution to the current is negligible.) The effective dielectric constant of an electron gas is not difficult to calculate from Maxwell's equations. (We will see that this dielectric constant becomes imaginary below a certain critical frequency that depends on the electron density; below this frequency, then, electromagnetic waves cannot propagate through the plasma.) No longer in a vacuum, we must use the general curl H equation, which includes real electric current, J, in addition to the displacement current:

$$\nabla \times H = J + \epsilon_0 \frac{\partial E}{\partial t}. \tag{29-7}$$

The electrons are in rapid thermal motion, but this motion is random and contributes nothing to the current. Yet all the electrons in any given region of the electric field of a wave are accelerated equally and move together to produce a net current, $J = Nev$, where N is the electron density, e is the electron charge, and v is the component of velocity imparted by the electric field. We will neglect the weak $v \times B$ force from the magnetic field, so Newton's second law of motion is just $m\, dv/dt = F = eE$. For a sinusoidal time dependence, $e^{j\omega t}$, we can write this equation of motion as $m(j\omega v) = eE$. Substituting $J = Nev = Ne^2 E/(j\omega m)$ in Equation (29-7) gives

$$\nabla \times H = \frac{Ne^2 E}{j\omega m} + j\omega_0 E = j\omega\epsilon_0 \left(1 - \frac{Ne^2}{\epsilon_0 m\omega^2}\right) E. \tag{29-8}$$

Note that the term in parentheses is the square of the relative dielectric constant and that it becomes negative for low frequencies, in particular for $\omega^2 < \omega_p^2$ where $\omega_p^2 = Ne^2/(\epsilon_0 m)$. This happens because the conduction current (the electron current) becomes greater than the displacement current. The total current (conduction current plus displacement current) changes sign and has the wrong polarity to source the H-field of a traveling wave. This critical frequency, ω_p, is known as the *plasma frequency*. (If the local charge neutrality of an electron–ion gas is disturbed, the densities will oscillate at this frequency the way a spring and mass system oscillates at its resonant frequency.) For a wave to propagate in the plasma, the dielectric constant must be real; only waves with frequencies lower than the plasma frequency will be reflected. The free electron gas that gives metals their conductivity is dense enough to reflect visible light, but the alkali metals (lithium, sodium, etc.) have relatively lower electron densities and are transparent in the ultraviolet.

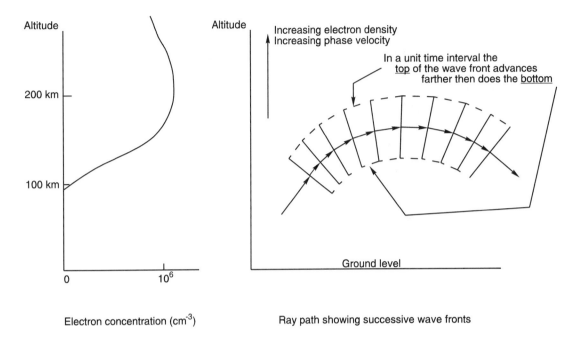

Figure 29-3. Ionospheric refraction.

REFLECTION OF WAVES FROM THE IONOSPHERE

The reflection of radio waves is normally a process of refraction because the waves are not vertically incident on the ionosphere. As they travel obliquely upward, the dielectric constant decreases, so the phase velocity increases, causing the perpendicular to the wave front to turn around gradually. This is shown in Figure 29-3.

For signal strength calculations over an ionospheric path the ionosphere can be considered a specular mirror, and field strengths can be calculated by taking the inverse square of the total path length of the ray.

DAYTIME VERSUS NIGHTTIME PROPAGATION

Short-wave broadcasts from distant transmitters on the higher frequency bands are not heard at night because the ionospheric electron density is too low for reflection, that is, the waves cannot get turned around sufficiently. On the other hand, low-frequency stations such as those in the AM broadcasting band are received from great distances *only* at night. During the day their energy is dissipated in the lower ionosphere through collisions between the accelerated electrons and the neutral particles. Why do higher frequency waves not suffer this daytime attenuation in the lower ionosphere? Consider a low-frequency and a high-frequency wave of equal power, that is, equal field strengths. Both cause the ionospheric electrons to execute synchronized sinusoidal motion as described above. But the low-frequency wave, because its period is long, will produce much higher electron velocities; we saw above that $v = Ee/(j\omega m)$. The average kinetic energy of each electron is therefore $m < v^2/2 > = m[Ee/(\omega m)]^2/4$, where E is the amplitude of the electric field. All the electrons in the lower ionosphere suffer collisions with the neutral particles (which are present since the ionization is not 100%). The average collision leaves the electron with a random velocity, that is, its share of the synchronized sinusoidal motion is lost. The frequency of collisions does not depend on the frequency of the electromagnetic wave (or on there being any electromagnetic wave present at all), so the rate of energy loss is inversely proportional to the square of ω, the wave frequency, and long-distance AM listeners have to wait for nighttime.

OTHER MODES OF PROPAGATION

Besides reflection by the ionosphere, there are a number of other ways that an electromagnetic wave can get around the curvature of the Earth. These include scattering from the ionized trails of meteors entering the Earth's atmosphere, scattering from irregularities in the ionosphere even when the ionosphere is otherwise not dense enough to turn the waves around by refraction, scattering from density irregularities in the neutral atmosphere (i.e. fluctuations in the index of propagation), and ducting beneath atmospheric temperature inversion layers.

BIBLIOGRAPHY
1. R. E. Collin (1985), *Antennas and Radiowave Propagation*. New York: McGraw-Hill
2. K. Davies (1966), *Ionospheric Radio Propagation*. New York: Dover.
3. M. C. Kelley (1989), *The Earth's Ionosphere: Plasma Physics and Electrodynamics*. San Diego: Academic Press.

PROBLEMS

1. The voltage at the terminals of a receiving antenna is proportional to the *E*-field of an incident electromagnetic wave. The "effective length" (or effective height if the antenna is vertical) is defined as the open-circuit voltage at the terminals divided by the incident E-field: volts/(volts/meter) = meters. Show that the effective length is given by

$$\text{effective length} = \sqrt{\frac{4RA_{\text{eff}}}{Z_0}}$$

where R is the real part of the antenna impedance, A_{eff} is the effective area ($A_{\text{eff}} = gain \times \lambda^2/4\pi$), and $Z_0 = 377$ ohms ($= \sqrt{\mu_0/\epsilon_0}$, the impedance of free space). Find the effective length of a half-wave dipole. (Gain= 1.6 and $R = 73$ ohms.)

2. Suppose we have a 1 W transmitter connected to a dipole antenna that is aligned to provide the maximum signal strength at a distant receiving antenna. Needing more signal strength, we obtain a second, identical dipole and, using a power splitter, feed each dipole with 1/2 W. We space the second far enough from the first so that they do not interact. We make sure that both antennas are aligned toward the receiver and we also make sure that the cables from the power splitter have equal length. At the receiving antenna, each transmitting antenna provides a field amplitude that is less than the original field by $1/\sqrt{2}$. But the two signals are in phase so the total amplitude is increased by $2/\sqrt{2} = \sqrt{2}$. Squaring this we see that the received signal strength is doubled. Have we gotten something for nothing? Could we repeat this process to increase the received power even more?

3. Let the individual antennas of Problem 2 be AM broadcast towers with omnidirectional patterns and vertical polarization. Suppose the spacing between these antennas is $\lambda/2$. As before, they are fed symmetrically, that is, with the same power and same phase. Find the radiation pattern in the horizontal plane, that is, make a polar plot of the relative field strength versus azimuth angle for a distance far from the antennas). *Hint*: At any observation point in the horizontal plane at a distance r from center of the line joining the two antennas, the total voltage is the sum of the contributions from the two antennas, $e^{j\phi_1}$ and $e^{j\phi_2}$. The phases ϕ_1 and ϕ_2 are the phase path lengths corresponding to r_1 and r_2, the distances from the observation point to the respective antennas. These phase paths are just $2\pi r_1/\lambda$ and $2\pi r_2/\lambda$. The field strength is given by $|e^{j\phi_1} + e^{j\phi_2}|^2$

AMPLIFIER NOISE II 30

NOISE MATCHING

In Chapter 21 we quoted a formula showing how the noise figure of any active two-port device depends on the impedance of the signal source:

$$F = F_{\min} + (R_n/G_s)|Y_s - Y_{opt}|^2 \qquad (30\text{-}1)$$

Here $F_{\min}$ is the minimum noise figure; $Y_s = G_s + jB_s$ is the source admittance; $Y_{opt} = G_{opt} + jB_{opt}$ is the source admittance that produces $F_{\min}$; and R_n is a real parameter known as the *noise resistance*. Equation (30-1) applies as well to audio and instrumentation amplifiers as to radio and microwave amplifiers, as long as the four noise parameters of the device, $F_{\min}$, R_n, G_{opt}, B_{opt}, are specified for the frequency in question. While this result will require some effort to derive, its use is simple, at least with RF (narrow-band) amplifiers. To design a low-noise RF amplifier, we get out the data sheet for the transistor, find the value Y_{opt} for the desired frequency, and design an input network to convert the impedance of the intended source (probably 50 ohms) into $1/Y_{opt}$. Figure 30-1 shows how a transformer and a reactor can form the matching network. (Of course, there are many equivalent networks.) With this input network the transistor is said to be "noise matched", because the noise figure of the overall amplifier will be $F_{\min}$. The input network that provides noise matching generally does not provide impedance matching, that is, it does not convert the source impedance into the complex conjugate of the input impedance of the amplifier. The resulting power mismatch means we will have less signal at the amplifier output than with a conjugate impedance match. But

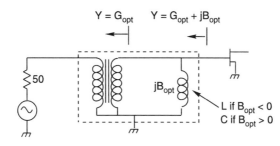

Figure 30-1. Input network to convert a 50 ohm source impedance to $G_{opt} + jB_{opt}$.

converting the source impedance into $1/Y_{\text{opt}}$, rather than into $1/Y_{\text{in}}^*$, will reduce the transistor noise more than the mismatch reduces the signal power. The net effect is to maximize the *signal-to-noise ratio*, that is, provide the minimum possible noise figure.

(Note that simultaneous noise matching and impedance matching might be achieved or at least approached. If the device has a nonzero reverse transfer coefficent, the input impedance of the amplifier is not fully determined by the input network. Depending on the transistor characteristics, a designer might find an output network (and termination) to make the input impedance of the amplifier approach Z_o while using an input network that provides noise matching. The use of feedback can provide additional degrees of freedom with which to pursue simultaneous matching.)

EQUIVALENT CIRCUITS FOR NOISY TWO-PORT NETWORKS

Equation (30-1) was derived by Rothe and Dahlke [2], who pointed out that a noisy two-port network (transistor, amplifier, etc.) can be modeled, for small signals, as an ordinary passive linear two-port with two external frequency-dependent noise sources. These noise sources can be series voltage generators at the input and output, shunt current generators at the input and output, a current source at one end and a voltage source at the other end, or, finally, both a voltage source and a current source at the same end. Figure 30-2 shows these equivalent circuits. Note that when a current generator and a voltage generator appear on the same side they can be interchanged without changing their parameters, so Figures 30-2e and f (or g and h) are actually the same circuit.

NOISE FIGURE OF THE EQUIVALENT CIRCUIT

Figure 30-2e (or f) is the easiest equivalent circuit to analyze for noise figure. Since both noise sources are placed at the input, the transfer characteristics (Z, Y, or S parameters) of the two-port will not come into play. In Figure 30.3, the circuit to be analyzed, the amplifier block can represent

Figure 30-2. Equivalent representations of a noisy transistor as a noiseless transistor with two external noise generators.

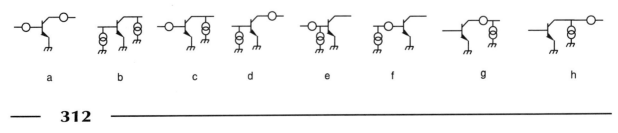

a　　　b　　　c　　　d　　　e　　　f　　　g　　　h

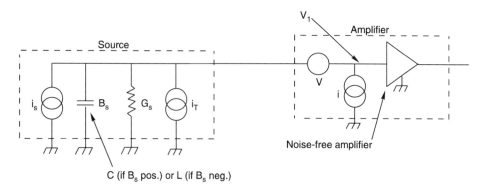

Figure 30-3. Signal source and noisy amplifier – equivalent circuit.

a transistor, a tube, or a complete amplifier. The two external noise sources, v and i, are equivalent to the internal sources of noise of the device, whatever their number and physical nature. The other current source, i_T, represents the thermal noise of the source, that is, $\langle i_T \rangle^2 = 4kTG_s$. Since there are four noise parameters to be determined, we need more than just $\langle i^2 \rangle$ and $\langle v^2 \rangle$. The other two parameters come from the fact that i and v are generally not independent but are correlated. The part of i that is correlated with v defines the "correlation admittance," Y_c, through the relation $i_c = Y_c v$. Thus

$$\langle iv^* \rangle = \langle (i_c + i_u)v^* \rangle = \langle i_c v^* \rangle = \langle Y_c |v|^2 \rangle = Y_c \langle |v|^2 \rangle = (G_c + jB_c)\langle |v|^2 \rangle. \tag{30-2}$$

To find an expression for the noise figure, we must find the voltage v_1, at the input of the device and use the definition of noise figure to find the ratio of the average of $\langle |v_1|^2 \rangle$ to the part of $\langle |v_1|^2 \rangle$ attributable to the source. The result is independent of the input impedance of the device because, using superposition, we can find v_1, considering the noise sources one at a time. They all have the same effective source impedance, $1/Y_s$. For convenience, then, we can take the device to have an infinite input impedance. It follows immediately that

$$v_1 = (i + i_T)/Y_s + v. \tag{30-3}$$

To get the equivalent input *power* we use the average square magnitude of v_1:

$$\langle |v_1|^2 = ((\langle |i|^2 \rangle + \langle |i_T|^2 \rangle))/|Y_s|^2 + \langle |v|^2 \rangle + (\langle iv^* \rangle / Y_s) + (\langle i^* v \rangle / Y_s^2). \tag{30-4}$$

Here we used the fact that there is no correlation between i and i_T or between i_T and v. The noise figure is given by this total power divided by the part due to the noise current of the source:

$$F = \frac{((\langle |i|^2 \rangle + \langle |i_{\mathrm{T}}|^2 \rangle)/|Y_{\mathrm{s}}|^2 + \langle |v|^2 \rangle + (\langle iv^* \rangle / Y_{\mathrm{s}} + \langle i^* v \rangle / Y_{\mathrm{s}}^*)}{\langle |i_{\mathrm{T}}|^2 \rangle)/|Y_{\mathrm{s}}|^2} \qquad (30\text{-}5)$$

or

$$F = 1 + (\langle |i|^2 \rangle / \langle |i_{\mathrm{T}}|^2 \rangle) + (\langle |v|^2 \rangle / \langle |i_{\mathrm{T}}|^2 \rangle)|Y_{\mathrm{s}}|^2 (1 + Y_{\mathrm{c}}/Y_{\mathrm{s}} + Y_{\mathrm{c}}^*/Y_{\mathrm{s}}^*). \qquad (30\text{-}6)$$

The noise current of the source is given by $\langle |i_{\mathrm{T}}|^2 \rangle = 4kT_o G_{\mathrm{s}}$. The *noise conductance*, g_{n}, and *noise resistance*, R_{n}, are defined by

$$\langle |i|^2 \rangle = 4kT_o g_{\mathrm{n}} \text{ and } \langle |v|^2 \rangle = 4kT_o R_{\mathrm{n}}. \qquad (30\text{-}7)$$

(The lower-case g_{n} follows the original notation of Rothe and Dahlke, who reserved G_{n} for the part of i that is correlated with v.) In terms of these parameters we have

$$F = 1 + g_{\mathrm{n}}/G_{\mathrm{s}} + (R_{\mathrm{n}}/G_{\mathrm{s}})|Y_{\mathrm{s}}|^2 (1 + Y_{\mathrm{c}}/Y_{\mathrm{s}} + Y_{\mathrm{c}}^*/Y_{\mathrm{s}}^*). \qquad (30\text{-}8)$$

Replacing Y_{s} by $G_{\mathrm{s}} + jB_{\mathrm{s}}$ and Y_{c} by $G_{\mathrm{c}} + jB_{\mathrm{c}}$, this becomes

$$F = 1 + g_{\mathrm{n}}/G_{\mathrm{s}} + (R_{\mathrm{n}}/G_{\mathrm{s}})(G_{\mathrm{s}}^2 + B_{\mathrm{s}}^2) + 2(R_{\mathrm{n}}/G_{\mathrm{s}})(G_{\mathrm{c}}G_{\mathrm{s}} + B_{\mathrm{c}}B_{\mathrm{s}}). \qquad (30\text{-}9)$$

Differentiating with respect to B_{s} and setting the derivative set to zero, we find the optimum source susceptance:

$$B_{\mathrm{opt}} = -B_{\mathrm{c}}. \qquad (30\text{-}10)$$

Differentiating next with respect to G_{s} and setting that derivative equal to zero gives the optimum source conductance:

$$G_{\mathrm{opt}} = \sqrt{g_{\mathrm{n}}/R_{\mathrm{n}} - B_{\mathrm{c}}^2}. \qquad (30\text{-}11)$$

After a few more steps we have

$$F = 1 + 2R_{\mathrm{n}}(G_{\mathrm{c}} + G_{\mathrm{opt}}) + (R_{\mathrm{n}}/G_{\mathrm{s}})[(G_{\mathrm{s}} - G_{\mathrm{opt}})^2 + (B_{\mathrm{s}} - B_{\mathrm{opt}})^2] \qquad (30\text{-}12)$$

which is the desired relation, the Rothe–Dahlke formula. The first term on the right is just $T_{\min}$. Summarizing these relations between the equivalent external noise generators and the standard noise parameters we have

$$R_{\mathrm{n}} = \frac{\langle |v|^2}{4kT_0} \qquad (30\text{-}13a)$$

$$B_{\mathrm{opt}} = \frac{-\mathrm{Im}\langle iv^* \rangle}{\langle v^2 \rangle} \qquad (30\text{-}13b)$$

$$G_{\text{opt}} = \sqrt{\frac{\langle |i|^2 \rangle}{\langle |v|^2 \rangle} - \left(\frac{\text{Im}\langle iv^* \rangle}{\langle |v|^2 \rangle}\right)^2} \qquad (30\text{-}13c)$$

$$F_{\text{min}} = 1 + \frac{2\langle |v|^2 \rangle}{4kT_0}\left(\frac{\text{Re}\langle iv^* \rangle}{\langle |v|^2 \rangle} + G_{\text{opt}}\right). \qquad (30\text{-}13d)$$

These inverse relations are

$$\langle |v|^2 \rangle = 4kT_o R_{\text{n}} \qquad (30\text{-}14a)$$

$$\langle |i|^2 \rangle = 4kT_o (G_{\text{opt}}^2 + B_{\text{opt}}^2) \qquad (30\text{-}14b)$$

$$\text{Im}\langle iv^* \rangle = -B_{\text{opt}}\langle |v|^2 \rangle \qquad (30\text{-}14c)$$

$$\text{Re}\langle iv^* \rangle = 4kT_o R_{\text{n}}\left(\frac{F_{\text{min}} - 1}{2R_{\text{n}}} - G_{\text{opt}}\right). \qquad (30\text{-}14d)$$

DEVICES IN PARALLEL

Noise matching allows us to get the best noise figure, F_{min}, from a transistor. We saw earlier that multiple devices in series (an amplifier cascade) result in an unavoidable increase in the noise figure. What about amplifiers in parallel? Identical transistors might be literally paralleled, emitter to emitter, base to base, and collector to collector. Or amplifiers might have their inputs directly connected and their outputs combined with a hybrid coupler. The result for two devices in parallel (see Problem 4) is that the resulting amplifier, compared to the individual amplifiers, has half as much noise resistance, twice as much noise conductance, and twice as much correlation admittance. This produces an unchanged F_{min}. It will, however, double Y_{opt}, making it easier to noise match to a low-impedance source, and can be a useful technique, for example, at very low frequencies where transformer matching is impractical.

NOISE MEASURE

We pointed out earlier that an amplifier with a good noise figure is of little value if its gain is low (a piece of wire has a perfect noise figure, unity, but has no gain). The application of negative feedback may lower the noise figure of a given transistor but at the same time it will reduce the gain. To recover the lost gain we can cascade more amplifiers. We saw earlier the expression for F_∞, the noise figure of a cascade of identical amplifiers

$$F_\infty - 1 = (F - 1)/(1 - 1/G). \tag{30-15}$$

This quantity, called the *noise measure*, is a better figure of merit than the noise figure in that a low-gain amplifier scores low, even if its noise figure is good. The optimum input network should minimize F_∞ rather than F. But the gain G, depends also on the output network. Haus and Adler [1] showed that the *minimum noise measure* is a fundamental invariant when the gain, G, is taken to be the "available gain" of the device (the maximum gain that can be obtained by varying the output match while leaving the input untouched). Minimum noise measure, like minimum noise figure, is therefore determined by the input network. Haus and Adler proved that the minimum noise measure of a device is left unchanged when the device is embedded in an arbitrary network of lossless reactances. It had been found earlier that the noise figure of a device could sometimes be lowered by using feedback. Haus and Adler's results showed that any improvement in noise figure must be accompanied by a decrease in gain such that the minimum noise measure remains constant. A corollary is that changing the orientation of the device (common emitter, common collector, or common base) does not change the minimum noise measure. Circuit techniques, then, cannot produce a breakthrough low-noise amplifier. Any standard configuration can have the best possible noise performance. Circuit techniques, however, can improve bandwidth, stability, and so on. If we know the characteristics of a given device, we can calculate the minimum noise measure (best possible noise performance) at every frequency. There is, however, no theory available to tell us how well we can do over a given band. Fano showed how well impedance matching can be done over a given band [3], but no one has done the same for noise matching.

BIBLIOGRAPHY
1. H. A. Haus and R. B. Adler (1957), "Optimum noise performance of linear amplifiers," *Proc. IRE* 46: 1517–33.
2. H. Rothe and W. Dahlke (1956), "Theory of noisy fourpoles," *Proc. IRE* 44: 811–18.
3. R. M. Fano (1950), "Theoretical limitations on the broadband matching of arbitrary impedances" *Journal of the Franklin Institute* 249: 57–83 and 139–54.

1. A transistor with noise parameters F_{min}, G, B_{opt} and R_n is provided with a noise-matching network to produce an amplifier with $F'_{min} = F'_{min}$, $B'_{opt} = 0$, and $G'_{opt} = 1/50$. What is the value of R'_n?

2. Can a transistor be fitted with an input network such that the correlation admittance, Y_c, is zero for the overall amplifier?

3. The portion of the noise current i_c, that is correlated with the noise voltage cannot exceed the total noise current, that is $\langle |i_c|^2 \rangle \leq \langle |i|^2 \rangle$. Use this to prove the following inequality:

$$F_{min} - 1 \leq 4R_n G_{opt}.$$

4. Show that when two identical transistors are paralleled, the minimum noise figure of the combination is the same as the minimum noise figure of the individual transistors.

5. A standard amplifier for use in 50 ohm systems has a gain G and a noise figure F. This amplifier is combined with a 50 ohm directional coupler to form the feedback amplifier shown below. The 180° coupler has a power gain of α in its "main channel" and therefore a power gain of $1 - \alpha$ in the adjacent arms. The 180° path of the coupler provides the negative feedback. (Assume that the amplifier has zero phase shift.)

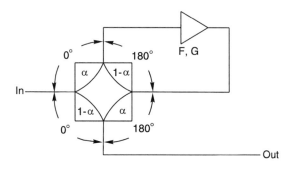

(a) Find the gain, G', of the feedback amplifier.

(b) Find the noise figure, F', of the feedback amplifier.

(c) Combine the results of Problems 5(a) and (b) to show that the noise measure of the feedback amplifier is the same as the noise measure of the embedded amplifier, that is, show that

$$(F' - 1)/(1 - 1/G') = (F - 1)/(1 - 1/G).$$

This problem provides a simple demonstration of noise measure remaining invariant when an amplifier is embedded in a lossless netowork. The negative feedback reduces the noise figure but it also reduces the gain, and the noise measure stays the same. (What would positive feedback do – short of causing oscillation?)

6. It is certainly possible (and common) for an amplifier to have a noise temperature less than its physical temperature. Is it possible to build an *electronic cold load*, that is, an active one-port circuit whose impedance is resistive but which generates less noise power than a resistor at ambient temperature?

7. A differential amplifier such as an operational amplifier (op amp), is a three-port device (output and two inputs). Find an equivalent circuit which represents an op amp as a noiseless device with three external noise sources. Arrange for these noise sources to be on the input side of the amplifier.

OSCILLATOR NOISE 31

The frequency stability of an oscillator is determined by changes in ambient temperature and humidity, mechanical vibration, component aging, power supply variations, load variations, and, finally, random noise generated by the circuit components – primarily the active element (transistor, tube, operational amplifier (op amp), etc). Except for noise, these effects are systematic, that is, nonrandom, and can, in principle, be eliminated by using constant-temperature chambers, thermal compensation schemes, shock mounting, and so forth. But even when these instabilities have been reduced to an acceptable level, random variations remain and determine the power spectrum of an oscillator ("lineshape" in spectroscopy). An ideal oscillator would have a power spectrum that is a delta function, that is, an infinitely narrow unmodulated carrier. Noise broadens the spectrum of any real oscillator. In some applications the concentration of power is important; a specification might state, for example, that at least 90% of the power must be contained in a band within 0.01 Hz of the nominal frequency. More often it is only necessary that the spectral density beyond some distance from the center frequency be very low. In such a case it might be specified that "the noise power in a 1 Hz bandwidth centered 1 kHz from the carrier must be at least 60 dB below the carrier power." Spectral energy outside the narrow band containing most of the power is referred to as "sideband noise".

In a superheterodyne receiver the noise sideband power of the local oscillator can mix with an incoming signal that is nominally outside the IF pass band to produce noise that falls within the pass band as shown in Figure 31-1. An undesired signal that is translated to a frequency outside the IF pass band will normally be rejected by the IF bandpass filter. But the noisy local oscillator (LO) will add noise sidebands that extend into the bandpass. If the undesired signal is near the bandpass and if it is much stronger than the desired signal, this noise can override the desired signal.

A similar situation occurs in MTI (moving target indicator) radar, where the Doppler effect is used to distinguish moving objects from background "clutter" (the ground and stationary objects). Moving targets can be identified as spectral lines shifted away from the frequency of the strong clutter echo. But, in as much as the master oscillator of the transmitter or

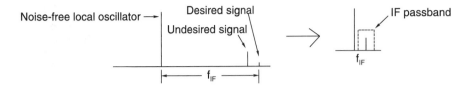

(a) Only the desired signal falls within the IF passband

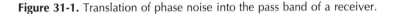

(b) Oscillator noise sidebands on the undesired
signal fall within the IF passband

Figure 31-1. Translation of phase noise into the pass band of a receiver.

the local oscillator of the receiver is noisy, moving targets with small velocities will be buried in the broadened clutter return. Good *subclutter visibility* therefore requires low-noise oscillators.

POWER SPECTRUM OF A LINEAR OSCILLATOR

In the simplest analysis, an oscillator is considered to be a linear system with just enough positive feedback to sustain oscillation. We have already used this approach to design a feedback oscillator. Here we will present a simple analysis [1,2] that predicts the shape of the spectral line. This shape does not depend on the circuit configuration, so here we will consider the simple phase shift oscillator presented earlier. This circuit, shown in Figure 31-2, uses a noninverting voltage amplifier having a voltage gain, A, (independent of frequency), infinite input impedance, and zero output impedance. We will put an equivalent generator at the input to account for the noise produced by the amplifier. As long as the amplifier is linear, we need only to consider its noise spectrum in the vicinity of the oscillation frequency. The feedback voltage is taken from the load resistor forming the

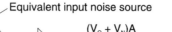

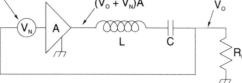

Equivalent input noise source

$(V_o + V_N)A$

V_o

V_N

A

L

C

R_L

Figure 31-2. Prototype oscillator for phase noise analysis.

bottom leg of a voltage divider whose top leg is the series LC circuit. By inspection of Figure 31-2 we can write the loop equation

$$V_0 = (V_0 + V_N)A\frac{R}{R + jX} \tag{31-1}$$

where A is the voltage gain of the amplifier and $X = \omega L - 1/(\omega C)$. Note that V_0 and V_N are spectral densities, that is, their units are volts/$\sqrt{\text{Hertz}}$. Solving for V_0 we have

$$V_0 = \frac{V_N R}{R(1 - A) + jX}. \tag{31-2}$$

Expanding X around the resonant frequency, we let $\omega = \omega_0 + \omega'$, where $\omega_0^{-2} = LC$. The reactance, X, then simplifies to $X \approx 2\omega' L$ and Equation (31-2) becomes

$$V_0(\omega') = \frac{-V_N}{(A - 1) - 2j\omega' L/R}. \tag{31-3}$$

The value of A is determined by the condition $\int V_0^2(\omega')/R\, d\omega' = P$ where P is the total output power:

$$P = \frac{1}{R}\int V_0^2(\omega')\, d\omega' = \frac{1}{R}\int_{-\infty}^{\infty} \frac{V_N^2\, d\omega'}{(A - 1)^2 + 4\omega'^2 L^2/R^2} = \frac{\pi V_N^2}{2(A - 1)L}. \tag{31-4}$$

Solving $(A - 1)$ we have

$$A - 1 = \frac{\pi V_N^2}{2PL}. \tag{31-5}$$

Note that $A - 1$ can be very small; the voltage gain is only slightly greater than unity. Substituting Equation (31-5) back into Equation (31-3) gives the power spectrum:

$$S(\omega) = \frac{V_0^2(\omega)}{R} = \frac{4P^2 Q^2 R^2}{\pi^2 V_N^2 \omega_0^2} \frac{1}{1 + (\omega'/\Delta)^2} \tag{31-6}$$

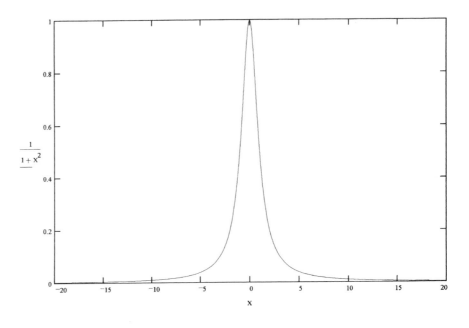

Figure 31-3. Lorentzian lineshape.

where $\Delta\omega'$, the half-width of the line, is given by

$$\Delta = \frac{\pi V_N^2 \omega_0^2}{4PQ^2R}. \tag{31-7}$$

This $1/(1 + x^2)$ lineshape ("Lorenztian") of Equation (31-6) gives the bell-shaped curve shown in Figure 31-3.

SIDEBAND SHAPE

As explained earlier, it is often important that the far-out sideband noise be low compared to the carrier power. If we assume $\omega' \gg \Delta$, then Equations (31-6) and (31-7) reduce to

$$\frac{S(\omega')}{P} = \frac{\omega_0^2 V_N^2}{\omega'^2 4Q^2RP} \tag{31-8}$$

(in reciprocal Hertz), and we see that the far-out sideband noise falls off as $1/\omega'^2$.

PHASE NOISE

So far the oscillator noise has the same character as if it had been produced by amplifying white noise and passing it through a very narrow filter. The

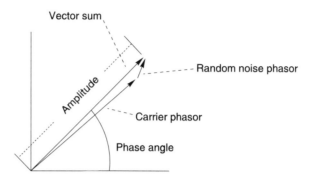

Figure 31-4. Vector diagram showing oscillator noise.

oscillator output voltage is equivalent to a noiseless phasor, that is, the carrier, with a random noise component added to its tip as shown in Figure 31-4.

When the noise component is parallel to the carrier, the vector sum has the correct phase but an altered amplitude. When the noise is perpendicular to the carrier, the result is mostly just a phase error. The signal can be passed through an amplitude limiter to remove the amplitude fluctuations. This leaves only the perpendicular noise components, that is, the phase noise. The parallel and perpendicular noise signals are uncorrelated, so if the amplitude variations are removed, the total noise power is cut in half. The symbol $\mathscr{L}(\omega')$ is often used to represent the phase noise power in a 1 Hz bandwidth at an offset of ω' from the nominal carrier frequency:

$$\mathscr{L}(\omega') = \frac{\omega_0^2 V_N^2}{\omega'^2 8 Q^2 R P} \tag{31-9}$$

(in reciprocal hertz). We can express the noise power $V_N^2(\omega')/R$ as FkT, where F is the noise figure of the amplifier, k is Boltzmann's constant, and T is the reference temperature, 290 K:

$$\mathscr{L}(\omega') = \omega_0^2 \frac{FkT}{\omega'^2 8 Q^2 P} \tag{31-10}$$

(in repricocal hertz). This result shows that best performance is obtained by using a low-noise device, by running it at high power, and by maximizing the loaded Q (high-Q resonator with light loading).

EFFECT OF NONLINEARITY

The analysis given above actually gives a lower limit for the noise. Very close to the carrier, oscillators depart from the predicted Lorenztian lineshape and, as the offset frequency approaches the carrier, the phase noise increases faster than ω'^{-2}. This additional noise is usually attributed to nonlinearity in the active device. We have made the assumption that the device is linear, but some nonlinearity *must* be present in any actual oscillator or its amplitude would either decay or increase without limit. The nonlinearity can translate baseband $1/f$ noise up to radio frequencies.

BIBLIOGRAPHY
1. D. B. Leeson (1966), "A simple model of feedback oscillator noise spectrum," *Proc. IEEE, 54: 329–30.*
2. W. P. Robins (1984), *Phase noise in Signal Sources.* London: Peter Peregrinus (on behalf of the Institution of Electrical Engineers, London).
3. G. D. Vendelin, A. M. Pavio and U. L. Rohde (1990), *Microwave Circuit Design Using Linear and Nonlinear Techniques.* New York: John Wiley.

PROBLEMS

1. Which of the following statements best describes the operation of an oscillator?

(a) Noise is necessary to get an oscillator started. Once running, the oscillation is self-sustaining and noise plays no role.

(b) The output of an oscillator consists of amplified noise; the very narrow lineshape derives from a resonator whose high loaded Q is effectively increased by the positive feedback of the circuit.

2. Discuss the consequences (sound or picture quality) of a noisy oscillator used as a local oscillator in (a) an AM receiver, (b) an FM receiver, and (c) a television receiver.

3. Sketch a design for an oscillator that somehow regulates its own gain in order to stay as linear as possible, that is, just enough to maintain oscillation.

4. Consider the relaxation oscillator shown in Figure 13-2. That circuit uses an op amp integrator together with a pair of voltage comparators. The integrator ramps up or down. When it hits the upper or lower limit it reverses direction. Suppose that noise in the op amps and comparators causes the average ramp time, T, to have a jitter. Suppose this jitter is a random variable with zero mean and an r.m.s. value of δT. Estimate the spectral width of this oscillator. *Hint*: Use the fact that the spectrum is the transform of the autocorrelation function of the output voltage. Estimate the width of the autocorrelation function; its reciprocal gives the width of the spectrum. Think about how you would do an exact calculation of the autocorrelation function.

32 RADIO AND RADAR ASTRONOMY

THE DISCOVERY OF COSMIC NOISE

Radio astronomy was discovered accidentally in the early 1930s by Karl Jansky, a physicist at Bell Telephone Laboratories. Jansky had been assigned to identify the sources of noise encountered in a newly installed transatlantic short-wave radio telephone service. Using a directional receiving antenna on 20.5 MHz, he observed that one component of the noise, a wide-band hiss, had a diurnal variation that reached a maximum intensity on average 4 minutes earlier each day. Jansky knew that the stars advance in just this way, and deduced that the source of the hiss must be outside the solar system. His observations showed that this "cosmic noise" came from the galactic plane and was strongest from the direction of the galactic center (in the constellation Sagittarius).

After Jansky, the second pioneer of radio astronomy was a radio engineer, Grote Reber, who in 1937 built a 30 foot (9 m) parabolic reflector beside his house in Wheaton, Illinois. This was maybe the first modern dish antenna. Reber began his observations using a receiver at 3 GHz (which pushed the high-frequency state of the art) because he assumed that cosmic radio noise was the low-frequency end of the thermally generated radiation spectrum from white-hot stars. The intensity of this radiation would increase as the square of the frequency, so using the highest practical frequency would make detection easier and would also make his antenna more directive. Detecting nothing at 3 GHz, he worked his way down to 160 MHz, where he was able to make contour maps of the cosmic noise intensity. The radiation he and Jansky observed is now known to be *synchrotron radiation*, caused by the centripetal acceleration of fast, that is, nonthermal, electrons spiraling in a magnetic field. By the end of World War II, the sun (an ordinary thermal source under low sunspot conditions) had been detected at microwave frequencies. After the war, a previously predicted spectral line at 21 cm (1,420 MHz) was quickly detected. This famous neutral hydrogen line corresponds to the energy difference between the parallel and antiparallel orientations of the magnetic moment of the nucleus (the proton) with respect to the magnetic moment of the spinning electron. Many radio telescopes have been built in the decades following the war, the largest being the 1,000 foot (305 m) diameter dish built by

Cornell University at Arecibo, Puerto Rico. Discoveries in radio astronomy include some 100 atomic recombination and molecular lines, pulsars, natural masers, and the isotropic 3 K black body cosmic background radiation.

RADIOMETRY

Most of the radio sources found in nature emit wide-band noise; their radiation comes from a great number of individual radiators whose contributions add randomly to produce Gaussian noise. (A histogram of voltage samples from the antenna terminals forms a Gaussian curve centered on zero.) Since such a signal is itself a random process, we can only measure its average properties. The most important of these is the average of the square of the voltage, that is, the power. If we take, say, several n minute averages of the power, these averages will be scattered around the true average. If n is made larger, the averages will be distributed more tightly about the true average. It might not take long to measure the receiver output power to a precision of, say 10 percent or even 1 percent. But that is almost never long enough to measure the power of a radio source to the same precision or, for that matter, even to detect a source. The problem is that the power from the source is masked by other sources of noise including receiver noise, antenna noise, and cosmic background noise. When an astronomer is trying to detect a source in a certain direction, the first step is to average the power received from that direction, the on-source direction. The next step is to measure the power from a nearby, but off-source, direction. Finally, the latter "off" power is subtracted from the former "on" power. For weak sources (most sources), these powers are almost identical and might correspond to a system temperature[*] of, say, 100 K. Yet the astronomer may need to detect a source that raises the system temperature by only 10 mK. This requires that the "on" and the "off" powers both be measured to an accuracy of, say, $3.3/\sqrt{2} = 2.3$ mK for a 3-sigma detection. The fractional accuracy of the "on" and "off" power measurements must therefore be $0.0023/100 = 2.3 \times 10^{-5}$ or about 1 part in 50,000 (see Problem 1).

   If we average N samples of the squared voltage, the relative standard deviation, $\delta P/P$, will be $\sqrt{2/N}$ if the voltage has a Gaussian distribution

---

[*]A *system temperature* of 100 K, for example, means that the equivalent noise power at the receiver input is the same as the noise power from a resistor at 100 K. This equivalent noise power is the sum of the actual noise power (sky noise from the main lobe, ground noise picked up from back lobes, some thermal noise if the antenna is lossy, plus a contribution representing the noise generated in the receiver).

and if all the samples are independent (see Problem 2). A signal from a channel of bandwidth B can furnish $2B$ independent samples evey second. Integrating for a time T we can therefore collect $2BT$ independent samples, and the relative standard deviation of the power measurement will therefore be

$$\frac{\delta P}{P} = \sqrt{\frac{1}{BT}}. \tag{32-1}$$

Usually a radio astronomer expresses power in terms of antenna temperature and would write $\delta T/T_{\text{system}}$ rather than $\delta P/P$. Equation (32-1) is commonly known as the "radiometer equation." Note that sensitivity increases with bandwidth, contrary to many communication situations where increasing the bandwidth just increases the noise and decreases the signal-to-noise ratio. We arrived at Equation (32-1) by considering digital (discrete) signal processing. It is also common to use an analog square law detector followed by an analog low-pass filter (which does the averaging). When the low-pass filter is just a simple RC integrator, an analog signal analysis shows that the radiometer equation can be written as

$$\frac{\delta T}{T_{\text{sys}}} = \sqrt{\frac{1}{2BRC}}. \tag{32-2}$$

SPECTROMETRY

Many sources produce "colored" noise rather than white noise, that is, the flux density from these sources varies with frequency. Instead of just measuring the total power, the signal is divided into adjacent frequency "bins" and the power in each bin is measured. When a spectrum of bandwidth B is divided into n frequency bins, each of bandwidth B/n, the radiometer equation shows that the integration time must be increased by n. It is therefore especially important in radio astronomy, where weak sources may require many hours of integration, to measure the n individual spectra simultaneously rather than sequentially. Simultaneous analysis is done with multichannel radiometers ("multiplex" spectrum analyzers). The simplest multiplex spectrometer is just a bank of n filters, each followed by its own square law detector and averager. Such an instrument is called a filter bank, but might better be called a radiometer bank. Today, most radio spectrometry is done digitally, often using digital *autocorrelators* (discussed in Chapter 33).

INTERFEROMETRY

The classic single-dish radio telescope such as Reber's backyard dish or the Arecibo dish has an angular resolution that is diffraction limited; the angular size of the beam is about λ/D radians, where D is the diameter of the dish. Such a telescope can make maps by scanning the vicinity of a source and doing radiometric averages at each point, but the resolution of the map is limited to λ/D. Better resolution requires a larger antenna. Interferometry uses more than one antenna to form a beam whose width corresponds to the *spacing* between the antennas rather than their diameters. The simplest interferometer has just two elements. The beam formed by the two antennas together has a series of narrow lobes in one dimension. If a point source moves through the beam the radiometer will trace out the lobes, just as a point source moving across a single dish antenna traces out the beam pattern. (Of course the fine lobe structure of the interferometer is multiplied by the broader beam pattern of the individual antennas.) The multi-lobed pattern of a two-element interferometer is unsuitable for the intensity mapping described above, but it does serve to set an upper limit on the angular size of a source.

IMAGING INTERFEROMETRY

By using data from multiple antennas it is possible to synthesize a beam that is small in both dimensions, that is, a beam that would correspond to a filled-aperture antenna whose diameter is the size of the interferometer array. The VLA (Very Large Array) at Socorro, New Mexico is an example of such a system. Signals from twenty-seven antennas are transmitted up to 21 km through low-loss circular waveguide to the central processing laboratory. The VLBA (Very Long Baseline Array) has ten antennas with spacings as large as the distance from Hawaii to St Croix in the Virgin Islands. Its signals are recorded on tape at each station, and the tapes are sent to New Mexico for after-the-fact combining and processing. Imaging interferometers work essentially as follows. Suppose we have an array of antennas such as shown in Figure 32-1. Assume the antennas are all pointed in the same direction, say up. If the voltages from the antennas are all added together in phase, squared, and averaged, the result is a narrow beam pointing straight up. This process can be written as follows:

$$P_0 = \langle |\sum_n V_n|^2 \rangle = \sum_{n,m} \langle V_n V_m^* \rangle. \tag{32-3}$$

Now suppose we add the voltages, multiplied by a progressive phase shift, to form a beam tilted slightly off the vertical. If we do this addition, followed by squaring and averaging, we get the kth beam:

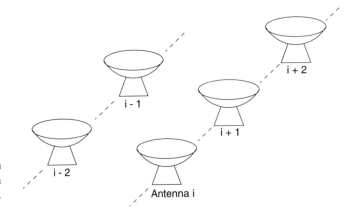

Figure 32-1. An interferometric antenna array.

$$P_k = \left\langle \sum_{n,m} V_n\, e^{(j\phi_{n,k})} V^*\, e^{(-j\phi_{m,k})} \right\rangle = \sum_{n,m} e^{[j(\phi_{n,k} - \phi_{m,k})]} \langle V_n V_m^* \rangle. \qquad (32\text{-}4)$$

This is the general case; for the straight-up beam $\phi_{n,k}$ and $\phi_{m,k}$ are both zero. The important thing to notice is that we can multiply the signals from every possible antenna pair and average the products first, independent of what beam we wish to form. After averaging these pairs we can *then* form each beam by weighting the averaged products with the appropriate phase factors and performing the double sums. Note also that the averaged products do not have to be measured simultaneously; we could even use one fixed antenna and one mobile antenna and measure the average product for one baseline after another. The VLA has twenty-seven antennas, so it can measure $(27 \times 26)/2$ baselines simultaneously. But, as the earth turns, these baselines become different baselines so, over time, it is possible to collect averaged products from a huge number of baselines. This set of baselines is virtually the same as all the baselines one could form between pairs of points on a dish with a radius of 21 km, and the final synthesized beam can be as sharp and clean as if it had come from a filled dish. An array can thus have the resolution of an enormous antenna even if it does not have the sensitivity (collecting area).

RADAR ASTRONOMY

The first attempts to detect radar echoes from the Moon were made at the Naval Research Laboratory in 1924. Further attempts were made as new search radars were developed. Ionized meteor trails produced echoes on World War II radars, but echoes from the Moon were not observed until after the war. Early in 1946 two efforts were successful. In the USA, an

Army Signal Corps group used modified World War II radar equipment. A group in Hungary [1] used a modest transmitter together with an interesting method of signal integration. (When appropriate signal processing can be done, detection depends only on the total *energy* received.) The Hungarians used electrolytic cells as stable analog integrators. By the end of their experiment the on-target cell had built up significant quantities of gas compared to the off-target cells. Many large radars were built in the late 1950s, and radar echo detections from the Sun, Venus, Mercury and Mars were made in the 1960s. Radar was used to measure the rotation rates of Venus and Mercury. The rings of Saturn were detected in the 1970s, as were the large moons of Jupiter, many asteroids, and several comets.

THE MOON

The distance of the Moon from the Earth, R, is 3.8×10^5 km, and it has a *radar cross-section*, σ, of about 6.6×10^5 km². This means that the power density of the echo, S_R (in watts per square meter), received back on the radar antenna will be given by $\sigma/(4\pi R^2)$ times the power density incident on the target, S_{inc}. (Note that this defines the *radar cross-section* as the area of an equivalent target that isotropically scatters the intercepted incident power.) The incident power density is just $S_{inc} = P_T G/4\pi R^2$, where P_T is the transmitter power and G is the antenna gain. The power into the receiver, P_R, is equal to $S_R A_{eff}$, where A_{eff} is the effective collecting area of the antenna. We have seen earlier that G is equal to $4\pi A_{eff}/\lambda^2$. The radar cross-section of the Moon is about 7% of its geometric cross-section, that is, it is 7% as reflective as a metal sphere of the same size. An echo from the Moon will therefore produce a power at the receiver of

$$P_R = P_T \frac{(4\pi A_{eff}/\lambda^2)}{4\pi R^2} \frac{\sigma}{4\pi R^2} = A_{eff} \frac{P_T A_{eff}^2 \sigma}{4\pi R^4 \lambda^2}. \tag{32-5}$$

Suppose we have a radar with a 1 kW transmitter and a modest 5 m diameter dish antenna with an aperture efficiency of 50%. The effective area of this antenna is therefore half of its geometric cross-section or $A_{eff} = 0.5 \times \pi(5/2)^2 = 9.8$ m². It is an experimental fact that σ for a planet is approximately independent of wavelength, so we see from Equation (32-5) that the received power is inversely proportional to λ^2 (because of the fundamental antenna relation between gain and effective area). Suppose that the wavelength is 30 cm, that is, the frequency is 1 GHz. With this radar system, the power received from the Moon will be

$$P_R = \frac{1000 \text{ W } (9.8 \text{ m}^2)^2 6.6 \times 10^{11} \text{ m}^2}{4\pi(3.8 \times 10^8 \text{ m})^4 (0.30 \text{ m})^2} = 2.7 \times 10^{-18} \text{W}. \tag{32-6}$$

Now whether this amount of power is easy to detect or not depends on the noise it competes with. Let us assume that the noise from the antenna (sky noise from background cosmic radio sources plus some "spill-over" noise from the surrounding ground) has an equivalent noise temperature $T_{ant} = 50\,\text{K}$. Let us assume that our receiver has an equivalent noise temperature $T_{rcvr} = 35\,\text{K}$. The total equivalent input temperature is therefore given by $T_{sys} = 85\,\text{K}$, and the noise power will be $kT_{sys}B$. If we make the bandwidth, B, very small we decrease the noise power, and hence detection becomes easier. But if we make B very small we will need a very accurate prediction of the Doppler shift caused by the relative motion of the Earth and the Moon in order to tune the return echo into the pass band of the filter. Finally, the rotation of the Moon causes a Doppler broadening of the return echo; if we make B less than 25 Hz we will begin to exclude some of the broadened signal. Let us compromise and use a bandwidth of 100 Hz. The signal-to-noise ratio (SNR) at the receiver output will then be given by

$$SNR = \frac{P_R}{kT_{sys}B} = \frac{2.7 \times 10^{-18}\text{W}}{1.38 \times 10^{-23}\text{W/Hz/K} \times 85\text{K} \times 100\,\text{Hz}} = 23.0.$$

(32-7)

This SNR of 23 is large enough that the signal would be visible on an oscilloscope connected at intermediate frequency (IF) voltage output; no signal averaging would be needed. The modest radar system assumed here could be assembled for a few thousand U.S. dollars.

VENUS

Venus is not nearly so easy to detect, being some 280 times more distant than the Moon. Its radar cross-section is about twenty times that of the Moon, so with the radar system described above, the SNR of the return echo would be lower by a factor $280^4/20$ or 307 million! This requires a *much* larger radar. The Arecibo radar, however, has an effective antenna diameter of 200 m and a power of 1 MW at 2.380 GHz. It would out-perform our Moon example radar by a factor of $(10^6/10^3)(2.38/1)^2(200^2/5^2)^2$ or 1.4×10^{10}. Thus, with the same assumed bandwidth and system temperature, the Venus echo from the Arecibo radar would have an SNR of 23 $(1.4 \times 10^{10})/(307 \times 10^6) = 1,090$. This is overkill for a simple detection, but is needed for high-resolution mapping.

DELAY-DOPPLER MAPPING

In most situations, planetary targets have angular sizes much smaller than the radar beam. Nevertheless, images of photographic quality can be made by a technique known as delay-Doppler mapping. This method uses short pulses (or equivalent wide-band signals) to obtain adequate range resolution. The relative motion between the radar and the target provides resolution in the transverse direction via the Doppler effect. This was the method used to get the first surface images of Venus, whose cloud cover kept its features hidden to optical telescopes. The technique is essentially the same as *side-looking* or *synthetic aperture radar* by which photographic-like images of ground targets are made from an airborne radar. Figure 32-2 shows how delay-Doppler mapping works. This is a view of the planet as seen from the radar. When a short pulse is transmitted, the first echo to return comes from the front cap or sub-earth point. At subsequent times the echo signal corresponds to loci of equal range that are rings on the planetary surface, centered on the sub-earth point. Because the planet is rotating, surface points on the right-hand side are moving away from the radar, and their echoes are Doppler shifted to lower frequencies. Likewise, points on the left-hand side are up-shifted. The magnitude of the Doppler shift is proportional to the cross-range distance, that is, the distance in the direction perpendicular to both the axis of rotation of the planet and the line between the radar and the planet. Loci of constant Doppler shift are rings parallel to the rotation axis of the planet and to the line from the planet to the radar. The two shaded surface elements have the same range and the same Doppler shift. This fundamental ambiguity can be removed, at least for Venus maps made at Arecibo, by tilting the radar beam slightly to illuminate only the northern or only the southern hemisphere of Venus. Obviously this technique requires the narrow beam provided by a very large antenna. Data taking is as follows. A series of IF voltage samples is read and stored separately as each return

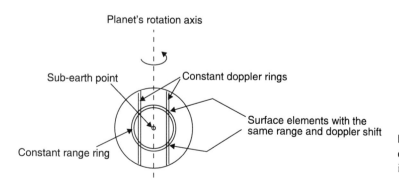

Figure 32-2. Geometry for delay-Doppler radar imaging.

pulse arrives. After many pulses have been received, we have a time series for each range ring. Each of these sequences is Fourier analyzed to get the Doppler spectrum. The magnitude of a given Doppler component corresponds to the reflectivity of the surface element at the intersection of the range ring and the Doppler ring.

OVERSPREADING

The basic delay-Doppler method will not work when the planet is too large and/or spins too fast. This is because the Nyquist sampling therorem requires that the (complex) sampling rate be at least equal to the bandwidth of the signal. Sampling at a lower rate will cause the high-frequency components to "fold" and appear as lower-frequency components. The bandwidth (twice the highest Doppler shift) is given by $2\Delta f = 3326$ $2f_0 2V_{max}/c = 4f_0 r\Omega/c$, where r is the radius of the planet, Ω is its angular velocity, and f_0 is the frequency of the radar. But if the time between samples, $1/\Delta f$, is less than the "depth" of the planet (the round-trip time for light to travel a distance r) then there will be more than one illuminated range ring on the planet at any time, and the range data will be ambiguous. Putting this together gives an upper limit for the radar frequency.

$$f_0 \leq \frac{c^2}{8r^2\Omega}.$$

(32-8)

This upper frequency is 1.55 GHz for Mercury and 1.0 GHz for Venus. These are quite practical frequencies for conventional microwave radar technology. For Mars, however, the upper frequency is only 13.7 MHz (caused by a high rotation rate). Equation (32-5) shows that such a low frequency would require an extremely large antenna. And even if terrestrial interference were avoidable, the sky noise at 13 MHz is orders of magnitude higher than it is at microwave frequencies.

BIBLIOGRAPHY

1. Z. Bay (1946), "Reflections of microwaves from the moon," *Hung. Acta Phys.* 1: 1–22.
2. J. D. Kraus (1986), *Radio Astronomy*. 2nd Edition, Powell, OH: Cygnus-Quasar Books.
3. P. Goldsmith (ed.) (1988), *Instrumentation and Techniques for Radio Astronomy*. New York: IEEE Press. (This book is prefaced with a 12-page historical overview.)
4. S. J. Ostro (1993), "Planetary radar astronomy," *Reviews of Modern Physics*, 65: No. 4.

1. Use the radiometer equation to find how much time is needed to make a 3-sigma detection of a 10 mK radio source if the system temperature of the radio telescope is 100 K. Assume the predetection bandwidth, B, is 10 MHz. A "3-sigma detection" requires that the 0.01 K contribution from the source be $3\sqrt{2}$ times the fluctuation predicted by the radiometer equation. (The factor $\sqrt{2}$ takes account of the increased fluctuation when the "off" power is subtracted from the "on" power, assuming an equal time T is spent measuring each.)

2. Verify that if a noise power estimator is defined as the average of N independent samples of the squared noise voltage, the standard deviation of this estimator will be $\sigma\sqrt{2/N}$. Assume that the probability distribution of the noise voltage is Gaussian with zero mean and variance σ^2. *Hint*: For a zero-mean Gaussian distribution, the expected value of V^2 is σ^2, and the expected value of V^4 is $3\sigma^4$.

3. Pulsars (rotating neutron stars) are radio sources that turn on and off with very regular periods ranging from about 2 ms to about 2 s. Given a radio telescope at your disposal, how would you go about searching for pulsars, that is, what kind of data-processing scheme would you use to find these periodic sources?

33

RADIO SPECTROMETRY

Spectrometry or spectral analysis is the statistical characterization of random (*stochastic*) signals such as the intermediate frequency (IF) voltage in a radio astronomy receiver. Most often the signal is Gaussian; if a fine-grained histogram of samples of the amplitude of the signal is scaled to make the area below the curve equal to unity, the average curve will be the Gaussian probability density function, $f(V) = (2\pi\sigma^2)^{-1/2}\exp(-V^2/2\sigma^2)$. You can verify (see Problem 1) that σ is the rms value of V, that is σ^2, is the power. But the total power does not completely characterize the signal. A complete description is contained in the *power spectral density function*, $S(\omega)$, the distribution of power versus frequency. A set of bandpass filters and power meters, that is, a set of radiometers, serves to measure points on the spectral density function (usually simply called the power spectrum). The signal is equally well characterized by specifying its *autocorrelation function*, $R(\tau)$, a function of time delay. The value of the autocorrelation function for a given time delay τ is defined as the average value of the "lagged product" $V(t)\,V(t+\tau)$, that is, $R(\tau) = \langle V(t)\,V(t+\tau)\rangle$. For a signal whose characteristics are unchanging (a *stationary process*), $R(\tau) = R(-\tau)$. The normalized correlation function, $\rho(\tau)$, is defined as $R(\tau)/R(0)$ or $R(\tau)/\sigma^2$. We will see that the Fourier transform of the autocorrelation function is the power spectral density so, directly or indirectly, we are describing the signal in terms of sine wave basis functions, that is, finding the magnitudes (but here not the phases) of a set of sine waves whose superposition has the same power spectrum as the original signal. Why are sine waves preferred over other sets of basis functions? Often the process under study is a spectral line that shows up clearly in a few adjacent Fourier coefficients (frequency bins) or a Doppler shift that just displaces the spectrum. The frequency spectrum is clearly a natural way to represent such signals. Of course we deal with sine waves all the time as the characteristic functions of linear systems; a sinusoidal signal of a given frequency remains a sinusoid with the same frequency after passing through any arbitrary chain or network of linear elements such as amplifiers, filters, and transmission lines. A variety of equivalent instruments have been developed for spectrometry. In approximate historical order these include analog filter banks, autocorrelators, chirp-z spectrometers, Fourier transform spectrometers, and acousto-optical spectrometers.

FILTERS AND FILTER BANKS

We have already discussed the analog filter bank (radiometer bank). Usually the signal is converted down to a convenient IF, for example tens of megahertz, and the filters are the various *LC* bandpass filters discussed earlier or crystal filters or even digital filters. If the spectrum under study is very wide, the filters must be in the UHF or microwave range. Here the *LC* or crystal resonators used in the radio range are replaced by resonant cavities, coupled transmission line elements, or other microwave resonators. A traditional disadvantage of an analog filter bank spectrometer has been that it is not "elastic," that is, the filters have a fixed width; if narrower filters are needed, another filter bank must be built. With high-speed digital-to-analog conversion and fast memory, this restriction can be lifted by using an interesting spectral expansion technique. Sequences of data are read into the memory at whatever slower rate is needed and then played back at a higher rate into a single wide-band filter bank.

AUTOCORRELATION SPECTROMETRY

Autocorrelation spectrometers estimate the power spectrum by taking pairs of samples with a given time separation, multiplying them, and averaging these "lagged products." (In direct Fourier spectral analysis the samples are multipled by sine waves but the resulting coefficients are squared before averaging. In both cases, powers, rather than voltages, are averaged.) When the time separation is zero, the average is just the total power. But if we measure other lagged products as well (where the time separation is not zero), we can indirectly find the power spectrum because the autocorrelation function (ACF) and the power spectrum are a Fourier transform pair:

$$S(\omega) = \int_{-\infty}^{\infty} R(\tau)e^{-j\omega\tau} \, d\tau \tag{33-1a}$$

$$R(\tau) = \frac{1}{2\pi} \int_{-\infty}^{\infty} S(\omega)e^{j\omega\tau} \, d\omega. \tag{33-1b}$$

This relation, the Wiener–Khinchin theorem, is often used to *define* the power spectrum. To see that $S(\omega)$ as defined by Equation (33-1a) agrees with our intuitive ideas of the power spectrum, consider the following (nonrigorous) argument. If the filter in the standard radiometer of Figure 33-1 is made very narrow about its center frequency, ω_0, the output voltage V_{out} will be a good measure of the value of the power spectrum of the input signal at ω_0. And if the filter shape becomes a delta function, the radiometer output will be exactly $S(\omega_0)$. Let us calculate the output of this radiometer and show that it is indeed equal to the ω_0 point on the Fourier

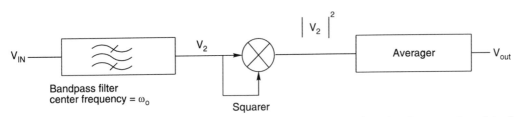

Figure 33-1. Narrow-band radiometer to measure one spectral point. (The Filter Response is a delta function at ω_o.)

transform of $R(\tau)$. The voltage at the filter output is related to the input voltage through the impulse response of the filter $H(\tau)$,

$$V_2(t) = \int_0^\infty V(t - \tau)H(\tau) \, d\tau. \tag{33-2}$$

The output of the squarer is therefore given by

$$|V_2(t)|^2 = \int_0^\infty \int_0^\infty V(t - \tau)V(t - \tau')H(\tau)H(\tau') \, d\tau \, d\tau' \tag{33-3}$$

and the output of the averager is

$$\langle |V_2(t)|^2 \rangle = \int_0^\infty \int_0^\infty \langle V(t - \tau)V(t - \tau') \rangle H(\tau)H(\tau') \, d\tau \, d\tau'$$
$$= \int_0^\infty \int_0^\infty R(\tau' - \tau)H(\tau)H(\tau') \, d\tau \, d\tau'. \tag{33-4}$$

Making the change of variable $\tau'' = \tau - \tau'$ we have

$$\langle |V_2(t)|^2 \rangle = = \int_0^\infty R(\tau'')\left(\int_0^\infty H(\tau)H(\tau'' - \tau) \, d\tau \right) d\tau''. \tag{33-5}$$

Invoking the convolution theorem, the inner integral is equal to the Fourier transform of $|H|^2$, that is, the transform of the power response function. For our filter, with its delta function response at $\pm\omega_0$, this transform is just $\exp(j\omega_0\tau'')+\exp(-j\omega_0\tau'')$, and we have the desired result,

$$\langle |V_2(t)|^2 \rangle = \int_0^\infty R(\tau'')(e^{j\omega_0\tau''} + e^{-j\omega_0\tau''}) \, d\tau'' = \int_{-\infty}^\infty R(\tau)e^{j\omega_0\tau} \, d\tau. \tag{33-6}$$

HARDWARE AUTOCORRELATORS

A hardware autocorrelator is a special-purpose parallel signal processor that calculates the averaged ACF (usually in real time). A single Fourier transform operation in a computer then turns the ACF into the power spectrum with the same number of frequency bins (points on the spectrum)

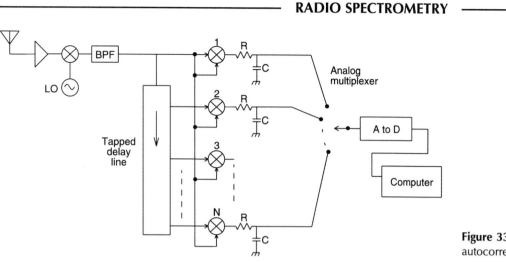

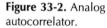

Figure 33-2. Analog autocorrelator.

as the number of points on the ACF. An autocorrelator is easy to build; the hardware has an extremely simple and expandable architecture. Figure 33-2 shows an analog version of the autocorrelator.

Figure 33-3 shows a digital autocorrelator. The analog delay line is replaced with a digital shift register. The analog multipliers and averagers are replaced by digital multipliers and accumulators.

Figure 33-3. Digital autocorrelator.

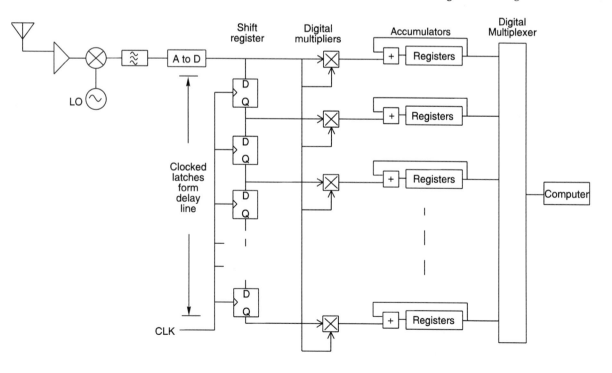

ONE-BIT AUTOCORRELATION

One-bit autocorrelation is a technique that greatly reduces the complexity of the digital circuitry. In the 1-bit technique, the input analog voltage is fed to a comparator whose output (1 bit) indicates the sign, that is, the polarity, of the voltage. The continuous range of analog voltages is compressed down to just two values: $+1$ and -1. (In the digital hardware, a digital "1" indicates a value of $+1$ and a digital "0" indicates a value of -1.) The delay line shift register of the autocorrelator needs a width of only 1 bit and the 1 bit $\times$ 1-bit multipliers can be exclusive NOR gates (the output of an exclusive NOR gate is $+1$ when the input signals are the same and -1 when they are different). The integrators that follow the multipliers are simple counters. The surprising thing about the 1-bit technique is that, while the output is indeed a distorted version of the actual correlation function, the distortion can be entirely corrected. Simply because the input signal has Gaussian amplitude statistics, the output of the 1-bit correlator turns out to be

$$\rho_{1-\text{bit}} = \frac{2}{\pi}\sin^{-1}\rho \tag{33-7}$$

where ρ is the normalized autocorrelation function, R/σ^2. This functional relation was derived during World War II by J. H. van Vleck during a study of the spectrum of clipped noise. (Jammers can use efficient final amplifiers if they do not need to be linear.)

In practice, the averaged values from the 1-bit autocorrelator are just inverted to get the true correlation function:

$$\rho = \sin\left(\frac{\pi}{2}\rho_{1-\text{bit}}\right). \tag{33-8}$$

A straightforward proof of the van Vleck relation (see Problem 4) proceeds from the Gaussian bi-variate probability density,

$$f(x_1, x_2) = \frac{1}{2\pi\sigma_1\sigma_2(1-\rho^2)^{1/2}}\exp\left[\frac{-1}{2(1-\rho^2)}\left(\frac{x_1^2}{\sigma_1^2} - 2\rho\frac{x_1 x_2}{\sigma_1\sigma_2} + \frac{x_2^2}{\sigma_2^2}\right)\right]. \tag{33-9}$$

Here x_1 and x_2 are the voltages $V(t)$ and $V(t+\tau)$, so $\sigma_1 = \sigma_2$. One gives up some radiometric sensitivity for the simplification of 1-bit processing; the integration time has to be increased by a factor of $\pi^2/4$ compared to a correlator using multibit quantization.

Several generations of digital correlators have been used in radio astronomy. A recently developed CMOS chip using the architecture of Figure 33-3 contains 1024 stages, each with a 32-bit accumulator and a 32-bit output buffer register. The chip runs at a clock rate of 130 MHz. A variety of serial and parallel multiplexing techniques are used to combine these chips to obtain many thousands of stages and overall input data rates of several hundred megahertz. Besides the all-analog and all-digital designs, there have also been numerous designs for hybrid (combination analog and digital) correlators.

FOURIER TRANSFORM SPECTROMETRY

Spectrometry can be done directly via Fourier transformation of the digitized IF signal. This processing became practical with the discovery (actually rediscovery) of the fast Fourier transform (FFT) and with advances in high-speed digital circuitry. The voltage is digitized and successive data strings are individually transformed. The squared magnitudes of the transform coefficients from successive transforms are averaged. Ultimately, Fourier transform spectrometry should displace autocorrelation spectrometry, simply because processing a sequence of n data points requires of the order of $n \log(n)$ operations for an FFT versus n^2 operations for autocorrelation. The simple architecture of the correlator, especially with the simplification of 1-bit arithmetic (and other not-quite-so-coarse digitization schemes) keeps this method competitive. Many "FFT boxes" are available as low-frequency laboratory spectrum analyzers, usually used for vibration analysis of machinery. With the very narrow frequency bins used in low-frequency spectrometry, the analysis time becomes long because the signal must be sampled over a time interval at least as long as the inverse of the resolution. In the case of random signals, even more time is needed to make a good average, that is, to satisfy the radiometer equation. The FFT analyzer, being a multiplex analyzer, at least processes all the frequency bins simultaneously.

ACOUSTO-OPTICAL SPECTROMETRY

The acousto-optical spectrometer (AOS) uses an optical crystal that becomes a diffraction grating by virtue of mechanical waves (sound waves at radio frequencies) propagating through it. These waves produce corrugations in the refractive index along the length of the crystal. The waves are launched in this *Bragg cell* by an electrical-to-acoustic transducer (piezoelectric crystal) driven by the IF signal to be analyzed. Figure 33-4 shows how the crystal is illuminated by a laser and how a multielement charged-coupled device (CCD) array detects (and averages) the diffraction pattern. The linear CCD array, like those used in supermarket checkout scanners, accumulates charge along its length at a rate given by the inci-

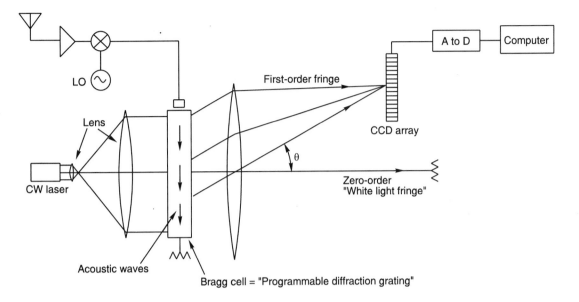

Figure 33-4. Acousto-optical spectrometer.

dent light intensity. For read-out, the charge packets are clocked down the length of the CCD in a bucket brigade fashion. The voltage at the end of the array, which is proportional to the charge at that position, is digitized and made available to the data-taking computer.

A figure of merit for any spectrometer is its number of frequency bins, that is, its analysis range divided by its resolution. For the AOS, the effective number of channels is given by the time–bandwidth product of the Bragg cell. To see this we note that the angular size (in radians) of the light incident on the CCD is given by λ_{laser}/L_{cell}, the ratio of the laser wavelength to the length of the cell (just as the beam size for an aperture antenna with diameter D is given by λ/D). The first-order diffraction condition (so that all the rays arrive in phase at the CCD) is given by $\lambda_{grating} \sin \theta = \lambda_{laser}$, where $\lambda_{grating}$ is the corrugation wavelength in the cell. Now, $\lambda_{grating}$ can be written as v_{cell}/f, where v_{cell} is the sound velocity and f is the input RF frequency. Approximating $\sin\theta$ by θ, we find the total angular range over the total frequency range given by $\Delta\theta/\Delta f = \lambda_{laser}/v_{cell}$. The number of channels is then given by the range over the resolution

$$N = \frac{\lambda_{laser}}{v_{cell}\lambda_{laser}} \frac{\Delta f \, L_{cell}}{} = \frac{L_{cell}}{v_{cell}} \Delta f = T \, \Delta f \qquad (33\text{-}10)$$

where T is the propagation time through the length of the cell. The quantity $T \, \Delta f$ is known as the *time–bandwidth product* of the cell. Materials used for Bragg cells include quartz, lithium niobate, and glass. A spectrometer for radio astronomy might use cells with a 200 MHz bandwidth and a time–bandwidth product of 1,000.

CHIRP-Z SPECTROMETRY

The chirp-z Spectrometer is shown in Figure 33-5. Also known as the microscan or compressive receiver, it uses a dispersive filter and a swept local oscillator (LO). While it appears to be a swept spectrum analyzer, it is actually a multiplex analyzer. The filter has a group delay characteristic that is linear with respect to frequency. In Figure 33-5, the delay is greatest for frequencies at the low end of the band of the filter. The frequency of the LO is given a linear sweep or *chirp*, repeated in a sawtooth fashion with period T. The ramp rate of the LO (megahertz per second) is made to be the reciprocal of the group delay slope of the filter (seconds per megahertz).

Consider a continuous wave (CW) input signal. Since the LO is swept, the mixer output will also sweep in frequency. In the case of a high-side mixer (LO frequency above the input signal frequency), the mixer output sweeps up in frequency at the same rate as the LO sweep. As the LO sweeps upwards, the first signals entering the filter are at the low-frequency end of the filter. These signals, as they travel through the filter, are delayed the most. Frequencies entering the filter farther up the sweep are delayed less. But all frequencies arrive simultaneously at the filter output, producing a single sharp output pulse. If the input signal consists of two CW tones, there will be two output pulses per sweep. The position of the pulses in time will indicate the frequencies of the CW input signals. Of course this

Figure 33-5. Chirp-z spectrometer.

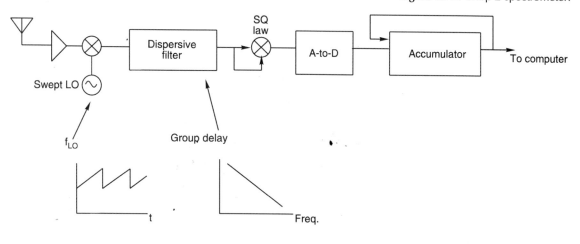

linear system can handle any combination, that is, spectrum, of input signals. The output data rate from the squarer is equal to the input data rate, so additional circuitry (the digital accumulator in Figure 33-5) is required for signal averaging.

The principle application of chirp-z spectrometers has been in electronic warfare – as surveillance receivers to sense radar or communication signals over a wide spectrum and with a high probability of interception. To find the effective number of channels in this spectrometer we can first consider the output pulse produced by a single CW input signal. This output pulse is produced by a coherent superposition of frequencies over the entire bandwidth B, of the filter. The width of this pulse is therefore given by $1/B$. Dividing the total output time T by the pulse width, the effective number of channels is given by TB, the time–bandwidth product, just as with the Bragg cell spectrometer.

Dispersive filters used for chirp-z spectrometry are usually surface acoustic wave devices. Piezoelectric transducers convert the input signal into mechanical surface waves and then back again to an electrical output signal. In one design, diffraction gratings on the surface of the crystal are arranged to make the input-to-output path length longer for low-frequency waves than for higher-frequency waves. Surface wave dispersive filters can have bandwidths of hundreds of megahertz and time-bandwidth products of several hundred. Dispersive filters are also made from CCDs. Generically these filters are all-pass networks, and the first dispersive filters were cascades of many second-order all-pass sections. (The number of sections required is equal to the time–bandwidth product.). Note: the filters are not just all-pass networks; they are usually given some amplitude "taper," that is, the amplitude response is made to roll off at the band edges, in order to eliminate the "side lobes" of the instrumental function that result from uniform amplitude response. The same is done with the AOS; the transmission of the Bragg cell is reduced near the ends.

RADAR PULSE COMPRESSION

A similar application of dispersive filters is in radar *pulse compression*. Good range resolution requires the use of short pulses. But, for coherent radar systems, sensitivity is a function of the total energy radiated. If the pulse is shortened to improve range resolution, its power must be proportionally increased to maintain sensitivity, that is, the average power of the transmitter must be left unchanged. Eventually peak power is limited in any transmitter by voltage breakdown or current limiting. In pulse compression radar the pulse is chirped, that is, the frequency is swept, just as the LO of the chirp-z spectrum analyzer is swept. The pulse can be long enough for the transmitter to achieve its maximum average power. An echo from a target returns with the same chirp. If the receiver IF amplifier includes a dispersive filter, the swept frequency echo will exit the filter as a short pulse, just as if the transmitter had produced a short pulse. Variations on this pulse compression scheme use other waveforms, such as pseudo-random phase shifting, and the corresponding matched filters.

BIBLIOGRAPHY

1. R. B. Blackman and J. W. Tukey (1959), *The Measurement of Power Spectra.* New York: Dover Publications.
2. D. B. Childers (ed.) (1978), *Modern Spectrum Analysis.* New York: IEEE Press.
3. M. A. Jack, P. M. Grant and J. H. Collins (1980), "The theory, design, and applications of surface acoustic wave Fourier-transform processors," *Proc. IEEE* 68: 450–68.
4. S. B. Kesler (ed.) (1986), *Modern Spectrum Analysis, II.* New York: IEEE Press.
5. J. B. Thomas (1969), *An Introduction to Statistical Communication Theory.* New York: John Wiley.
6. A. R. Thompson, A. Richard, J. M. Morgan and G. W. Swenson, Jr (1991), *Interferometry and Synthesis in Radio Astronomy.* Malabar, FL: Krieger.

PROBLEMS

1. For the Gaussian probability density $f(V) = (2\pi\sigma^2)^{-1/2} \exp(-V^2/2\sigma^2)$ verify that

$$\int_{-\infty}^{\infty} f(V)\, dV = 1 \text{ and } \int_{-\infty}^{\infty} f(V)V^2\, dV = \sigma^2.$$

2. (a) The noise-like signals observed in radio astronomy can be modeled as a comb of delta functions in frequency space – a picket fence of sine waves. Use your computer to generate 50 sine waves spanning the angular frequency range 1 to 1.10. Let these frequencies be somewhat random, for example, $\omega_n = 2\pi[1 + 0.002n + 0.05\,\text{rnd}(1)]$, where rnd(1) is a random variable in the range 0 to 1. Give each sine wave a random starting phase, $\phi_n = 10\,\text{rnd}(1)$. Plot the sum of the sine waves as a function of time for $t = 0$ to $t = 20$ using a time interval of 0.1. This narrow-band spectrum should look quite sinusoidal but with a slowly varying phase and amplitude.

(b) Repeat this exercise but with the angular frequency range extending from 1 to 2, that is, $\omega_n = 2\pi[1 + 0.05n + 0.05\,\text{rnd}(1)]$. This wide-band spectrum should look like random noise.

(c) In either case, Problem 2(a) or (b), the resulting voltage (sum) behaves like a random variable with Gaussian statistics. Plot a histogram of voltage samples to verify this.

3. Suppose a random signal is produced by applying an ideal low-pass filter to white noise. The resulting power spectrum is flat from dc to the filter cutoff frequency ω_c. Show that the normalized ACF is given by $\rho(\tau) = \sin(\omega_c\tau)/(\omega_c\tau)$.

4. Derive the van Vleck relation (Equation (34-4)) that gives the 1-bit autocorrelation function in terms of the (normalized) ACF. Use the bivariate Gaussian probability density function (Equation (33-6), where x_1 and x_2 represent the voltages $V(t)$ and $V(t+\tau)$ and $\sigma_1 = \sigma_2$). Since the product $\text{sign}(x_1)\,\text{sign}(x_2)$ is equal to +1 in the first and third quadrants and -1 in the second and fourth quadrants, the expectation of $\text{sign}(x_1)\,\text{sign}(x_2)$ is the integral of $f(x_1, x_2)$ over x_1 and x_2, weighted by +1 or -1 according to the quadrant. *Hint*: Change to polar coordinates and keep a table of integrals handy.

5. If the dispersive filter used in the chirp-z spectrometer or pulse compression radar is given an impulse (delta function in time), what sort of output waveform does it produce?

LABORATORY TEST EQUIPMENT 34

POWER MEASUREMENTS

As pointed out in Chapter 27, power measurements are best made with a square law detector. (If the waveform is unknown or noisy, a square law detector *must* be used.) Powers down to fractions of a microwatt can be measured directly using a diode detector. The dynamic range of the diode square law detector is limited because the voltage must be kept low enough that the fourth and higher even power terms in I versus V will be negligible. Thermistor (temperature-sensitive resistor) power meters are not as sensitive, but they do have more dynamic range. These instruments use a small "bead"-type thermistor. The power to be measured heats the thermistor, which is in a bridge circuit with a second thermistor. A feedback circuit supplies power to heat the second (reference) thermistor. When the bridge is balanced the two thermistors come to equal resistance values and the heater power is equal to the input RF power. The heater power, which is proportional to the heater current if the heater voltage is maintained constant, is displayed. Flow calorimetry can be convenient in high-power measurements. This method uses a liquid-cooled dummy load. The fluid temperature is measured on both sides of the load to get the temperature difference (a pair of thermocouples makes this easy). The flow rate must also be measured. Note that all the power measurements discussed above are "true r.m.s." measurements; they do not require that the unknown voltage be sinusoidal.

VOLTAGE MEASUREMENTS

An RF voltmeter that displays phase as well as amplitude is called a *vector voltmeter*. Since phase is relative, the instrument has an auxiliary "reference" input. A simplified circuit is shown in Figure 34-1. The unknown signal is fed to a diode peak detector, to measure the amplitude. The reference signal and the unknown signal are both fed to limiting amplifiers to provide a constant signal level to a multiplier (mixer), used as the phase detector. Without the limiters, the phase reading would depend on the amplitude as well as the phase.

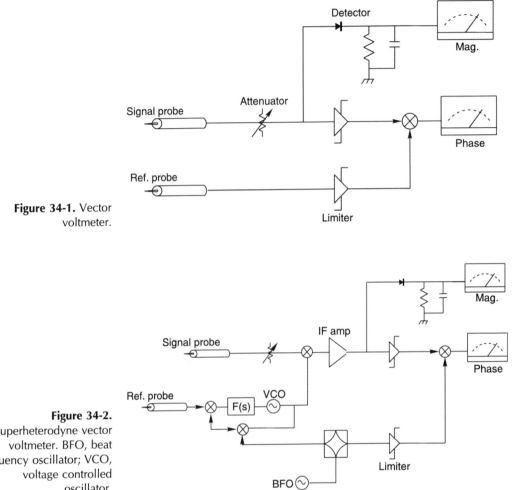

Figure 34-1. Vector voltmeter.

Figure 34-2. Superheterodyne vector voltmeter. BFO, beat frequency oscillator; VCO, voltage controlled oscillator.

To make a more sensitive vector voltmeter, the signals are mixed down to an intermediate frequency (IF) and then amplified. A phase lock loop circuit controls the local oscillator (LO) so that the proper IF frequency is produced. This superheterodyne instrument is shown in Figure 34-2.

IMPEDANCE MEASUREMENTS

At low frequencies (where lumped components are practical) impedance can be measured directly with a bridge such as that shown in Figure 34-3. If the device under test (DUT) is purely resistive, the variable capacitor, C, will be set to the same value as the fixed capacitor C_1. It is convenient to use a variable capacitor with a midpoint capacitance equal to C_1. The

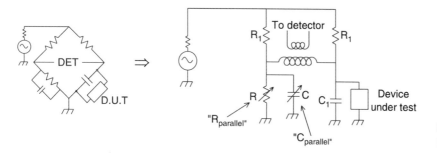

Figure 34-3. Vector impedance bridge.

front panel "C_{PARALLEL}" dial is offset so that this point is marked zero. If the load has an inductive component, the bridge can be balanced by setting C to less than C_1. The dial then indicates a negative capacitance, which the operator correctly interprets as a parallel inductance. The value of the variable resistor is indicated on its dial as "R_{PARALLEL}". This kind of bridge was made famous by the Boonton Electronics Corporation's "RX Meter."

The RX Meter uses an amplified detector, actually a superheterodyne circuit. A simple version of this bridge, the "noise bridge", is shown in Figure 34-4. This bridge uses a wide-band source and a frequency-selective detector (e.g. an external communications receiver). The source is a simple noise generator, usually an avalanche diode followed by 20 or 30 dB of gain. The instrument can be hand held and battery powered.

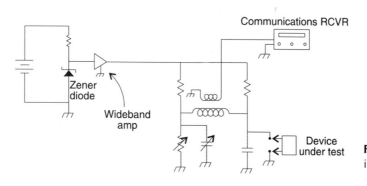

Figure 34-4. Noise bridge impedance meter.

The so-called *Q-meter*, shown in Figure 34-5, is a simple instrument for testing inductors. It reads both the reactive part and the resistive part of the impedance. A generator provides a constant voltage across R, a low-value resistor. The variable capacitor is tuned to maximize V. This maximum occurs when $X_c = X_L$ (series resonance). The operator can read the capacitance dial and, knowing the generator frequency, calculate the unknown inductance. The value of V gives the value of the equivalent series resistance, r. The current through the DUT is just V_R/r, so
$$V = X_c I = (X_c/r)V_r = Q_{\text{DUT}}V_r$$

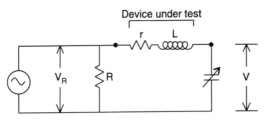
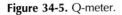

Figure 34-5. Q-meter.

SWEPT FREQUENCY IMPEDANCE MEASUREMENTS

A directional coupler or some equivalent is the heart of a *network analyzer*, usually a swept frequency instrument. In the simplified circuit of Figure 34-6, any power reflected from the DUT is divided between the generator port (where it is absorbed by the matched generator) and the output port (where it can be measured). A coupler plus a swept RF source (sweep generator) and power detector make up a *scalar network analyzer* (Figure 34-6). For calibration, a perfect reflector is substituted for the DUT. (This calibration reflector can be a short, an open, or any pure reactance.) With the reflector on, we note the power meter reading. We then put the DUT on the test port. If the power meter reading goes down *n* decibels we know the "return loss" of the DUT is *n* decibels. A 20 dB return loss, for example, indicates that 1/100 of the power is reflected from the DUT, that is, the magnitude of the (voltage) reflection coefficient is 1/10.

A *vector network analyzer* lets us measure the phase of the reflection coefficient as well as its magnitude. This can be done by combining a vector voltmeter with the above circuit. But it is easy to build a circuit that produces a polar display, that is, a Smith chart on a cathode ray tube (CRT) screen. Such a circuit is shown in Figure 34-7.

Just as with the vector voltmeter and the Q-meter, the vector network analyzer can be made more sensitive by equipping it with amplification. Again, a superheterodyne circuit is used. Figure 34-8 shows how a commercial network analyzer uses two RF conversions before the final conversion to baseband. Newer network analyzers digitize the baseband data

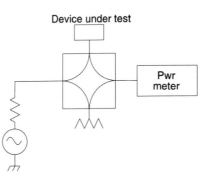

Figure 34-6. Scalar reflectometer.

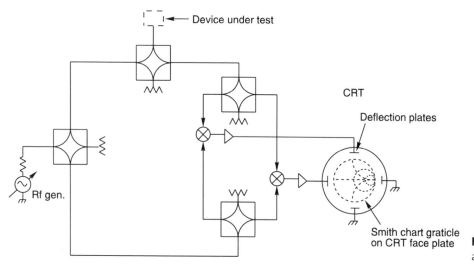

Figure 34-7. Vector network analyzer.

Figure 34-8. Superheterodyne vector network analyzer.

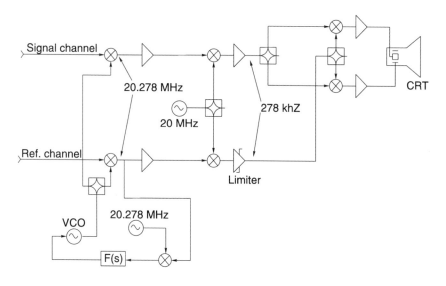

and apply corrections before displaying the reflection coefficient. Note that the vector network analyzer includes the functions of all the simpler instruments. It can measure power, vector volts, impedance, and transmission.

Spectrum analyzers are usually just superheterodyne receivers with swept local oscillators. A special nonlinear IF amplifier provides a logarithmic output voltage, that is, the detector output is proportional to the log of the input signal power. This output is displayed on the *Y*-axis of an oscilloscope while the *X*-axis is swept in synchronization with the LO. This produces a display of signal strength versus frequency. These analyzers usually have a wide range such as 10 MHz to 20 GHz. A block diagram for such an instrument is shown in Figure 34-9.

A swept frequency spectrum analyzer has objectionable flicker in situations where the sweep time must be greater than about 1/20 s. The sweeping filter must spend a time at least equal to the reciprocal of its bandwidth on each frequency to get the desired resolution. The total sweep time must therefore be greater than the effective number of points divided by the resolution bandwidth, which is the same as the frequency span divided by the square of the resolution bandwidth. For example, if it is necessary to analyze a span 20 kHz with 200 effective points, the resolution bandwidth would be 20,000/200 = 100 Hz, and the sweep time would have to be 2 s. Digital storage (or a very long persistence display screen) is a satisfactory solution to the flicker problem when the unknown spectrum is stable (not noise-like) and when the total sweep time (time to acquire the spectrum) is not objectionable.

Multiplex spectrum analyzers, as opposed to swept frequency spectrum analyzers, evaluate all the spectral points simultaneously. The probability of intercept is 100%, that is, a signal that pops up momentarily will not be lost because the instrument is off scanning a different part of the band. Spectrometry in radio astronomy requires very accurate spectral measurements in order to distinguish radio sources from background radiation. The accuracy requires long integration times for every spectral point, so it is necessary to do these integrations simultaneously rather than sequentially.

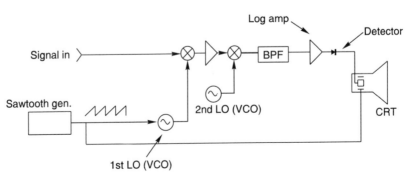

Figure 34-9. Spectrum analyzer. BPF, bandpass filter.

1. In testing a high-power transmitter with a water-cooled dummy load it is found that the water leaving the dummy load is hotter by $10°C$ than the water entering the load. If the flow rate is 10 liters/second, what is the output power of the transmitter? (1 cal=4.18 Joules.)

2. Design a phase meter that has a range of more than $\pm\pi$, that is, an instrument that counts complete turns as well as partial turns. Use the basic phase measurement method of the vector voltmeter and add digital circuitry – an up/down counter – to count turns.

INDEX